AF505453

THE PASSAGE OF THE THAMES

HOLOCENE ENVIRONMENT AND SETTLEMENT AT RUNNYMEDE

THE PASSAGE OF THE THAMES

HOLOCENE ENVIRONMENT AND SETTLEMENT AT RUNNYMEDE

Runnymede Bridge
Research Excavations, Volume 1

STUART P. NEEDHAM

with contributions by
Janet Ambers, Tony Clark, Mike Cowell, John Evans, Mary Evans,
Susan Limbrey, Mark Robinson and Rob Scaife

and illustrations by
Stephen Crummy and Phil Dean

THE BRITISH MUSEUM PRESS

© 2000 The Trustees of the British Museum

First published in 2000 by The British Museum Press
A division of The British Museum Company Ltd
46 Bloomsbury Street, London WC1B 3QQ

A catalogue record for this book is available from the British Library

ISBN 0 7141 2315 3

Designed by John Hawkins Book Design
Typeset in Ehrhardt 10/12pt
Printed in the United Kingdom at the University Press, Cambridge

This volume has been published with the
assistance of a grant from English Heritage

Contents

vi

List of Plates

List of Figures

List of Tables

Acknowledgements

To the British Museum must go the first credit; its support has been legion, including many hidden costs as well as the visible funding of the excavation campaign and post-excavation analysis by outside specialists. Dr Ian Longworth's backing throughout was crucial, as was that of the Trustees and Sir David Wilson at a higher level. Other forms of essential clearance were readily granted by the landowners, the Department of Transport (later the Department of Environment and Transport), and by English Heritage in giving consent to excavate this scheduled monument. I thank the various Inspectors involved there, but especially Roger Thomas for his long-abiding interest. We also remain very grateful to the Runnymede Hotel, onto whose grounds we encroached during one season; Michael O'Dwyer during his tenure as manager was a true friend in offering help and taking an interest in our operation.

Yet again, my debt to my illustrator colleagues, Stephen Crummy and Phil Dean, will be obvious from a mere cursory glance through the volume. Stephen deserves particular acknowledgement on this occasion for translating the field drawings into intelligible diagrams.

Help with the programme of excavation and post-excavation came from many sources. Those participating as diggers are too numerous to list, but do not go unremembered; others must be mentioned by name: Ray Caple, Julie Carr, Chris Evans, Jill Guthrie, Andrew Herne, J.D. Hill, Phil Mason, Josh Pollard, Cath Price, Marie-Louise Stig Sørensen, Martin Trott, but especially, in the context of this volume, Sue Wales and Sherrian Edwards for their ever-demanding role in environmental sampling. Caroline Cartwright subsequently played an important part in organising meetings of environmental specialists and keeping track of their progress; she also contributed to the discussions on interpretation of the sequence. Needless to say, the combined knowledge of the specialists actually working on material, whose reports appear below, made a phenomenal impact on the overall interpretation offered in Chapters 10–12; however, I have distilled these in my own way and must bear full responsibility for lapses, false deductions, or wilder speculations.

Thanks also go to others who visited the site to give advice or take samples: John Allen and Mark Macklin in relation to fluvial geomorphology; Ian Máté for taking soil samples; Nick Debenham and Eddie Rhodes respectively for sampling for thermoluminescence and optically stimulated luminescence. Janet Ambers's contribution went beyond her report, additionally supplying re-evaluations of earlier radiocarbon results from the site, and Tony Spence assisted with editing. I would also like to record here my grateful thanks to Dorli Williamson for her perseverance with the daunting task of marking a goodly proportion of the finds assemblage.

Identifications on a wide range of artefacts and on wood species were gratefully received from Stephen Castle, David Gaimster, Rowena Gale, Roger Jacobi, Phil Jones, Ian Kinnes, Beverley Nenk, John Orna-Ornstein and Val Rigby. Their respective parts are acknowledged in the tables in Chapter 3. Further assistance came in the form of unpublished information on local sites from David Barker, Mark Birley, Steve Dyer, Steve Ford, Phil Jones and Robert Whytehead. J. Sneed is thanked for his assistance with soil-phosphate analysis.

Help was always at hand when it counted most from Sheridan Bowman, whether in the finds hut, or in the tedium of compilation and editing. If sanity has prevailed and a worthwhile result been achieved, it is thanks to her.

Summary

This volume serves as the anchor in the Runnymede Research publication series. In reconstructing a substantial history of changing Holocene environment for the local catchment of the middle Thames valley, it provides context for the various phases of cultural activity on the Runnymede Bridge site (Surrey; formerly Berkshire). At the same time it sets out the evidence for the chronology of those phases in terms of stratigraphic position, archaeological comparanda and independent dating. The enormous quantity of data relating to the two main phases of occupation, in the Middle Neolithic and Late Bronze Age, is only summarised here, but material from other phases is given fuller attention.

Stratigraphic sequences representing various stages of the Holocene period were accumulated during the research excavations through a series of deep cuttings taken down to, and sometimes below, the modern water table. The physical structure of the site is set out in detail, taking an archaeological-context approach to alluvial members. These have been grouped in terms of a single site sequence of alluvial parcels using sedimentary correlations made across the site. This provides the framework for the detailed reconstructions of geomorphological and environmental change that are presented later in the volume.

Five key branches of environmental archaeology receive in-depth analysis using extensive column sampling of the alluvial deposits: soils and sediments involving particle size analysis and micromorphological techniques; inorganic phosphate analysis; the study of Mollusca, of insect remains and of pollen. In addition, plant macrofossil remains were thoroughly investigated from an Early Neolithic context. Charcoal and wood remains from contexts other than the main occupation horizons are summarised; they include material derived from the activities of both humans and beavers.

These diverse sources give rise to an extremely detailed picture of the environment at different scales; for example, contrasting the existence of very small openings on the site in the Atlantic period with a general background of dense alder forest along the valley floor. After the formation of marl in the Boreal period, probably due to extensive ponding in the Egham–Staines basin, there was a phase of marked incision, the creation of channels and a nascent island. The rise of alder woodlands locally is dated to the early/mid-seventh millennium BC. Growth of the island through subsequent clayey alluviation led to a stable land surface during the Late Mesolithic and Early Neolithic periods and attracted low-scale human activity. However, this resulted, inter alia, in the digging of a pit to hold a post early in the seventh millennium BC, and a ditch perhaps early in the fifth millennium BC. Signs of very early agriculture – cereal cultivation, possibly preceded by some woodland grazing by

domesticates – were recovered from Early Neolithic channel deposits and give an important prelude to the intensive Middle Neolithic occupation of the site. Deforestation during the earlier Neolithic, although still affecting only a minor portion of the land, nevertheless had a dramatic impact on the stability of the valley ecosystem. One effect was a massive flood involving the deposition of a large body of gravel over the banks late in the fourth millennium BC.

During the later Neolithic, vegetation compositions in the region fluctuated in response to a shifting patchwork of habitats involving episodes of woodland regeneration, but also some continued arable farming. A beaver presence seems more significant for site history than scant cultural debris. Human activity on the site itself remains exiguous during the Early to Middle Bronze Age, but there may have been a phase of cultivation, explaining the particular distribution of a small assemblage of Beaker and Early Bronze Age pottery. However, during this period the site bears witness to a much more substantial legacy of human interference with the landscape. A thick body of alluvium is argued to have derived directly from large-scale clearance and the cultivation of erosion-vulnerable ground in the catchment. Having covered the whole site, alluviation ceased around the middle of the second millennium BC, and may signal the widespread conversion of open land in the valley bottom to grassland, a landscape interpreted to continue into the Late Bronze Age phase of renewed intensive occupation.

The second millennium also saw a change in the nature of the river system. Some branches of the prior anastomosed network became choked, resulting in the loss of the original island on the site, the single channel remaining migrated, and eventually a new breakthrough channel created a quite different island. This was the setting for Late Bronze Age occupation, but only after a phase of woodland regeneration. During this sequence and into the later first millennium BC, channels are seen to be dynamic and to feature point-bar aggradation. The northern channel was quickly blocked again, perhaps during the occupation, leaving a southern meander to become more accentuated over the ensuing millennia. A small group of later Iron Age and Roman sherds implies cultivation on the site or nearby, while associated flood silts tie in with third-century AD alluviation elsewhere in the locality suggesting more widespread changes in agricultural practice.

It is uncertain whether the modern near-straight course of the Thames, making a true island – *Tyngeyt* – was re-established in Late Medieval times or later, but the southern loop ceased to carry current around the sixteenth to seventeenth centuries AD at a time when river-borne trade was on the increase. The succeeding centuries saw a number of modifications to channel and bank, some deliberate due to towpath definition and lock and weir construction, others a consequence of the altered patterns of flow. The lock complex gave rise to an associated settlement, gradually acquiring a more leisure-pursuit orientation. Eventually the river crossing, possibly on a long-standing route, was converted from ferry to bridge.

The concluding chapter extracts some significant findings or more general points relating to channel form and configuration, the make-up of river islands, flooding and alluviation and, finally, the nature and occurrence of occupation. The different nature of the various occupations is highlighted and discussed in the context of the social milieus. The siting of the Late Bronze Age settlement is now seen as responding to both a system of regional social interaction and spiritual needs. Close association with the Thames is viewed as paramount in sanctioning a social order and the site as a sacred heart.

'... Athos learned that every river is a tongue of commerce, finding first geological then economic weakness and persuading itself into continents.'

(Anne Michaels, *Fugitive Pieces*, 1997)

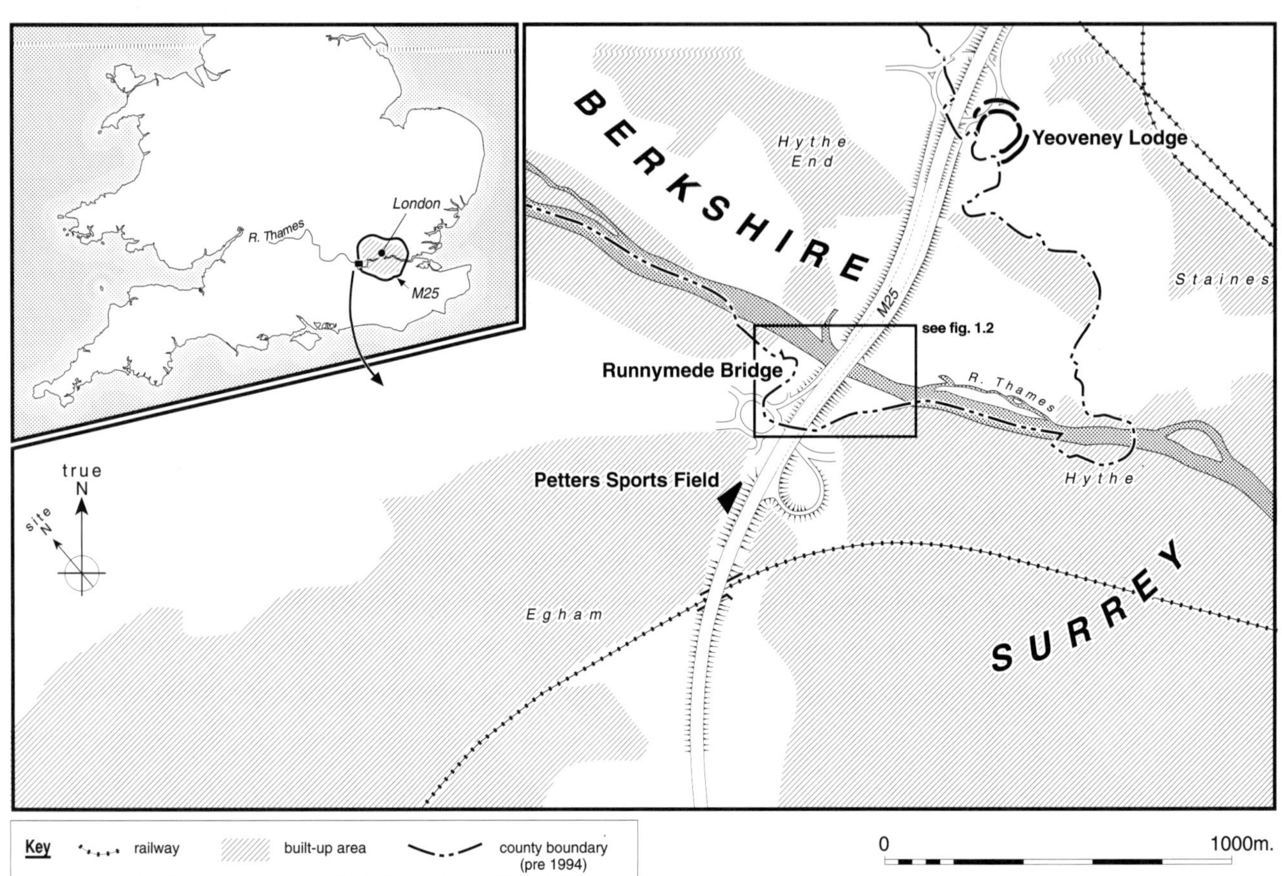

1.1 *Location map for Runnymede Bridge on the middle Thames.*

CHAPTER 1

Introduction

The passage of the Thames: providing a context

The 'passage' of the Thames, in all its connotations, is inherently absorbing, being Britain's premier river, artery of the capital and conduit to shores near and far. While this volume, based on one small segment of the valley, cannot profess to deal with the full temporal and spatial parameters of this great river, it may serve to illustrate the importance of interplay between host river in all its moods and ten thousand years of human settlement history alongside.

What follows is, at core, the landscape history of a small plot of land at Runnymede Bridge in the middle Thames (Fig. 1.1); indeed, in truth, it is as much a riverscape history. The intense enquiry into this plot of land and, by extrapolation, its environs stems directly from the nature of the project, to explore in depth its two major phases of human occupation and various further evidence for local human activity. As befits any serious archaeological study of a site nowadays, there was an associated desire to understand the direct physical circumstances of the different phases of activity. However, the potential for highly dynamic environments in valley bottoms, arising from river channel movement and the effects of other hydrological changes (notably flooding and mean water tables), invites close attention to the environmental scene at as fine a scale as possible. Such environmental factors bear very directly on the nature of occupation or land use feasible within any given psycho-framework of society, this being made up of such factors as tolerance thresholds, the prevailing cosmography, perceived risks and perceived rewards (e.g., Bell 1992; Brown 1997).

The valley-bottom setting of the Runnymede Bridge site may traditionally be described, geomorphologically, as part of a 'floodplain'. While it is hard to avoid this term, it is important to appreciate that the delimitation of the land that constitutes a floodplain can only be vague. Even within a period for which hydrological conditions barely altered, flood rates and flood levels would constantly vary according to climatic oscillations and vagaries. As a consequence, the frequency of inundation of a particular spot and the maximum area of land inundated in the local reach of the valley would also be subject to persistent variation. Floods are notorious for their considerable and unpredictable

oscillations over the short to medium term. Over greater timescales, changing hydrological conditions can alter the balance between wetness and dryness to the extent that it might be possible to recognise differing aggregate conditions from the archaeological and palaeoenvironmental evidence. These are of course a vital starting point for our central enquiry, but they cannot address the other, socially oriented issues raised above which can only be drawn from a wider consideration of the culture concerned.

Neither is the deposition of overbank alluvium helpful to the definition of floodplains, since flooding does not necessarily involve alluviation (e.g., Robinson and Lambrick 1984). Flood regimes, of whatever type, have the potential to alter the valley-bottom landform. Even when accretion is insignificant, there is potential for the erosion of soil from surfaces and river banks, and occasionally the more dramatic effects of avulsions, where a new channel is carved out through part of the floodplain.

Although the research project has necessarily focused on the specific site of human occupation at Runnymede Bridge, the diverse enveloping alluvial deposits give witness to changes in the environment on a much broader scale than the site itself. This provides opportunities for seeing how the activity phases on the site might interrelate with activity in its hinterland, both contemporary activity and that belonging to preceding or succeeding phases. When also tied in with knowledge of other sites in the immediate region, valuable insights into settlement and land-use histories may be gleaned. Only such integrated interpretation can hope to identify true cause and effect relationships: effects of human action on the ecosystem, or conversely externally driven environmental change which caused or contributed to changes in demography or land use. Typifying this inverse relationship is the debate on whether alluviation and floodplain change were primarily of climatic or anthropogenic causation. Brown has recently made the case cogently that this is a false question and talks of '. . . the reality of the combined nature of causality . . .' (1997, 312 ff). Broad patterns of Holocene environmental change are generally now well understood, although this is still less true of alluviation and channel change; but it might be argued that the specific conditions affecting early communities were created at a relatively local scale, even though broader factors may have governed some underlying trends.

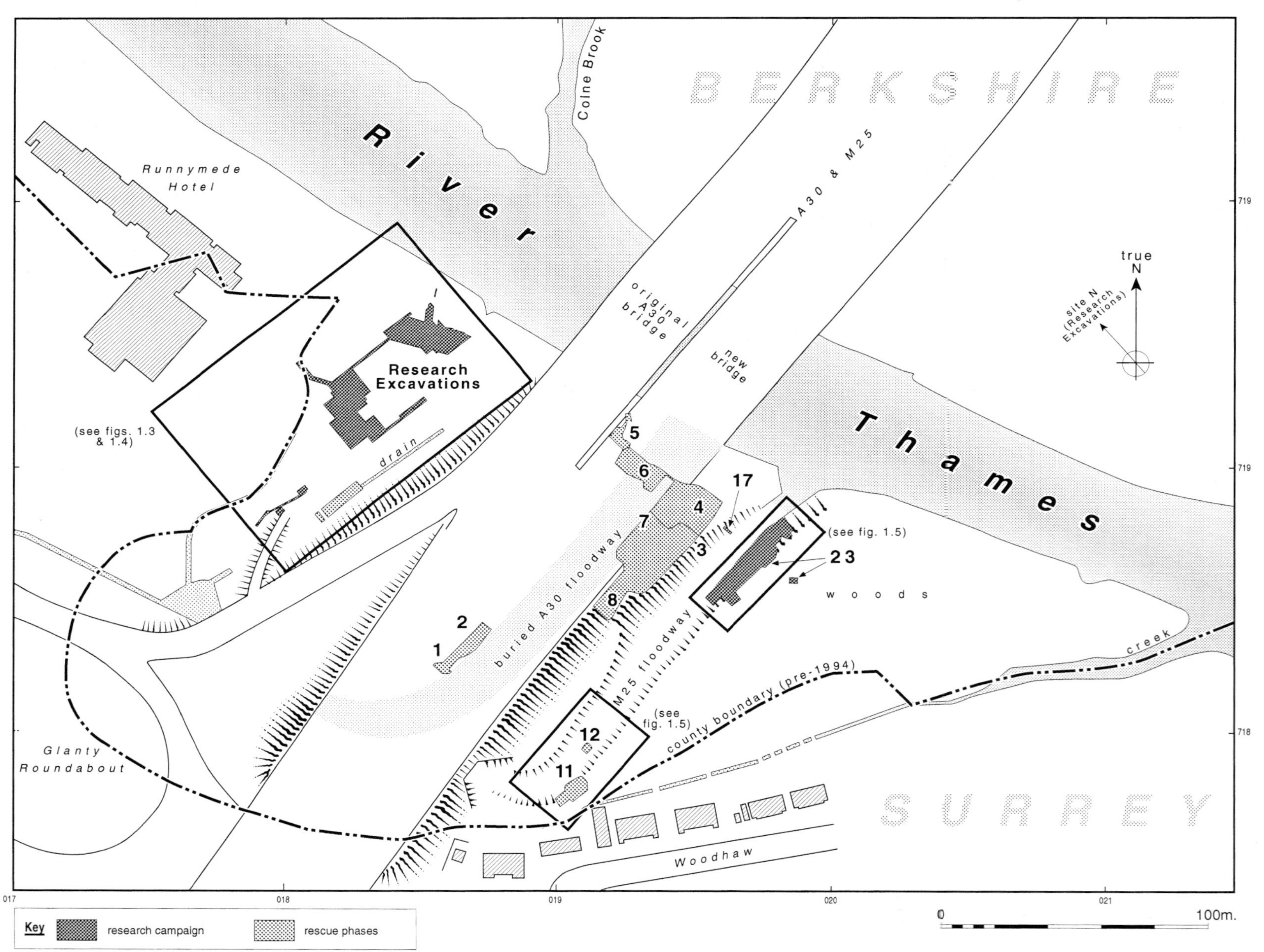

1.2   *Site plan showing the excavations of the rescue and research phases in relation to the modern built landscape.*

The research excavations: strategy, sequence and notations

Six seasons of research excavation took place at Runnymede Bridge from 1984 to 1989 (Fig. 1.3). This campaign followed intermittent rescue archaeology on the site between 1975 and 1980, most of which has been published in full (Longley 1980; Needham 1991). The last rescue season, in 1980, involved the excavation of Area 10 by the author with David Longley (Late Bronze Age layers) and Rob Poulton and Martin O'Connell of the Surrey Archaeological Unit (Neolithic layers). This demonstrated for the first time that the in situ prehistoric deposits previously encountered on the line of the M25 extended to the north-west of the A30/M25 bridge embankment (Fig. 1.2). Consequently, it was considered likely that much or all of the triangular plot of rough ground trapped between the bridge embankment and Runnymede Hotel grounds would contain undisturbed deposits. In response English Heritage agreed to schedule the plot as an ancient monument. The research excavations were undertaken having obtained scheduled monument consent and permission from the Department of Transport (the landowners; later the Department of Environment and Transport). The area specified remains scheduled.

The research campaign set out with the following objectives:

1. To obtain a better idea of the limits of the two settlement phases and their relationship to changing floodplain topographies.
2. To excavate some reasonably large areas in order to determine the structural layout of parts of the site; this had been precluded during rescue phases by limitations on the size of trenches.
3. Furthermore, to record artefact-rich surface deposits in relationship to site layout and thereby to comprehend better the organisation of the settlement and its constituent activities.
4. To seek useful stratigraphic sequences of occupation deposits which might document change in artefact assemblages at a resolution hitherto unknown on British early first millennium BC sites.
5. To exploit further the demonstrated environmental potential of the site in the hope of reconstructing a longer sequence of change through the post-glacial period; and to interrelate the changes with evidence for local human activity.

Objectives 1 and 5 are the key concerns of this volume.

Choice of excavation areas

During an earlier road construction phase the scheduled part of the site had been raised by laying a raft of rubble around 0.70 m thick (Fig. 2.1). This capping, although excellent from the point of view of protecting the archaeology, made boreholing extremely laborious and furthermore precluded any geophysical surveys. A further constraint between 1984 and 1987 was the presence on the site of a row of caravans, some occupied, just inside the western hedge line.

In the first season of excavation, 1984, two trenches were laid out to confirm the presence of in situ archaeological deposits across a wide part of the scheduled plot (Areas 13 and 14; Fig. 1.3). They were also sited in the hope of falling close to the north-western edge of the deposits. A few boreholes taken beforehand encouraged the view that these areas would yield worthwhile evidence. Most of the ensuing research campaign was built around these two trenches since they yielded the range of evidence that had been hoped for. Also in 1984 a small slot (Area 15) was dug into the modern river bank; it revealed only Post-Medieval deposits.

Thus in the following seasons two major zones of excavation were developed around Areas 13 and 14, with scattered outlying trenches and extension arms (Figs 1.2 and 1.3). The major zones are conveniently termed 'interior' and 'riverside'. A summary of the zoning of trenches, their seasons of excavation and the occurrence of the two main occupation horizons is presented in Table 1.1.

In detail the sequence of excavation was as follows. In 1985 Areas 13 and 14 continued to be excavated, both having extensions added to determine the limits of the site and the age of flanking palaeochannels. Area 16 was also started, being intended to give a transect through the deposits and underlying topography on a SW–NE alignment. In particular it was hoped to observe an expected fall in the Neolithic land surface in a north-easterly direction. Progress in Area 16 was slower than expected owing to the depth of LBA deposits at the eastern end. During the season it was therefore decided to concentrate on the two ends of the long trench in order to see full profiles as early as possible (Area 16 West, Area 16 East). Excavation was never resumed in the central part of Area 16, but a modern pit enabled easy recording of a deep section and sampling for environmental evidence (column 9).

A hand-dug slot, Area 17, was excavated through rubble cover at the foot of the M25/A30 embankment on its south-eastern side, i.e. the farther side from the scheduled area (Fig. 1.2). It was in close proximity to Area 4 of the rescue excavation. It was hoped to establish whether the LBA river channel and attendant structures seen in Areas 4 and 6 might continue undisturbed to the south-east. The result was inconclusive at the time, but subsequent research places Area 17 in the edge of a later channel. The difficulty of accessing the undisturbed deposits here prevented any expansion of this trench.

The first substantial additions to the core areas came in 1986. Areas 19 and 20 were opened to abut Area 13, where excavation continued, now probing the Neolithic deposits. Area 19 effectively swallowed up Area 16 West, where Late Bronze Age contexts had been fully excavated in 1985, but Neolithic deposits had been left untouched. In the riverside zone Area 18 was laid out to abut the southern side of completed Area 14. The trench was set for convenience

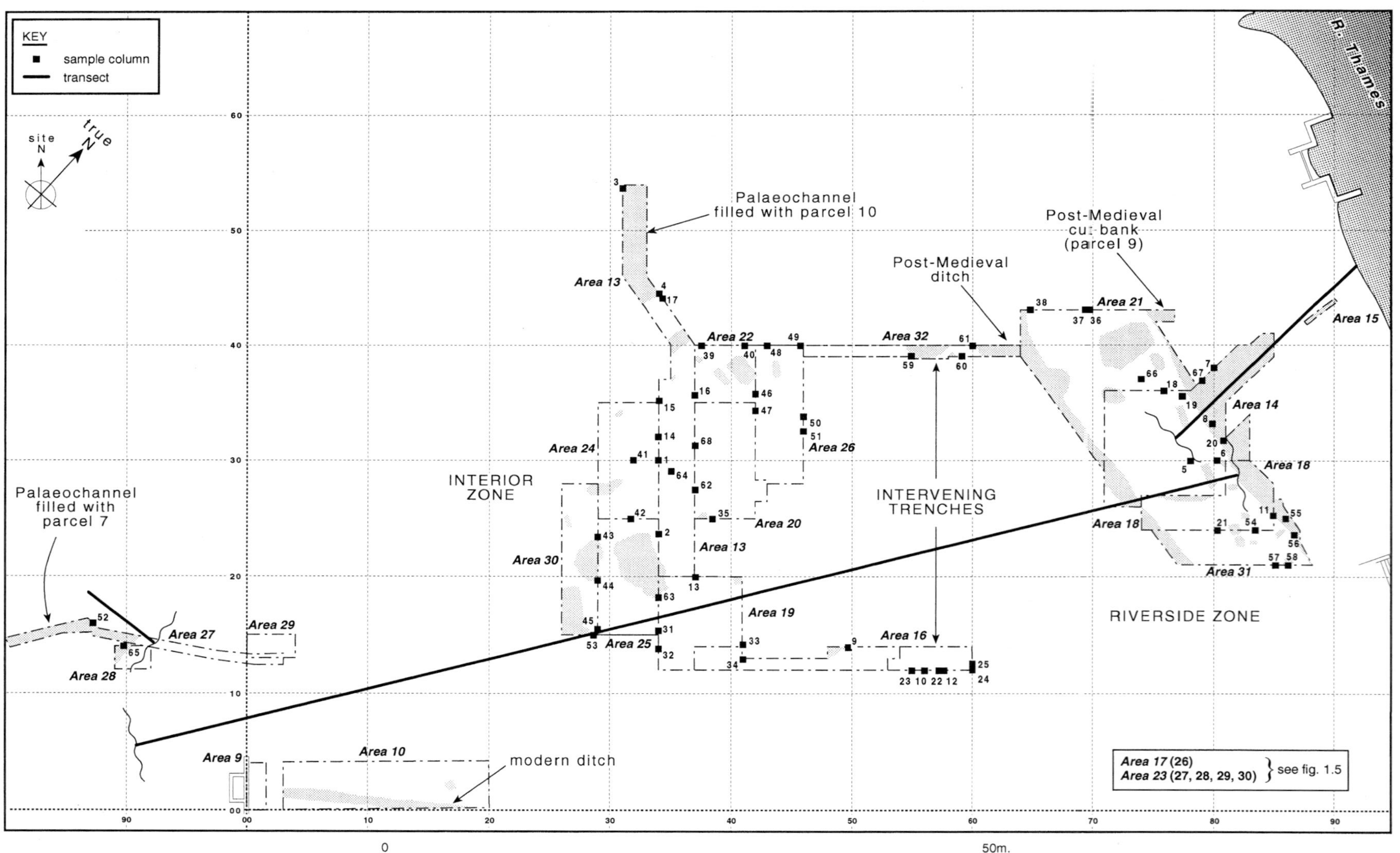

1.3   *Trench plan showing the location of environmental and dating sample columns in the main area of research excavations. The transects are the axes used for the composite section shown in Fig. 2.2.*

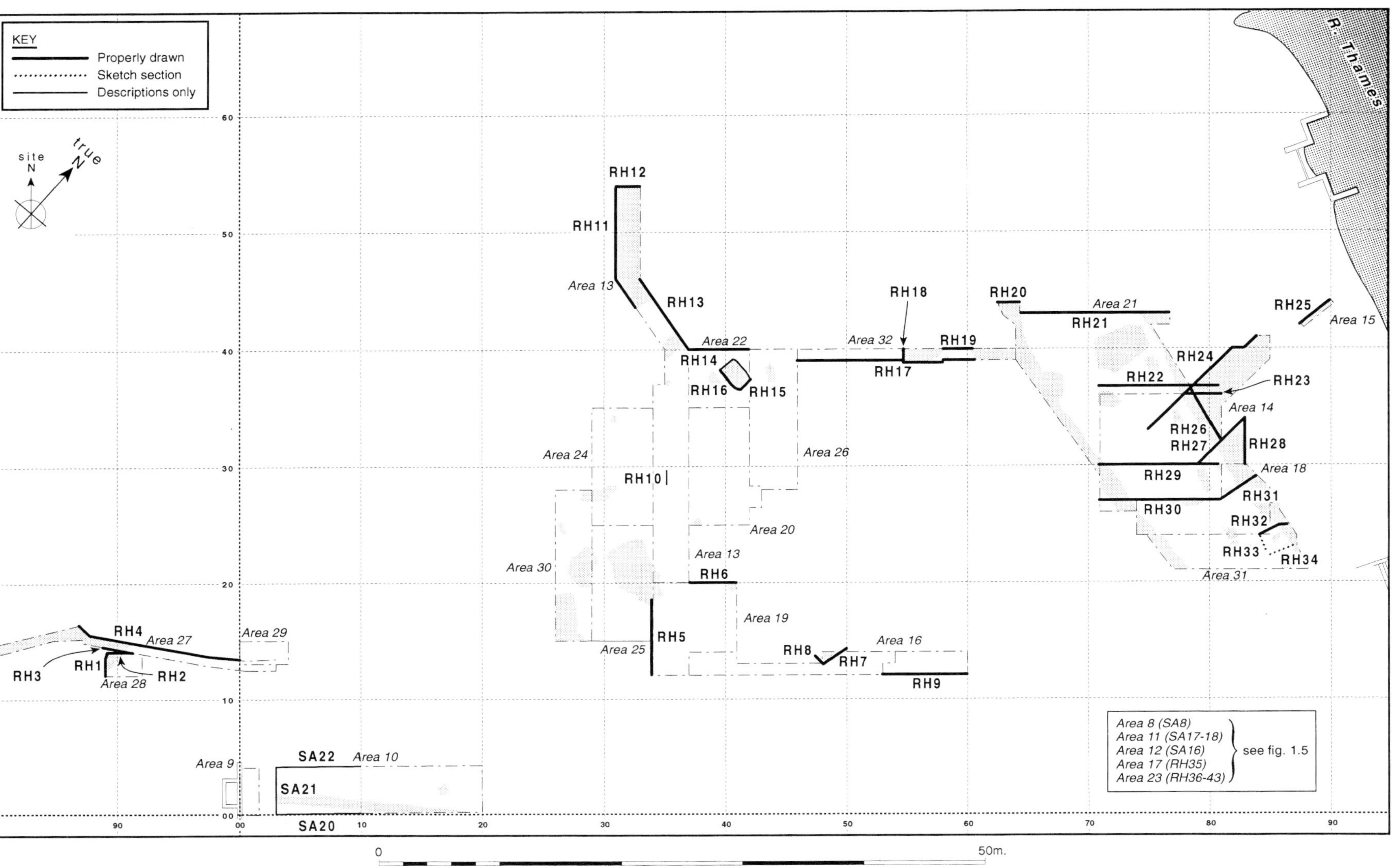

1.4   *Trench plan showing the locations of recorded deep sections in the main area of research excavations.*

Table 1.1 *Summary details of the excavated trenches of the research campaign, 1984–89.*

Area	Seasons	In situ Neolithic	In situ LBA
Interior zone			
13	ERB 84-87	excavated	excavated
16 West (→19)	ERB 85-89	excavated	excavated
19	ERB 86-89	excavated	excavated
20	ERB 86-89	excavated	excavated
22	ERB 87-89	excavated	excavated
24	ERB 87-89	excavated	excavated
25	ERB 88	unexcavated	excavated
26	ERB 88-89	unexcavated	excavated
30	ERB 89	unexcavated	excavated
Intervening zone			
16 Centre	ERB 85	unexcavated	partially excavated
16 East	ERB 85-86	partially excavated	excavated
32	ERB 89	partially excavated	excavated
Riverside zone			
14	ERB 84-85	none present	excavated
15	ERB 84	none present	none present
18	ERB 86	none present	excavated
21	ERB 87-89	none present	excavated
31	ERB 89	none present	excavated
Outlying trenches			
17	ERB 85	none present	none present
23	ERB 87	none present	none present
27	ERB 88	partially excavated	partially excavated
28	ERB 89	excavated	excavated
29	ERB 89	unexcavated	partially excavated

between two Post-Medieval linear cut features, and thus staggered relative to the prior trench.

In 1987 the interior zone saw excavation in five trenches. The last of the Neolithic was dealt with in Area 13 South, while the Neolithic deposits were begun in Area 19 (now actively incorporating the corner formerly excavated as Area 16 West). Late Bronze Age evidence was seen to completion in Area 20 and substantially excavated in two new trenches, Areas 22 and 24.

In the riverside zone, the LBA deposits having been worked out in one season in Area 18, a new larger trench was opened, Area 21. This was appended to the northern edge of Area 14, again set between the continuing Post-Medieval cut-lines. Boreholing had suggested that a natural trough containing stratified LBA deposits was becoming deeper in this direction, thus promising a deeper stratified sequence.

The final new addition in 1987 took place on the far side of the embankment, Area 23, as a follow-up to Area 17. This was difficult to access and the closest strip of ground had been cut into by the broad swathe of the M25 relief floodway. The Department of Transport gave permission for a 40 m length of the eastern floodway wall to be stripped of overlying vegetation and spoil to check that a suspected palaeochannel ran through this area. In addition to exposure of the natural deposits in the floodway wall, limited perpendicular transects were cut back into them to establish dip orientations. A discrete hole was machine dug behind the main exposure to check for any rapid lateral changes in sediment pattern.

In 1988 Neolithic deposits continued to be explored in Area 19 and were begun in Areas 20 and 22. Late Bronze Age cut features were excavated in Area 24, while Areas 25 and 26 were appended on the western and eastern sides respectively of the agglomerating block. Thin surface deposits in Area 25 allowed LBA contexts to be fully worked out in one season; it was not reopened for excavation of the Neolithic material. A long trench, Area 27, was largely machine-cut through the scrubwood to the west of Area 10 in order to investigate suspected channel deposits and hopefully also to delimit the intact prehistoric deposits. The latter were found towards its north-eastern end. In the riverside zone, Area 21 was reopened with the exception of its eastern corner, where concentrated excavation had worked out the Late Bronze Age evidence in 1987. However, a triangular extension (southern spur) was added on the south to fill in the Late Bronze Age land surface between Area 14 and the Post-Medieval ditch.

The 1989 season concluded the campaign. All unfinished trenches from previous seasons needed to be completed down to satisfactory horizons, effectively the base of Late Bronze Age or Neolithic occupation deposits. New trenches were, as ever, still desirable to clear up crucial questions. However, any new trenches also needed to reach a satisfactory horizon in a single season. For the most part this meant full excavation of the LBA deposits with no expectation of investigating any underlying occupation deposits. A fairly ambitious programme was set and an extra-long season granted to carry it out. Favourable weather in 1989 helped the programme to a good conclusion.

In the interior zone excavation continued in Areas 19, 20, 22 and 24 to see completion of the Neolithic levels and features. A small part of Area 13 was reinvestigated to check feature continuity with Area 20. Meanwhile, in Area 26 the Late Bronze Age deposits were bottomed, as also in a new trench, Area 30, appended to Areas 24/25 to extend their critical posthole plans. To the west, two small trenches were set out alongside parts of Area 27 to investigate specific issues: Area 29 to pursue a particular feature (which failed to materialise); Area 28 to examine more thoroughly the channel/land interface on this margin of the site.

In the riverside zone the LBA deposits across much of Area 21 (including its southern spur), were seen to completion, while a new trench, Area 31, was laid out to the south of Area 18 and excavated through to the base of LBA levels. This direction was followed to broaden the context of some substantial postholes and distinctive soil deposits recovered in the south and east respectively of Area 18. The final trench to be added was a 1 m wide link between the riverside and interior zones, Area 32. This trench (not possible before 1988 owing to the resident caravans) gave the prospect of obtaining a long profile of the topography for at least the Late Bronze Age ground surface. In the event it was also possible to excavate the Neolithic deposits in a 0.5 m wide transect and thus deepen the attendant section.

Alluvial deposits

Whenever opportunities arose, the alluvial deposits underlying and abutting the cultural horizons were examined. Circumstances normally dictated that access to underlying deposits occurred in any given area towards the end of its final season of excavation, that is only after all cultural layers and cut features had been excavated. In many cases a machine was used at this stage to cut deep trenches quickly; the recorded sections underpin the environmental and topographic sequence published in detail here. The locations of sample columns are shown in Figs 1.3 and 1.5 and those of the deep sections in Figs 1.4 and 1.5.

Sample columns were taken for varied reasons, predominantly for recovery of biological remains, sedimentological information and scientific dating. Most columns were taken through alluvial deposits as well as cultural horizons, to document as fully as possible the picture of environmental change, but others were geared to questions concerned specifically with one or other phase of human activity and were correspondingly shorter (Table 1.2). Such distinctions also determined analysis type, for example whether for phosphate measurement or extraction of ecofacts. The type of biological remains sought was determined by the level of the deposits in relation to the water table, only those below it being suitable for pollen, plant macros and insect remains. Molluscan analysis alone could be applied to longer sequences of alluvium as they rose above the water table. Overlap between the different kinds of analysis was often achieved, but no attempt was made to cover every viable branch systematically for every column. Even so, the quantity of material to process and analyse was large and in some cases decisions were made to sub-sample the full column. Details of the columns – the alluvial parcels and cultural horizons they span and the types of analysis intended/applied – are summarised in Table 1.2.

For the columns reported in this volume, samples of soil were almost invariably taken for processing in the laboratory, normally by the appropriate specialist. They were removed from site either as monolith segments in trays, or as loose soil carefully excavated into bags. Details of procedure and sample sizes are to be found in the relevant chapters. Plant remains and animal bones which had been introduced to the site by human agency, i.e. those from the cultural levels, are not reported on in this volume since they belong with the other detailed evidence for occupation. Recovery of these 'cultural ecofacts' was enhanced by extensive flotation and wet-sieving on site.

Cardinal point citations

The site grid of the research campaign is not only almost diagonal to the National Grid (Fig. 1.3), but also at variance with the site grid for Area 6 (Needham 1991). Area 23 also has a local site grid. In order to present a coherent discussion of the whole site, orientations in this volume are cited as cardinal points, relating to the Ordnance Survey's National Grid. Occasionally, parts of trenches are specified according to the local site grid – hence: 'Area 16 West', strictly speaking 'South-west'.

Citation of dates

All dates given as BC or AD are calendar dates, for the most part being based on calibrated radiocarbon measurements. Where uncalibrated radiocarbon dates, or date ranges, are cited, they are always expressed as BP.

County location

Historically, although the site has been landlocked on the southern, Surrey bank of the Thames for the past three centuries, its previous separation by a major meander of the Thames had placed it within the county on the northern bank. In recent times (and during the excavations) this was Berkshire, parish of Wraysbury; earlier it had been Buckinghamshire. Administrative revision has, however, in its wisdom and disregard for historical circumstance, now moved the county boundary up to the modern course of the Thames, taking the site into Surrey. Interestingly, the location is also only just beyond the western limit of Middlesex, prior to its erasure from the map!

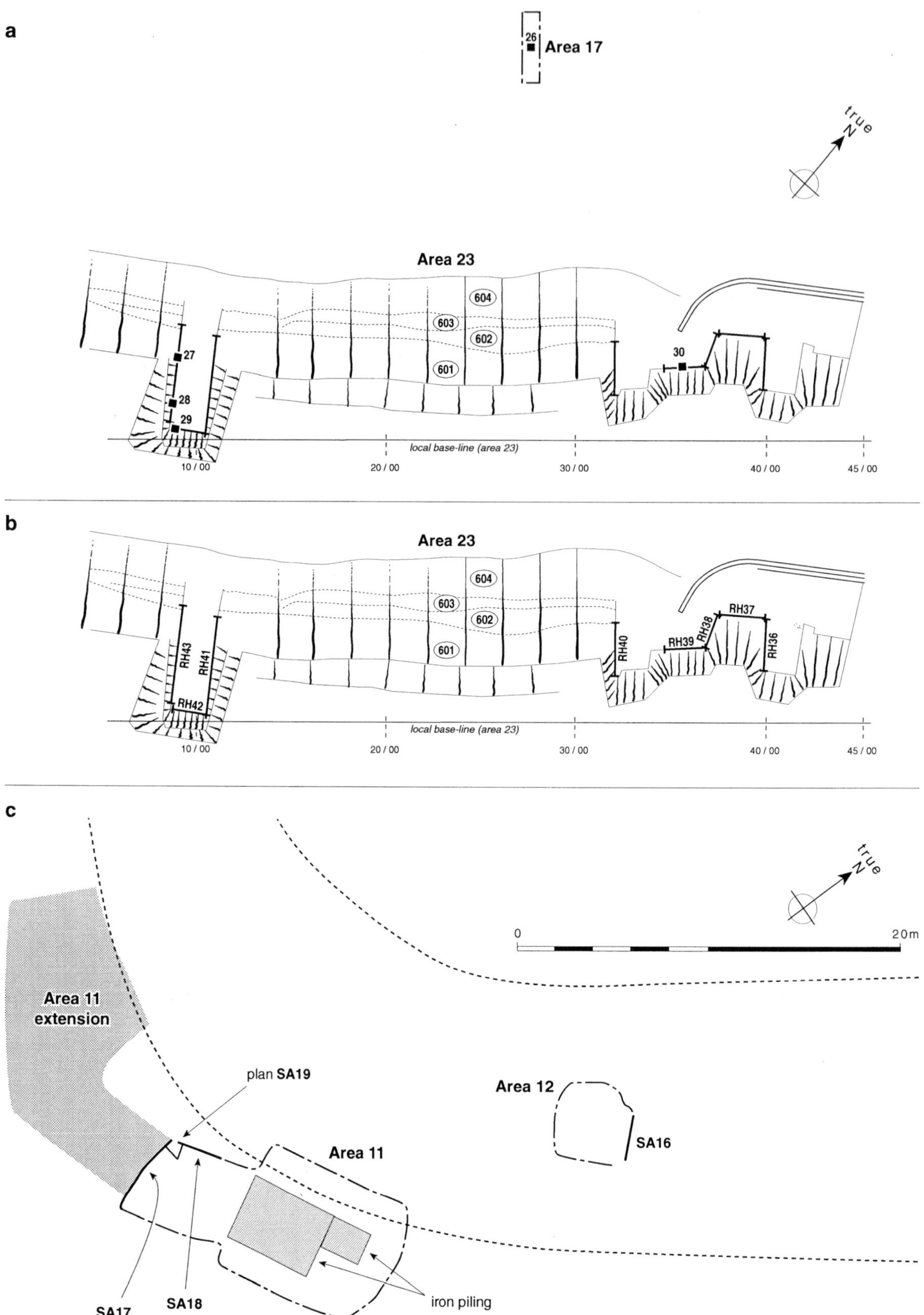

1.5 *Location of: a) sample columns in Areas 17 and 23; b) recorded deep sections in Area 23; c) recorded deep sections in Areas 11 and 12.*

Table 1.2 *Summary table of sample columns and spot samples for environmental and dating purposes.*

Col. no.	Season Area	Grid	Material	No. samples	Sample interval (cms)	Total column span (m OD); Sample nos	Column span reported on; Sample nos	Parcels/ periods represented
1a	85 A13	34.0/30.0	Mollusca	7	5*	14.98 - 14.65 Sn102 - 108	-	6D, LBA
			Pollen	7	5*	14.98 - 14.65 P70 - 76	None found	6D, LBA
			Phosphates	18	2	14.95 - 14.59	14.95 - 14.59	6D, LBA [Fig. 6.1]
1b	86 A13	34.0/30.0	Phosphates	7	2	14.54 - 14.40	-	2B, 2F
2	84 A13	34.0/24.0	Pollen	7	5*	14.90 - 14.59 P77 - 83	None found	6D, LBA
			Mollusca	7	5*	14.90 - 14.59 Sn109 - 115	-	6D, LBA
3	85 A13	31.0/53.5	Mollusca	13	10	13.90 - 12.60 Sn1 - 13	-	10 [Fig. 2.9]
4	85 A13	34.0/44.5	Mollusca	11	10	13.81 - 12.71 Sn36 - 46	13.81 - 12.91 Sn36 - 44	1K, 1E [Figs 2.10, 7.3 and 7.4] (possible root disturbance at top)
			Archaeo-magnetic dating	5 pairs + 3 at base	20	13.90 - 12.90 1 - 13	13.90 - 12.90 1 - 13	1K, 1E [Fig. 2.10] [Table 4.2]
			Pollen	5 overlapping 20 cm trays		13.51 - 12.71 P9 - 13	13.06 - 12.71 P12 - 13	1E [Figs. 2.10 and 9.2]
			Insects	5	10	13.21 - 12.71 I1 - 5	12.91 - 12.71 I4 - 5	1E [Figs 2.10 and 8.1] [Tables 8.3 and 8.4]
5	84 A14	78.0/30.0	Mollusca	37	5	14.34 - 12.49 Sn48 - 56, 58 - 85	14.34 - 12.49 Alternate samples	(LBA), 2B, 2E, 2D [Figs 2.23, 7.9 and 7.10]
			Pollen	8 x 24 cm trays		14.39 - 12.47 P58 - 65	13.07 - 12.75 P63 - 64	2E, 2D [Figs 2.23 and 9.7]
6	84 A14	80.3/30.0	Mollusca	14	5	13.18 - 12.48 Sn86 - 99	13.03 - 12.58 Intermittent samples	2E [Figs 2.23 and 7.11]
			Pollen	3 x 24 cm trays		13.18 - 12.46 P66 - 68	-	2E [Fig. 2.23]
(6)	84 A14	79/30	Insects	one bulk sample context 14.120		c. 12.80 - 12.60 A14 W1	12.80 - 12.60	2E [Fig. 8.1] [Tables 8.3 and 8.4]
7	85 A14	80.0/38.0	Mollusca	6	10	13.55 - 12.95 Sn47 - 52	-	4B [Fig. 2.19]
			Archaeomag-netic dating	3 (at one horizon)		13.97 - 13.92	13.97 - 13.92	9 [Fig. 2.19]
8	85 A14	80.0/33.0	Mollusca	14	5	13.86 - 13.16 Sn53 - 66	13.86 - 13.21 Intermittent samples	4B [Figs. 2.21, 7.14 and 7.15]
9	85 A16	49.5/14.0	Phosphates	19	4	14.70 - 13.94	14.70 - 13.94	6D, LBA, 2B, 2F, ?MN [Figs 2.7 and 6.1]
		49.7/14.0	Mollusca	22	5*	14.68 - 13.48 Sn14 - 35	-	6D, LBA, 2B, 2F, ?MN, ?1K [Fig. 2.7]

Col. no.	Season Area	Grid	Material	No. samples	Sample interval (cms)	Total column span (m OD); Sample nos	Column span reported on; Sample nos	Parcels/ periods represented
			Pollen	8 overlapping 20 cm trays		14.52 - 13.31 P1 - 8	None found	LBA, 2B, 2F, ?MN, ?1K [Fig. 2.7]
10	86 A16	56.2/12.0	Mollusca	19	5	14.67 - 13.72 Sn1 - 19	14.67 - 13.72 Sn1 - 19	6D, LBA, 2B, 2F, MN [Figs 2.8 and 7.13]
11	86 A18	85.0/25.5	Mollusca	7	5	14.20 - 13.85 Sn22 - 28	14.20 - 13.85 Intermittent samples	6D, LBA, 4B [Figs 2.26 and 7.20]
12	86 A16	58.0/12.0	Mollusca	2	5	13.76 - 13.66 Sn20 - 21	13.76 - 13.66	2B [Fig. 2.8]
13a	85 A13	37.0/20.0	Phosphates	10	3	14.86 - 14.56	14.86 - 14.56	6D, LBA, 2B, (?2F) [Figs 2.6 and 6.1]
13b	89 A19	37.2/20.0	Phosphates	25	2	14.72 - 14.22	14.72 - 14.22	6D, LBA, 2B, 2F, MN, 1K [Figs 2.6 and 6.1]
		37.1/20.0	Mollusca	13	2	14.51 - 14.25 Sn1 - 13	14.51 - 14.25 Alternate samples	MN, 1K [Figs 2.6 and 7.5]
			Soil micro-morphology	3 overlapping 10 cm trays		14.51 - 14.25 248 - 250 bags 462-464	14.51 - 14.25	MN, 1K [Fig. 2.6] [Table 5.1]
14	85 A13	34.0/32.0	Phosphates	3	-	-	-	(Test samples)
15	86 A13	34.0/35.3	Soil micro-morphology	7 overlapping 10 cm trays (one gap)		14.82 - 14.41, 14.33 - 14.15 207- 212, 233	14.82 - 14.41, 14.33 - 14.12	6D, LBA, 2B, 2F, MN, 1K [Table 5.1]
						14.79-14.12 bags 410-421, 453 - 454		
16a	85 A13	37.0/35.6	Phosphates	13	3	14.84 - 14.45	14.84 - 14.31	6D, LBA, 2B, 2F, MN [Fig. 6.2]
16b	86 A13			7	2	14.45 - 14.31		
17a	85 A13	34.3/44.0	Phosphates	9	4	14.40 - 14.04	14.40 - 14.00	(10), 6D, ?LBA, 2F, MN [Fig. 6.2]
17b	86 A13			1	4	14.04 - 14.00		
18	85 A14	75.9/36.0	Phosphates	11	3	14.49 - 14.16	14.49 - 14.16	6D, LBA [Fig. 6.2]
19	85 A14	77.4/35.5	Sediments	5 overlapping 20 cm trays		14.04 - 13.44	-	4B [Fig. 2.19]
20	85 A14	80.5/31.5	Archaeomag-netic dating	15	6	13.94 - 13.04 1 - 15	13.94 - 13.04	4B, 2B, 2E [Fig. 2.22] [Table 4.2]
21	86 A18	80.4/24.0	Soil micro-morphology	8 overlapping 9 cm trays		14.32 - 13.71 213 - 220	14.44 - 13.71	6D, LBA, 4B, 2B [Table 5.1]
						14.44 - 14.29, 14.19 - 13.71 bags 422-434		

Col. no.	Season Area	Grid	Material	No. samples	Sample interval (cms)	Total column span (m OD); Sample nos	Column span reported on; Sample nos	Parcels/ periods represented
22	86 A16	57.7/12.0	Soil micro-morphology	12 overlapping 9 cm trays		14.56 - 13.60 221 - 232 bags 435-452	14.56 - 13.60	6D, LBA, 2B, (2F), MN [Fig. 2.8] [Table 5.1]
23	85 A16	55.0/12.0	Phosphates	10	3	14.62 - 14.32	14.62 - 14.32	6D, LBA [Figs 2.8 and 6.2]
24	85 A16	60.0/12.0	Phosphates	10	3	14.53 - 14.23	14.53 - 14.23	6D, LBA [Figs 2.8 and 6.3]
25	86 A16	60.0/12.3	Phosphates	33	2	14.30 - 13.64	14.30 - 13.64	LBA, 2B, 2F, MN [Figs 2.8 and 6.3]
26	85 A17	-	Non-specific	7	layers	-	-	5
27	87 A23	10.0/04.1	Mollusca	7	10*	14.14 - 13.53 Sn1 - 7	14.14 - 13.53 Alternate samples	6A, 5 [Figs 2.28, 7.23 and 7.24]
28	87 A23	09.7/02.0	Mollusca	5	10*	13.95 - 13.47 Sn8 - 12	13.85 - 13.57 Sn9, 11	6A, 5 [Figs 2.28, 7.23 and 7.24]
29	87 A23	09.9/00.5	Mollusca	7	10*	13.84 - 13.18 Sn13 - 19	13.84 - 13.18 Alternate samples	6B [Figs 2.28, 7.23 and 7.24]
30	87 A23	36.7/03.8	Mollusca	10	10*	14.10 - 13.28	-	6A, 5 [Fig. 2.27]
31	89 A19	34.0/15.0	Archaeomag-netic dating	21	*c.* 5	14.42 - 13.37 1 - 21	14.42 - 13.37	1K, 1C [Fig. 2.6] [Table 4.2]
			Sediments	26	5	14.44 - 13.14 bags 1 - 26	14.44 - 13.14	MN, 1K, 1C [Fig. 2.6] [Table 5.1]
		34.0/15.3	Mollusca	26	5	14.44 - 13.14 Sn14 - 39	14.39 - 13.14 Alternate and intermittent samples	MN, 1K, 1C [Figs 2.6, 7.1 and 7.2]
		34.0/15.6	Pollen	2 overlapping 20 cm trays		13.18 - 12.88 P4 - 3	13.18 - 12.90	1C [Figs 2.6 and 9.1]
			Insects	(from pollen samples)		13.18 - 12.88	None found	1C
32	89 A19	34.0/13.9	Soil micro-morphology	1 x 10 cm tray		14.53 - 14.43 245 bag 475	14.53 - 14.43	MN [Fig. 2.6] [Table 5.1]
33	89 A19	41.0/14.2	Phytoliths	14	2	14.63 - 14.35 Ph1 - 14	-[1]	?6D, LBA, 2B, 2F, ?MN
34	89 A19	41.0/12.9	Soil micro-morphology	2 overlapping 10 cm trays		14.45 - 14.26 246 - 247	14.45 - 14.26	2F, MN [Table 5.1]
						14.35 - 14.26 bag 476		
35	89 A20	38.5/25.0	Soil micro-morphology	1 x 20 cm tray		14.11 - 13.91 bags 471-474	14.11 - 13.91	LM/EN ditch fill [Table 5.1]
36	89 A21	69.8/43.0	Mollusca	16	5	14.14 - 13.34 Sn68 - 83	13.99 - 13.34 Alternate samples	LBA, 4B [Figs 2.16, 7.18 and 7.19]

Col. no.	Season Area	Grid	Material	No. samples	Sample interval (cms)	Total column span (m OD); Sample nos	Column span reported on; Sample nos	Parcels/ periods represented
		69.9/43.0	Phosphates	35	2	14.34 - 13.64	14.34 - 13.64	6D, LBA, 4B [Figs 2.16 and 6.3]
37	89 A21	69.4/43.0	Soil micro-morphology	6 overlapping 10 cm trays		14.24 - 13.75 253 - 258 14.00 - 13.84 bags 480-481	14.24 - 13.75	LBA, 4B [Fig. 2.16] [Table 5.1]
38	89 A21	64.8/43.0	Soil micro-morphology	2 overlapping 10 cm trays		14.37 - 14.20 259 - 260 bags 467-468	14.37 - 14.20	LBA [Fig. 2.16] [Table 5.1]
39	89 A22	37.4/40.0	Insects	2	10	13.01 - 12.81	13.01 - 12.81	1D [Fig. 2.11] [Tables 8.3 and 8.4]
		37.5/40.0	Pollen	2 overlapping 20 cm trays		13.01 - 12.70 P5 - 6	12.97 - 12.70	1D [Figs 2.11 and 9.3]
40	89 A22	41.0/40.0	Pollen	2 overlapping 20 cm trays		13.02 - 12.68 P7 - 8	13.02 - 12.68	1F-1G [Figs 2.11 and 9.4]
		41.2/40.0	Insects	5	5	13.02 - 12.77	12.97 - 12.82	1F-1G [Figs 2.1 and 8.1] [Tables 8.3 and 8.4]
			Mollusca	2 (from insects)		13.02 - 12.87	13.02 - 12.87	1F-1G [Table 7.1]
			Archaeomag-netic dating	14	5	13.75 - 13.05 1 - 14	13.75 - 13.05	1G [Fig. 2.1] [Table 4.2]
41	89 A24	32/30	Mollusca	5	5	14.29 - 14.04 Sn84 - 88	14.29 - 14.09 Sn84-87	Neolithic pit fill, 1K (Sn88) [Fig. 7.6]
42	89 A24	31.8/25.0	Soil micro-morphology	1 x 10 cm tray		14.53 - 14.43 242 14.48 - 14.43 bag 477	14.53 - 14.43	2F, MN [Table 5.1]
43	88 A25	29.0/23.4	Phytoliths	18	1	14.76 - 14.58 Ph38 - 55	-[1]	6D, LBA, (2B)
44	88 A25	29.0/19.8	Phytoliths	15	1	14.72 - 14.57 Ph23 - 37	-[1]	6D, LBA
		29.0/19.5	Phosphates	6	2	14.71 - 14.59	14.71 - 14.59	6D, LBA [Fig. 6.3]
45	88 A25	29.0/15.9	Phytoliths	19	1	14.68 - 14.49 Ph56 - 74	-[1]	6D, LBA, 2B
		29.0/15.5	Phosphates	6	2	14.69 - 14.57	14.69 - 14.57	6D, LBA [Fig. 6.3]
46	88 A26	42.0/35.7	Phytoliths	11	2*	14.49 - 14.28 Ph1 - 11	-[1]	LBA, 2B
47	88 A26	42.0/34.3	Phytoliths	11	2	14.53 - 14.31 Ph12 - 22	-[1]	LBA, 2B
48	89 A26	43.0/40.0	Phytoliths	14	2	14.56 - 14.28 Ph30 - 43	-[1]	6D, LBA, 2B
49	89 A26	45.7/40.0	Phosphates	20	2	14.63 - 14.23	14.63 - 14.23	6D, LBA, 2B [Fig. 6.4]
50	89 A26	46.0/33.7	Phytoliths	15	2	14.64 - 14.34 Ph15 - 29	-[1]	6D, LBA, 2B

Col. no.	Season Area	Grid	Material	No. samples	Sample interval (cms)	Total column span (m OD); Sample nos	Column span reported on; Sample nos	Parcels/ periods represented
51	89 A26	46.0/32.6	Phosphates	16	2	c. 14.57 - 14.25	14.57 - 14.25	6D, LBA, 2B [Fig. 6.4]
			Soil micro-morphology	2 overlapping 10 cm trays		14.44 - 14.27 262 - 261 bags 470-469	14.44 - 14.27	LBA, 2B [Table 5.1]
52	88 A27	87.0/16.2	Mollusca	20	5*	14.48 - 13.42 Sn1 - 20	14.43 - 13.55 Intermittent samples	6C, 7 [Figs 2.5, 7.21 and 7.22]
53	89 A30	28.7/14.9	Soil micro-morphology	2 overlapping 10 cm trays		14.66 - 14.53 244 - 243 bag 478	14.66 - 14.53	LBA [Table 5.1]
54	89 A31	83.5/24.0	Soil micro-morphology	1 tray pressed into surface 31.597		14.14 - 14.09 241, bag 461	14.14 - 14.09	LBA sub-soil [Table 5.1]
55	89 A31	86.0/25.0	Pollen	3 overlapping 20 cm trays		13.01 - 12.51 P9 - 11	13.01 - 12.51	4B [Figs 2.26 and 9.8]
			Insects	10	5	13.01 - 12.51	13.01 - 12.61	4B [Figs 2.26 and 8.1] [Tables 8.3 and 8.4]
			Mollusca	(from insect samples)		13.01 - 12.61	13.01 - 12.61 1 - 8	4B [Figs 2.26, 7.16 and 7.17]
56	89 A31	86.7/23.6	Soil micro-morphology	1 tray pressed into surface 31.598		13.96 - 13.91 240, bag 460	13.96 - 13.91	LBA sub-soil [Table 5.1]
57	89 A31	85.2/21.0	Phosphates	10	2	14.34 - 14.14	14.34 - 14.14	6D, LBA [Fig. 6.4]
58	89 A31	86.2/21.0	Phosphates	12	2	14.33 - 14.09	14.33 - 14.09	6D, LBA [Fig. 6.4]
59	89 A32	55.0/39.0	Mollusca	16	5	13.72 - 12.92 Sn52 - 67	13.72 - 12.97 Alternate samples	2F, (?MN), 1H [Figs 2.13, 7.7 and 7.8]
			Sediments	16	5	13.72 - 12.92 39 - 54	13.72 - 12.92	2F, (?MN), 1H [Fig. 2.13] [Table 5.1]
			Pollen	1 x 20 cm tray		13.10 - 12.90 P2	13.10 - 12.95	1H [Figs. 2.13 and 9.5]
(59)	89 A32	c. 55/39	Insects and macro. plant remains	one bulk sample		13.00 -12.95 A32.040	13.00 - 12.95	1H [Fig. 8.1] [Tables 8.1 - 8.4]
60	89 A32	59.0/39.0	Pollen	1 x 20 cm tray		13.13 - 12.93 P1	13.13 - 13.00	2D [Figs 2.13 and 9.6]
			Mollusca	12	5	13.57 - 12.97 Sn40 - 51	13.57 - 12.97 Intermittent samples	2B, 2D [Figs 2.13 and 7.12]
			Sediments	12	5	13.57 - 12.97 27 - 38	13.57 - 12.97	2B, 2D [Fig. 2.13 [Table 5.1]
		59.3/39.0	Archaeomag-netic dating	32	5	14.60 - 13.00 1 - 32	14.60 - 13.00	6D, LBA, ?4B, 2B, 2D [Fig. 2.13] [Table 4.2]
61	89 A32	59.9/40.0	Phosphates	15	2	14.62 - 14.32	14.62 - 14.32	6D, ?LBA [Figs 2.14 and 6.4]
62	86 A13	37.0/27.5	Soil micro-morphology	1 x 10 cm tray		13.88 - 13.78 234, bag 455	13.88 - 13.78	MN-LN feature fill (13.669) [Table 5.1]

13

Col. no.	Season Area	Grid	Material	No. samples	Sample interval (cms)	Total column span (m OD); Sample nos	Column span reported on; Sample nos	Parcels/ periods represented
63	86 A19	34.0/18.2	Soil micro-morphology	7 overlapping 10 cm trays		14.77 - 14.19 200 - 206	14.77 - 14.19	6D, LBA, 2F, MN, 1K [Fig. 2.6] [Table 5.1]
						14.72 - 14.20 bags 400-409		
64	87 A13	*c.* 35/29	Sediments	6	layers	14.21 - 13.16 1 - 6	14.21 - 13.16	1K, 1C [Table 5.1]
65	89 A28	89.8/14.0	Archaeomag-netic dating	16	5.3 (aver-age)	14.67 - 13.82 1 - 16	14.67 - 13.82	6C, 7 [Fig. 2.3] [Table 4.2]
66	85 A14	*c.* 74/37	Archaeomag-netic dating	4 pairs	5	*c.* 14.50	Erratic results obtained	6D
67	85 A14	79/37	Sediments	5	layers	13.50 - 13.00	-	4B [Fig. 2.19]
68	86 A13	37.0/31.3	Soil micro-morphology	3 overlapping 9 cm trays		13.82 - 13.58 235 - 237 bags 456-457	13.82 - 13.58 (not particle size analysis)	LM-EN feature fill (13.774)

* Series includes occasional samples of different depth
[1] Selected samples tested; phytoliths found not to be preserved
- All altitudes and sample numbers listed in order from top to bottom
- 'Soil micromorphology' normally involves a set of monolith trays for the micromorphology and a parallel series of 'bagged' samples, separately numbered, for macroscopic inspection and particle size analysis; the respective sample numbers are given one above the other in the same entry
- MN and LBA refer respectively to the presence of in situ Middle Neolithic and Late Bronze Age occupation deposits within soil/sediment profiles

CHAPTER 2

Stratigraphic sequences through the alluvial deposits

The stratigraphic sequences are described below (Table 2.2) in geographical order working first through the main research area from south-west to north-east, then crossing to the south-eastern edge of the site. They are additional to the sequences recorded in the 1978 excavation campaign (Needham 1991 and summarised in Table 2.1) and include one 1978 profile (Area 8, section SA8) not previously published. Otherwise the profiles documented here were recorded in 1980 (Areas 10, 11 and 12) or during the research campaign. The locations of the sections revealing these profiles are shown in Figs 1.4 and 1.5, while the exact grid references of their end-points are given on the respective drawings (Figs 2.3–2.31). Sections deriving from the rescue/salvage campaign continue the 'SA' sequence (of Needham 1991, 18, table 1), whereas all those from the research campaign are united in a single sequence prefixed 'RH'; these are not the original section drawing numbers.

Layers within each sequence are listed in stratigraphic order from earliest to latest. Stratigraphic sequence is unequivocal, except where comments to the contrary are made in the table. The key parts of sequences are shown as matrices in Figs 2.32–2.39, which also illustrate the interpreted correlations between the locally established sequences. Any necessary clarification of correlations along with the discussion of chronological evidence is left to Chapter 10, where topographic, environmental and other implications are also drawn together. Soil and sediment descriptions were made by archaeologists (primarily the author) in the field under varied conditions of moisture and lighting. In one or two cases the continuation of deposits beneath the (reduced) water table follow the descriptions made by Rob Scaife on monoliths in the laboratory. More precise descriptions of character and colour may be found in Susan Limbrey's contribution (Chapter 5) for those sediments penetrated by her columns.

The stratigraphic sequences have been synthesised into a single framework of alluvial parcels covering the Runnymede Bridge site. Definition of the concept of alluvial parcels was given in Needham 1992 and the main points are repeated here for convenience. They represent blocks of sediment which may contain both horizontal and inclined layer members. They are intended as interpretative units, interpretation being based on the available knowledge of: a) dated horizons; b) sedimentary characteristics; c) changes in hydrological/depositional regimes, particularly those giving rise to major stratigraphical interfaces. Defined alluvial parcels have a temporal span which is largely exclusive of those for proximate parcels. When seen in juxtaposition, the interface will be either an oblique erosion line (palaeobank) against which one set of deposits is broadly later than that which it abuts, or, a more horizontal surface believed to represent a significant standstill with or without truncation.

Definition is non-genetic; it is not intended that a given parcel has a limited range of sediment types (although this may be the case) and the immediate environment is likely to change as the parcel builds up. Variations within a parcel may also include steady gradational changes along deposits, dependent on topographic undulations and sedimentation conditions. Obviously, the recognition of lateral gradation will depend on the length of exposures available; it may pass unrecognised when two or more sections are well separated. Where coeval blocks of sediment occupy discrete palaeochannels, they are treated as discrete parcels, even if sedimentary characteristics are very similar. Otherwise the main parcel numbers broadly reflect their position in the overall stratigraphic sequence (parcel 7 is, however, anomalous since it pre-dates parcel 6). Parcel 1 underlies Neolithic occupation on the dryish banks; parcel 2 spans Neolithic deposits and continues until a new pre-LBA land-form was created; parcel 3 is also pre-LBA, but fills a distinct channel; parcel 4 is another channel fill starting equally early, but active throughout and after the occupation phase; parcels 5 and 7 represent fills of channels cutting into different sides of the LBA island during and after occupation; parcel 6 is overbank alluvium sealing both those fills and the LBA-occupied surface with all its deposits; parcel 8 comprises channel fills in the south dated to the Post-Medieval period; parcel 9 is of sediments accumulating after an artificial Post-Medieval cut in the north, and parcel 10, the upper, Post-Medieval channel fill in the west. Sub-parcels are used to define important distinctions within this broad structure. There have been a number of changes to the parcel attributions previously published and the revised occurrence is summarised in Table 2.1. The full picture is presented in Fig. 2.1, while Fig. 2.2 shows most of the parcels in an idealised section through the main zone of research excavations from the south-west (Area 27) to north-east (riverside zone).

The main cultural horizons of the earlier Neolithic and the Late Bronze Age are only covered here in summary fashion, using representative excavated contexts. The full 'exploded' view of excavated contexts for these horizons is reserved for the relevant volume (e.g., Needham 1991; Needham and Spence 1996).

Table 2.1 *Summary of the alluvial parcels present in previously published Areas taking account of the revisions outlined in Chapter 10.*

Area	Parcel	Layers
1 and 2	3	1-2
	6	?9, 10-11
3		insufficient information
4	2A*	120b-123
	2C*	lower: 117-120a [118 ≡ 2F] upper: 115-116
	2B	114
	4A	104-113
	4C	98-103
5	1J	140-143, 145
	4A	126-139
6	2A	41c-42
	2C*	lower: 40-41b upper: 39
	2B	38
	4A*	8b, 10-30
	4C*	4-8a, 9
7 and 8	1A	88-92, 93 (NE end), 94-97
7	2A	73-77
	2C	lower: 66 (part), 68-72 [72 ≡ 2F] upper: 63, 65, 66 (part), 67
	2B	62, ?64
	4A	54b-66
	4C	54a
8	1L	85-87, 93 (SW end)
	2F	84
	2B	83
	3	78-82 and 154-156 (this volume)
	5	46-52 and 146-153 (this volume)
	6	44-45

* Parcels yielding results from analysis of environmental remains (columns A4, SS1 and WF1)

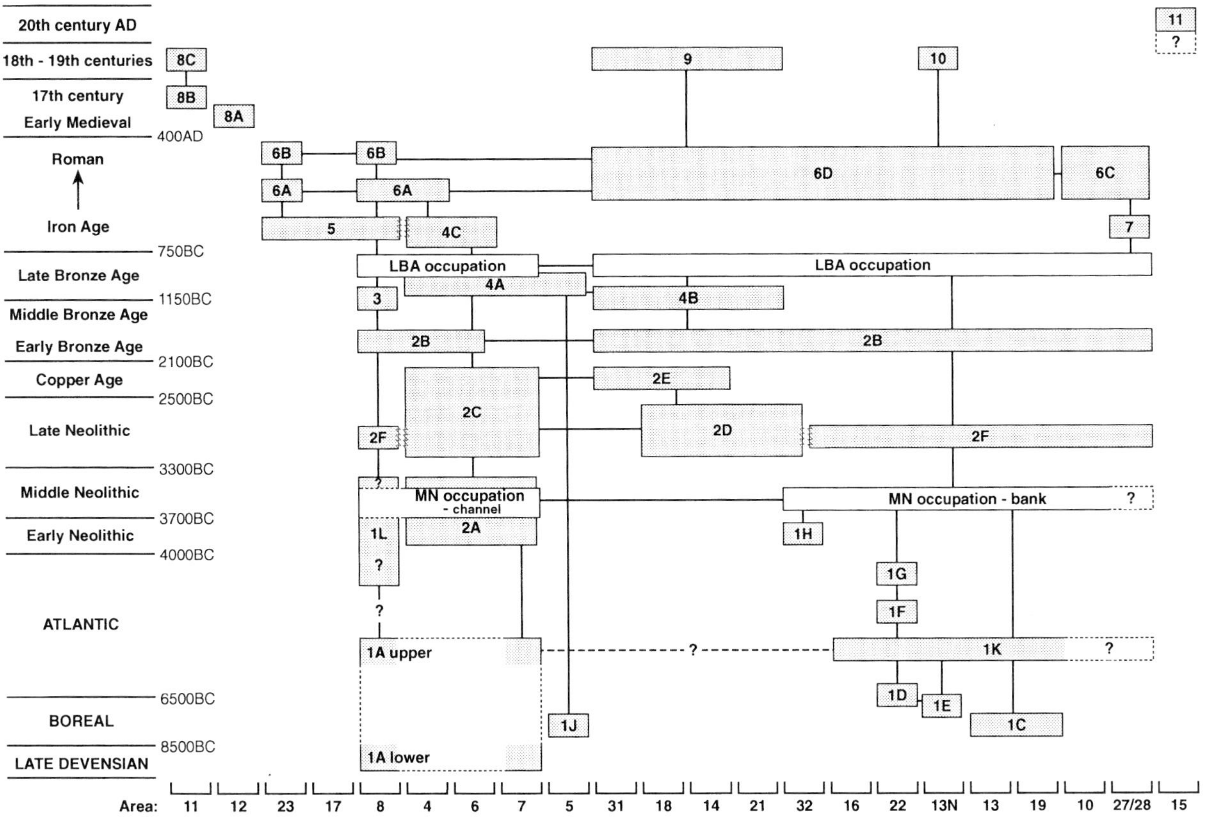

2.1 *Schematic representation of ages and distribution of alluvial parcels.*

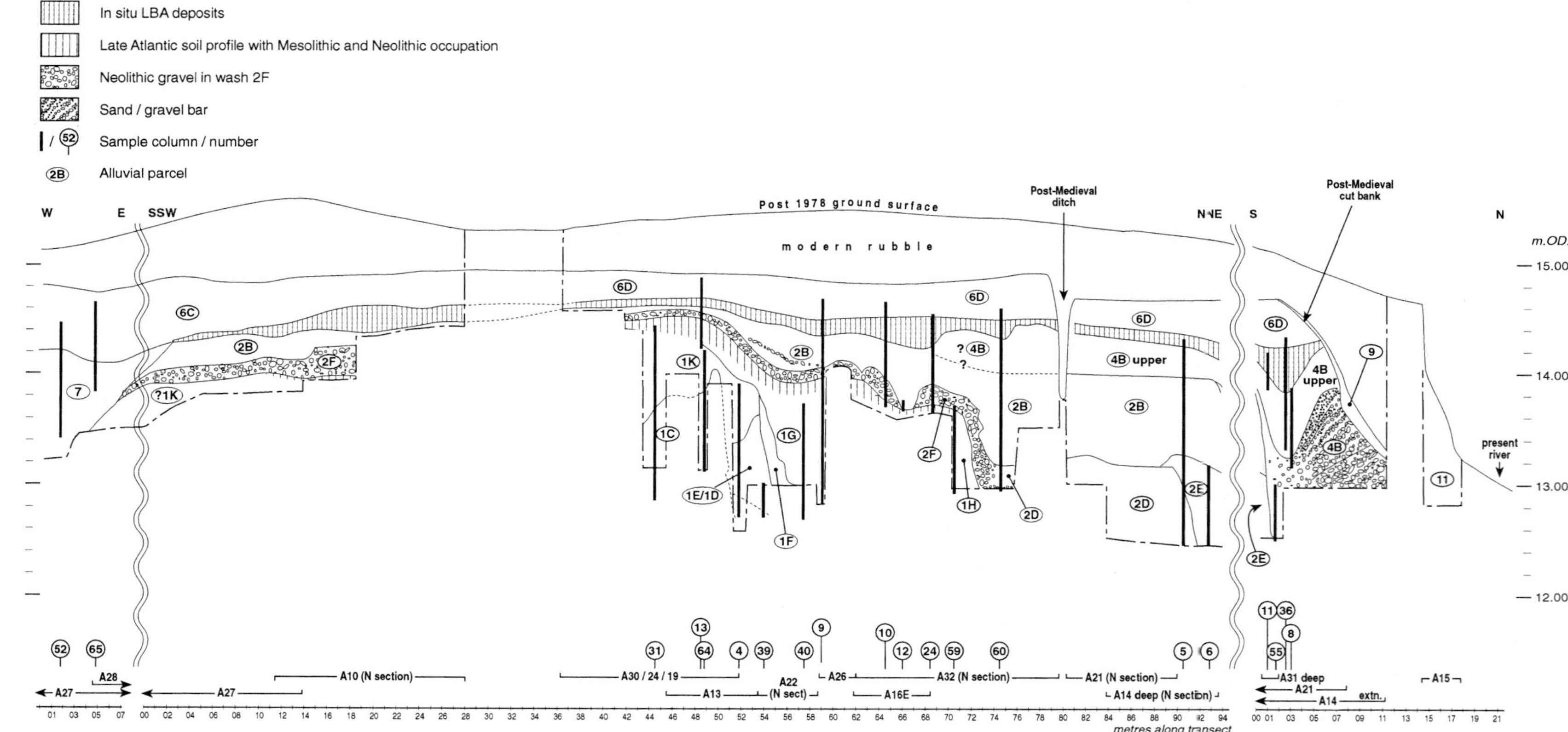

2.2 *Composite section through the alluvial deposits in the main zone of research excavations. The axes of the transects are shown on Fig. 1.3. Important sample columns and layer boundaries have been projected onto those axes orthogonally. Where separated trenches would give slightly divergent profiles, those yielding the illustrated columns are favoured.*

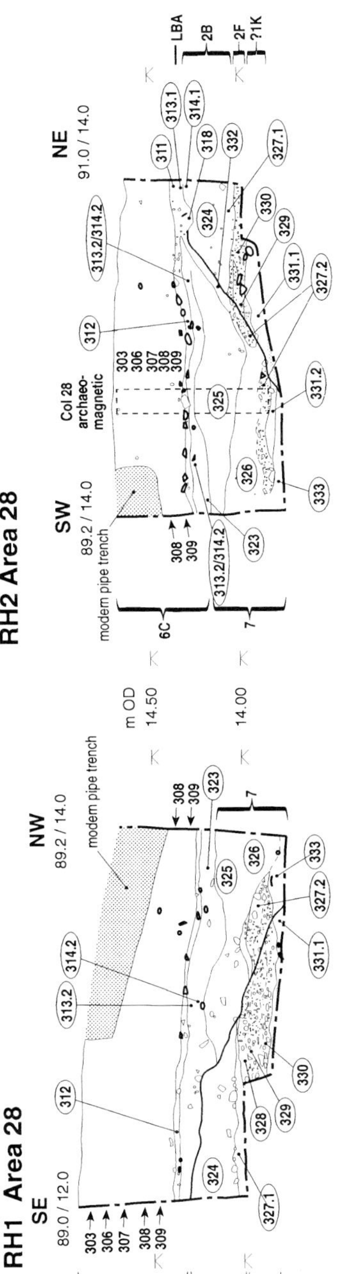

2.3   *Sections RH1 and RH2, Area 28 south-west wall and north-west wall (part) respectively.*

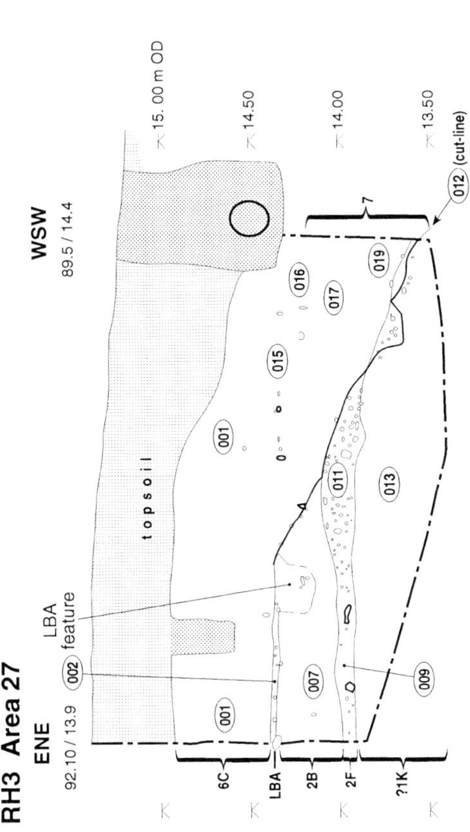

2.4   *Section RH3, Area 27 south-east wall (part).*

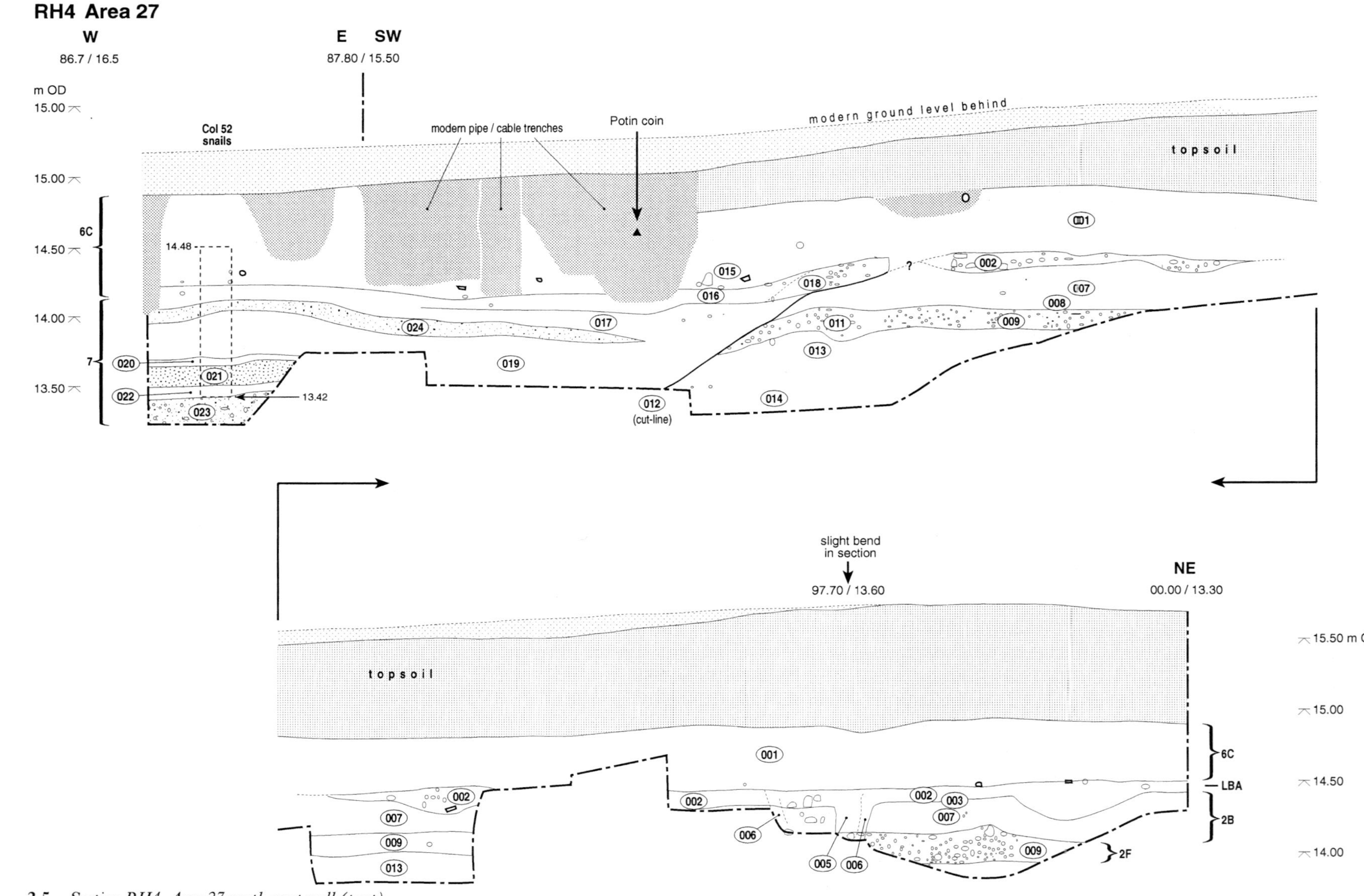

2.5   *Section RH4, Area 27 north-west wall (part).*

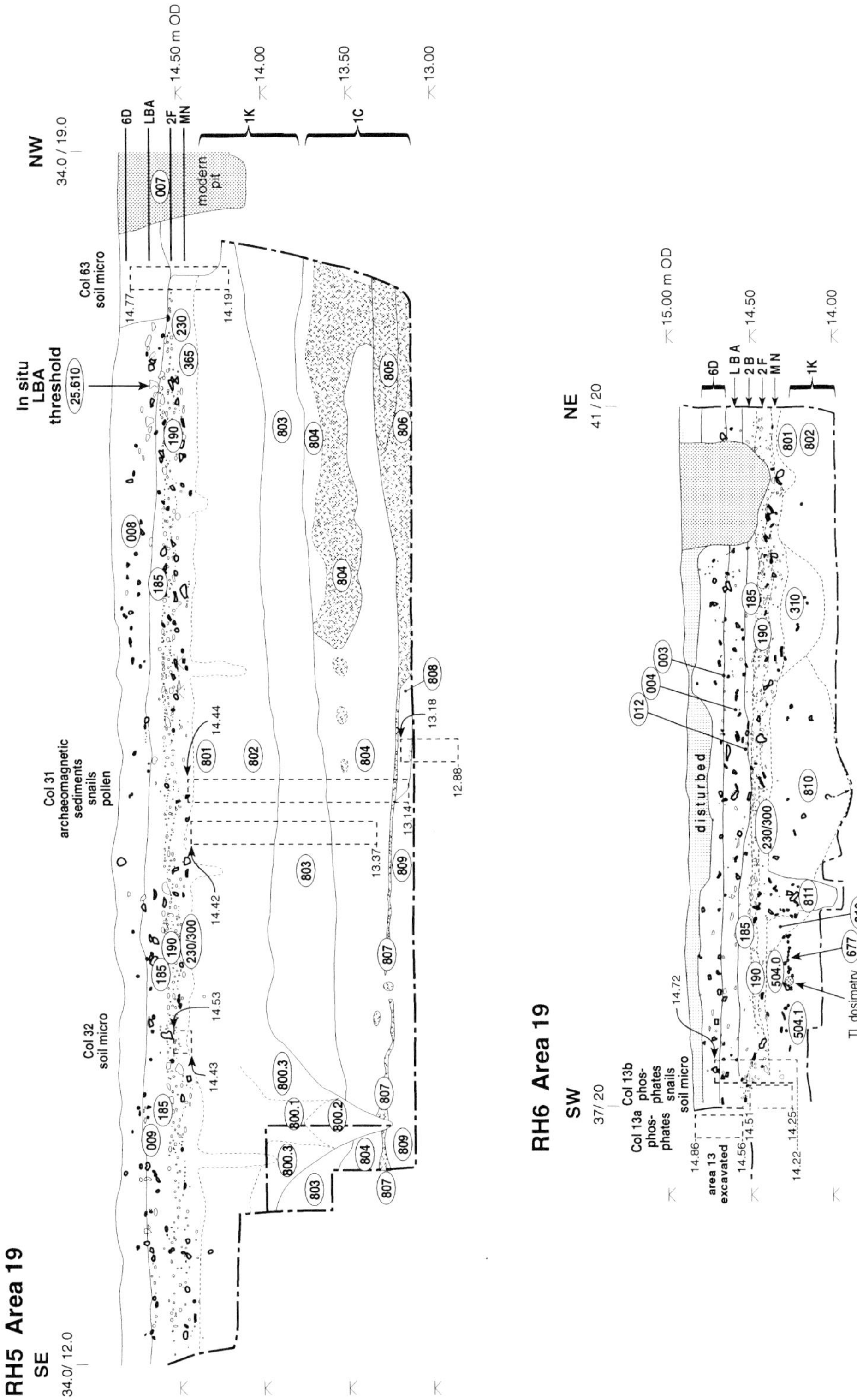

2.6   *Sections RH5 and RH6, Area 19 south-west wall (part) and north-west wall (part) respectively.*

21

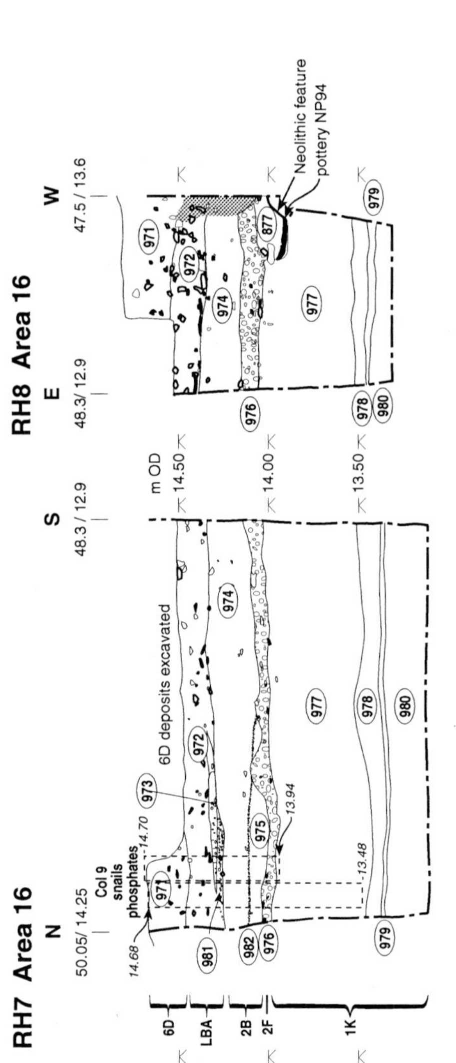

2.7   *Sections RH7 and RH8, Area 16 Centre, walls of modern pit 16.002.*

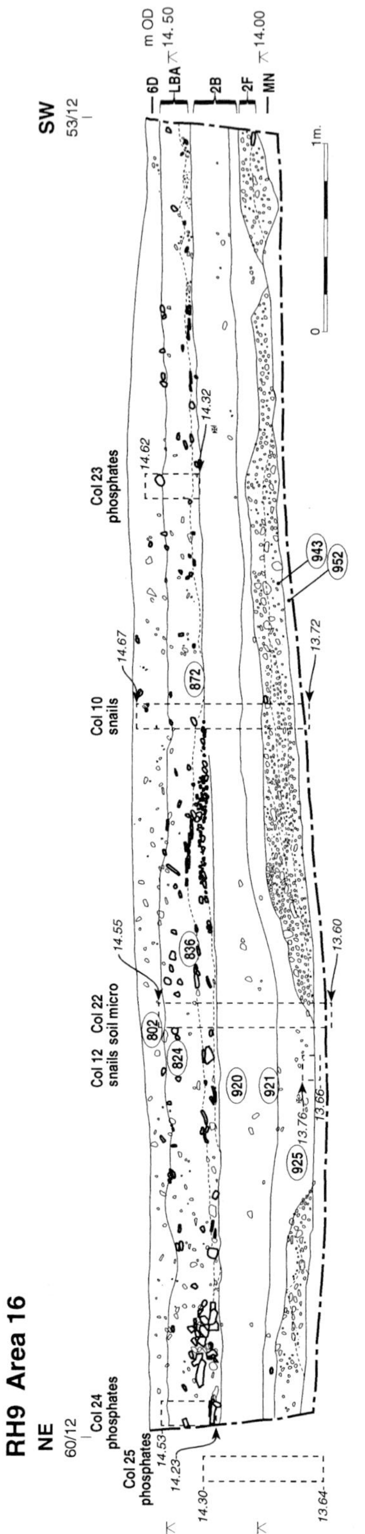

2.8   *Section RH9, Area 16 East, south-east wall.*

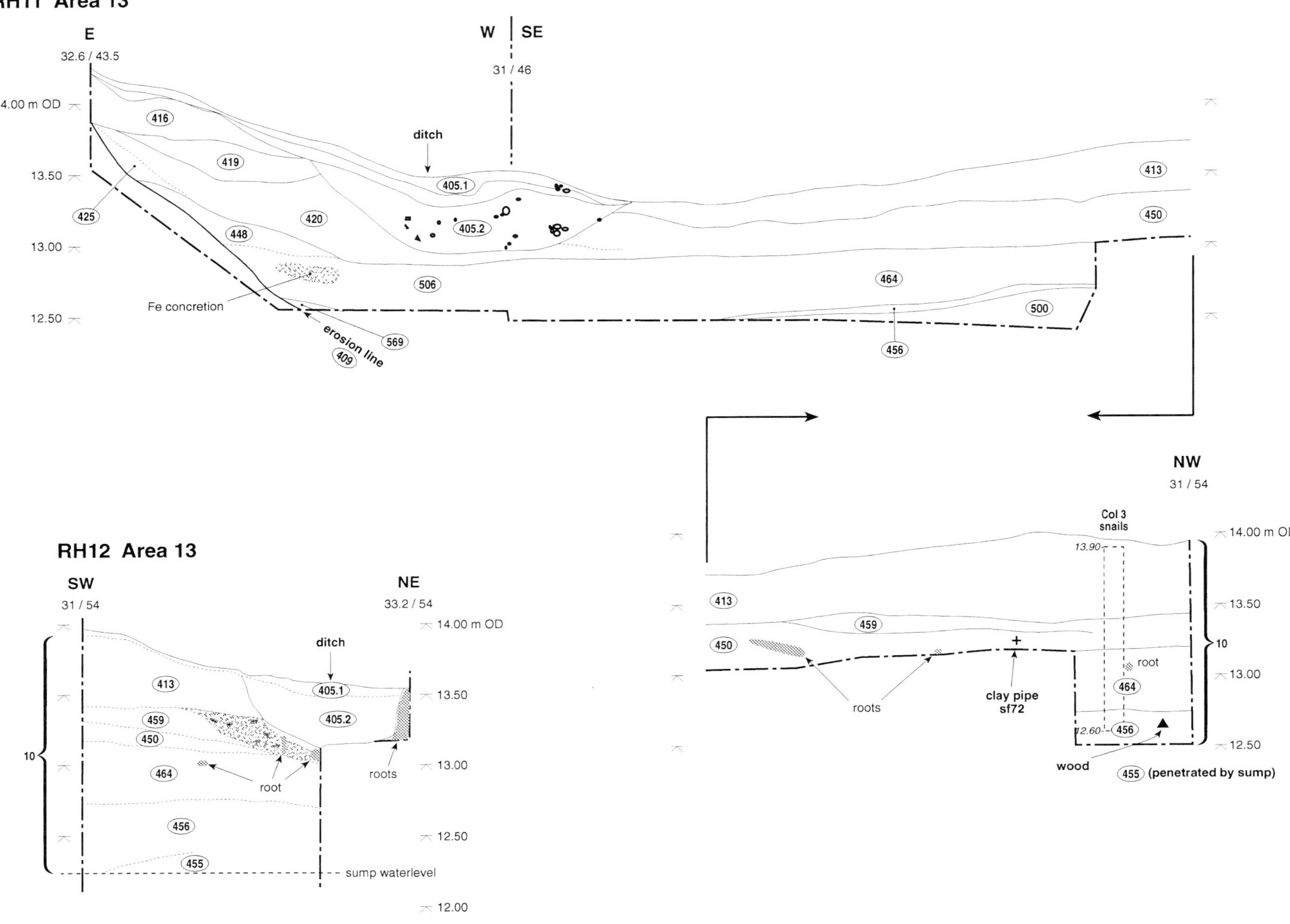

2.9 *Sections RH11 and RH12, Area 13 north extension, south-west and north-west walls respectively.*

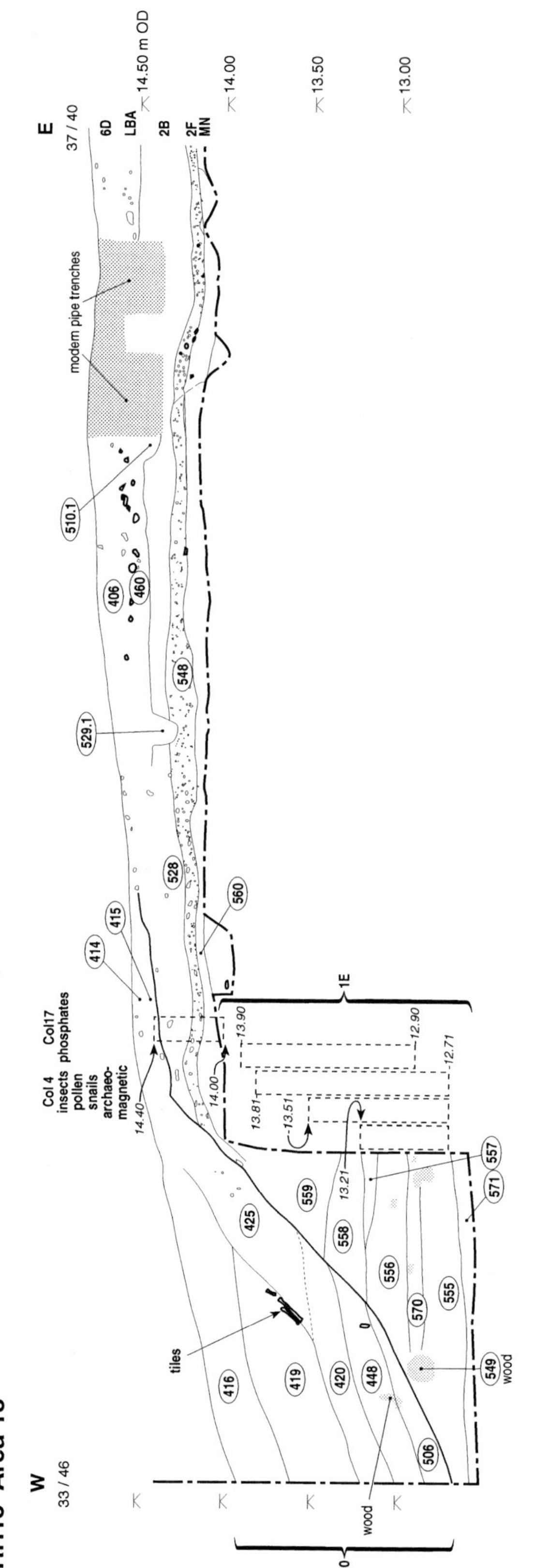

2.10   *Section RH13, Area 13 north extension, north wall.*

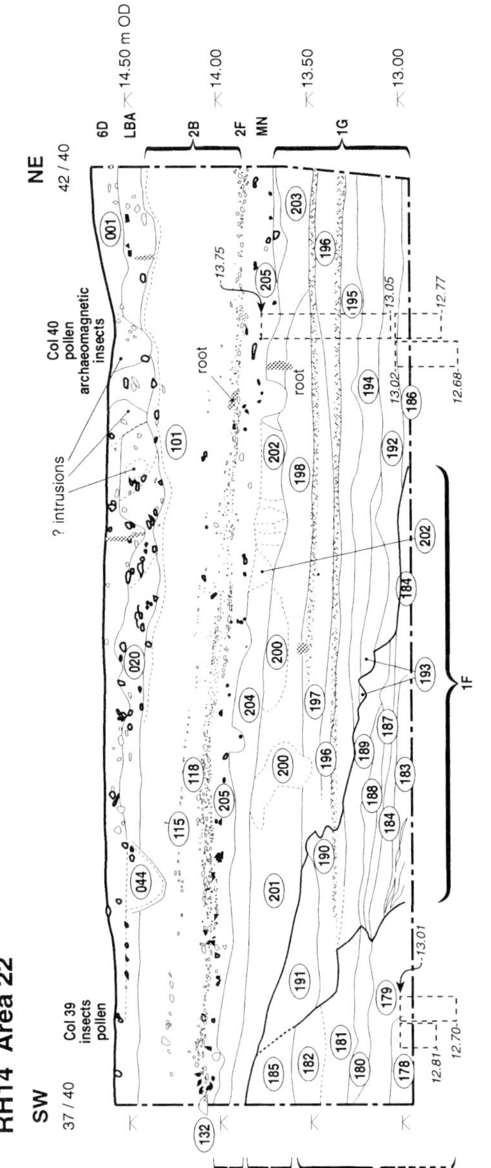

2.11 *Section RH14, Area 22 north-west wall.*

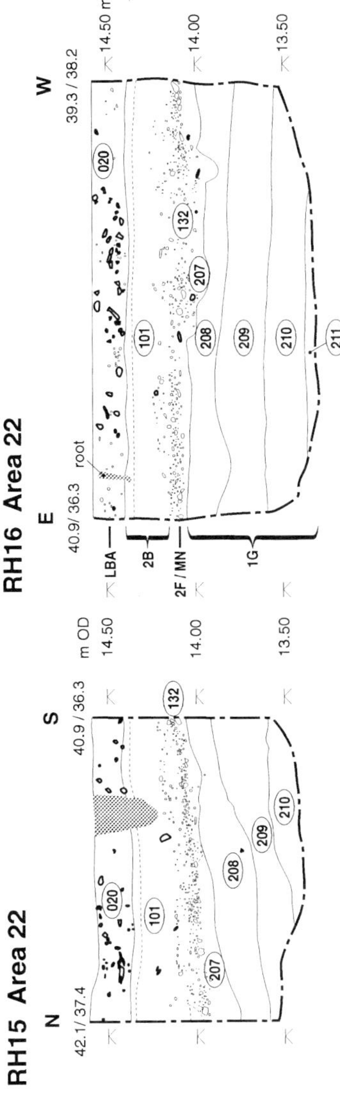

2.12 *Sections RH15 and RH16, Area 22 east and south walls respectively of modern pit 22.006.*

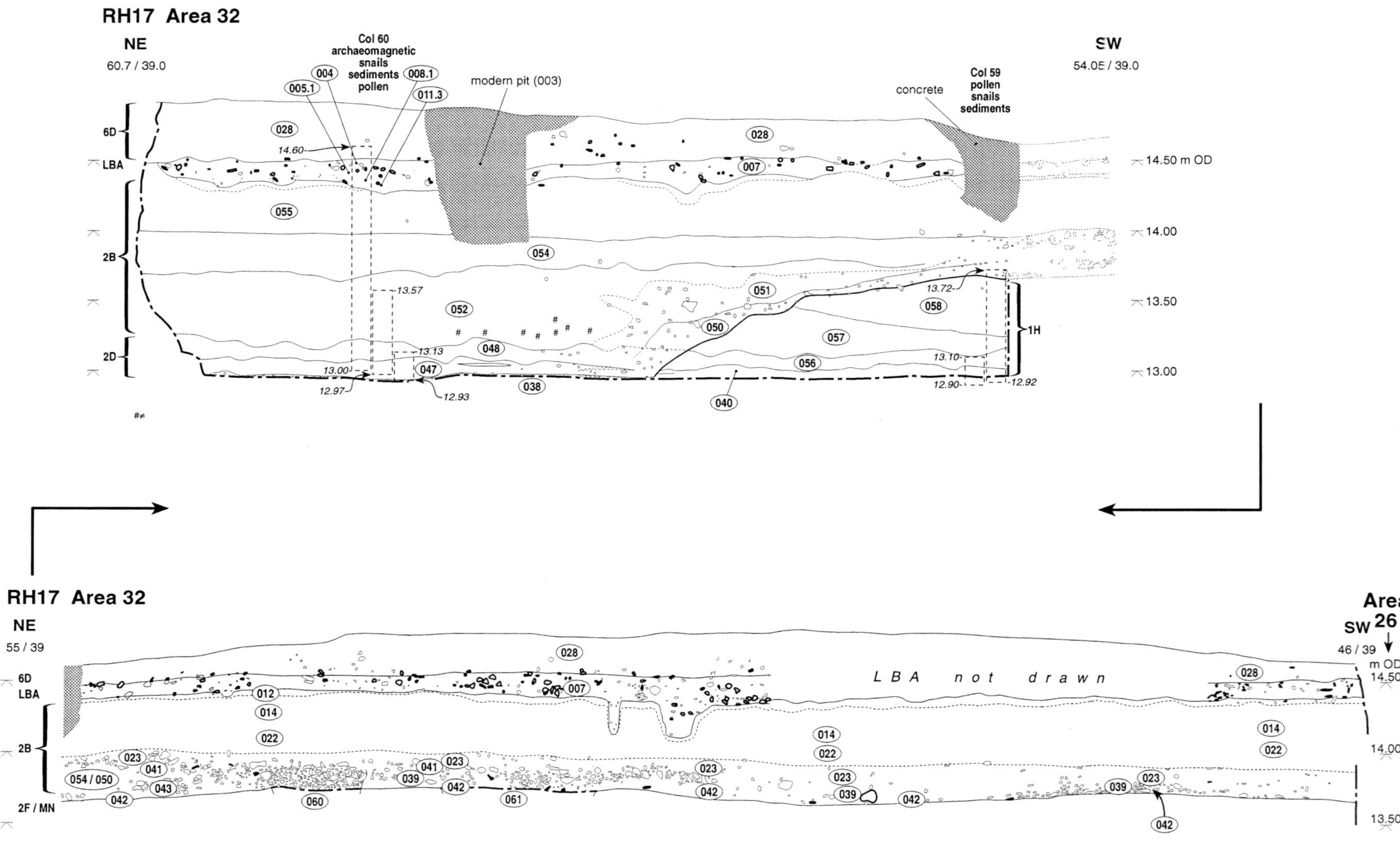

2.13   *Section RH17, Area 32 south-east wall (part).*

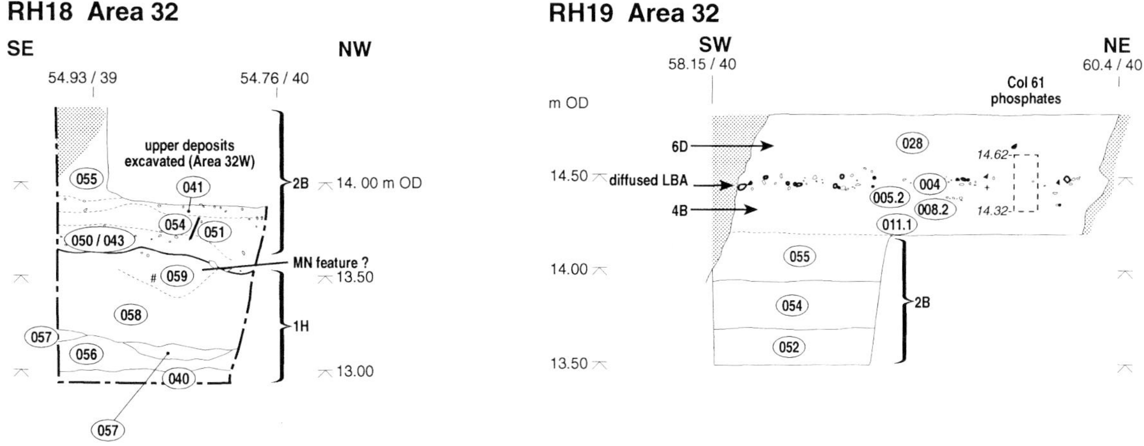

2.14   *Sections RH18 and RH19, Area 32 central transverse section and north-west wall (part) respectively.*

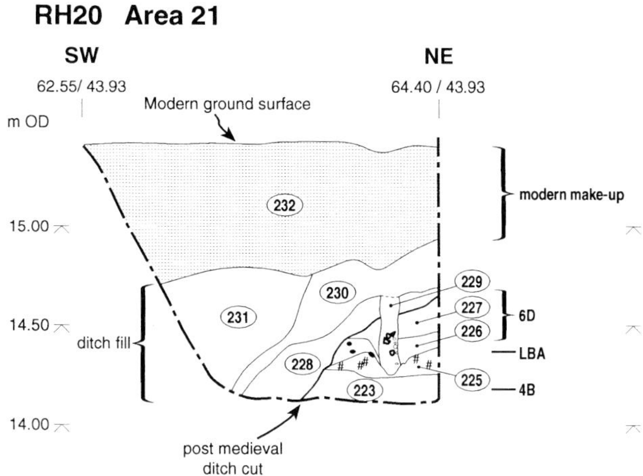

2.15   *Section RH20, Area 21 west corner, modern pit wall.*

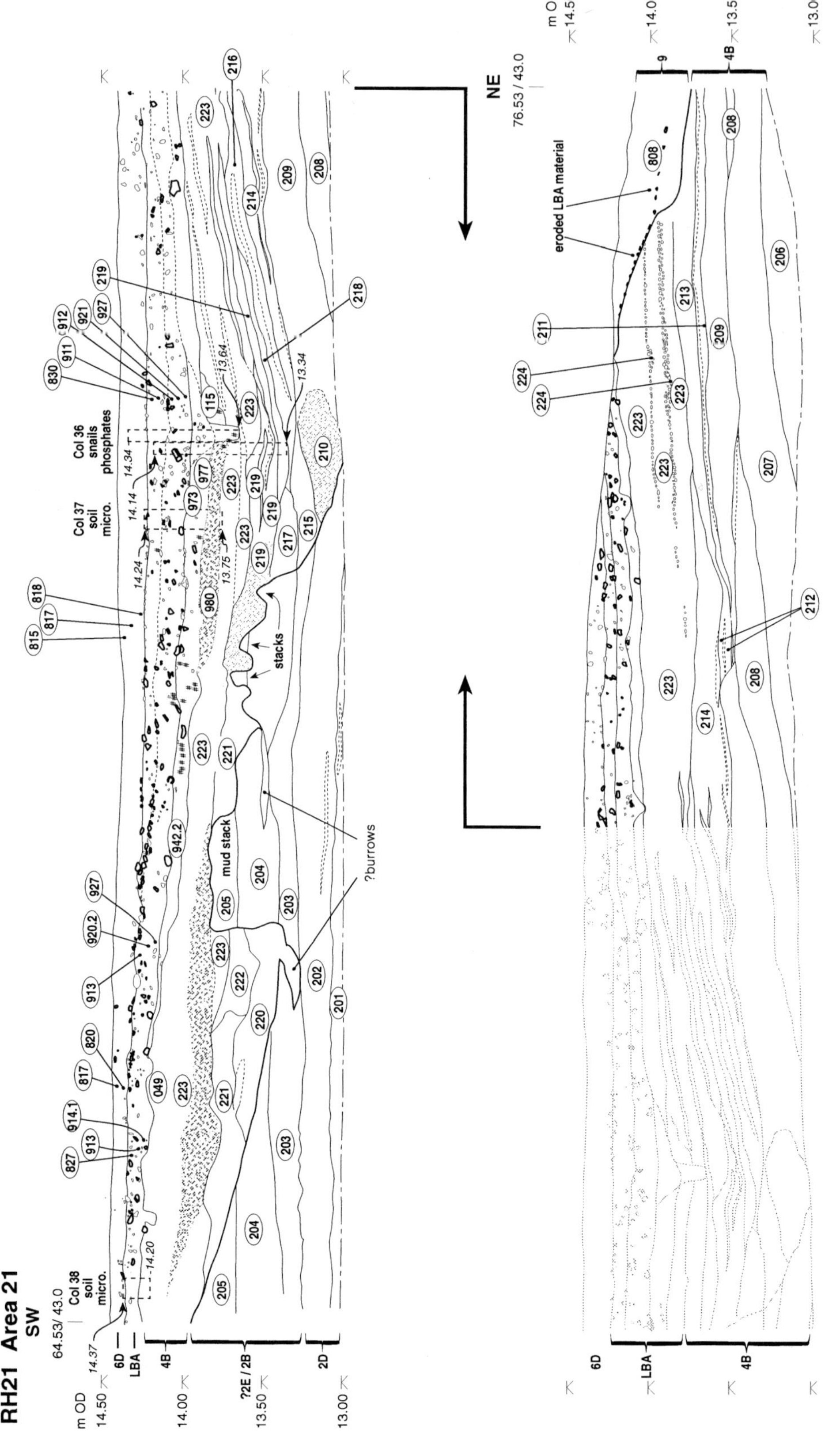

2.16 *Section RH21, Area 21 north-west wall.*

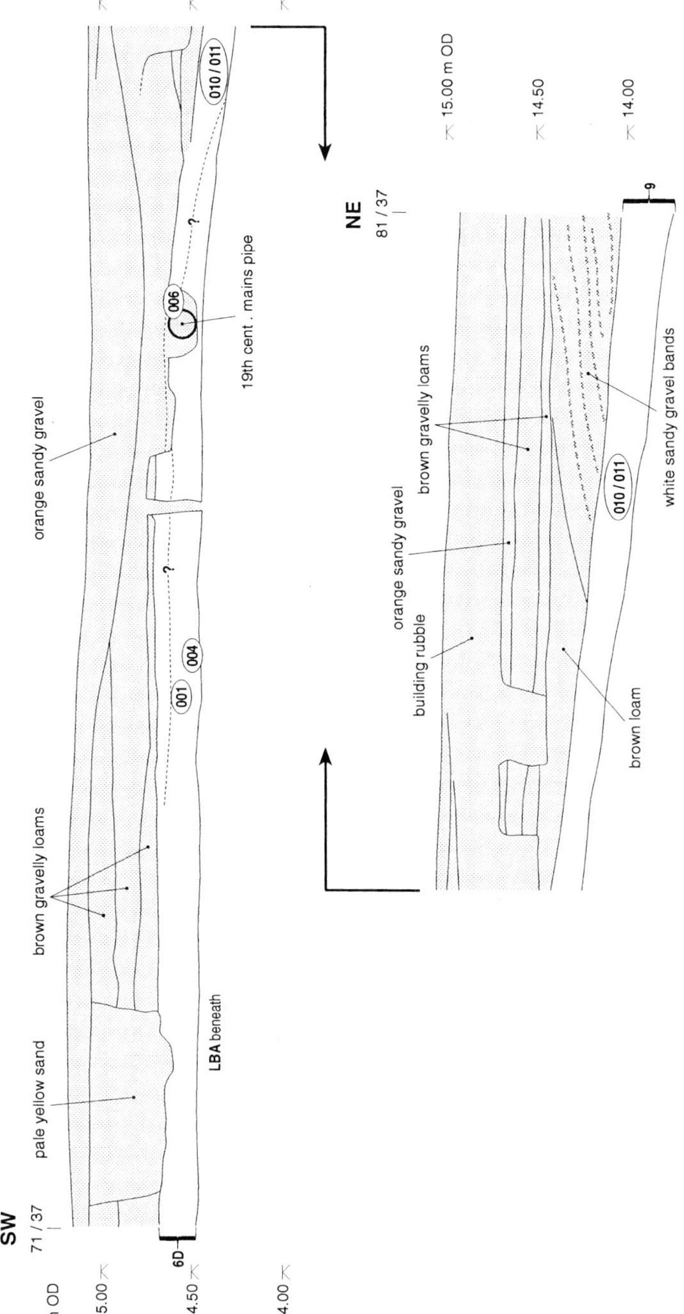

2.17   *Section RH22, Area 14 north-west wall (above walkway).*

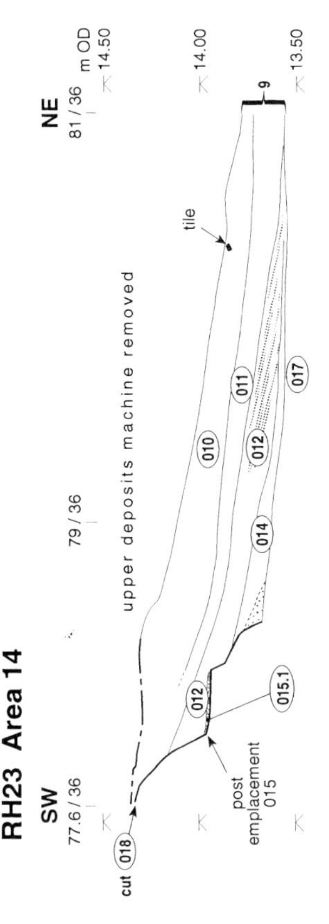

2.18   *Section RH23, Area 14 north-west wall (main trench; part).*

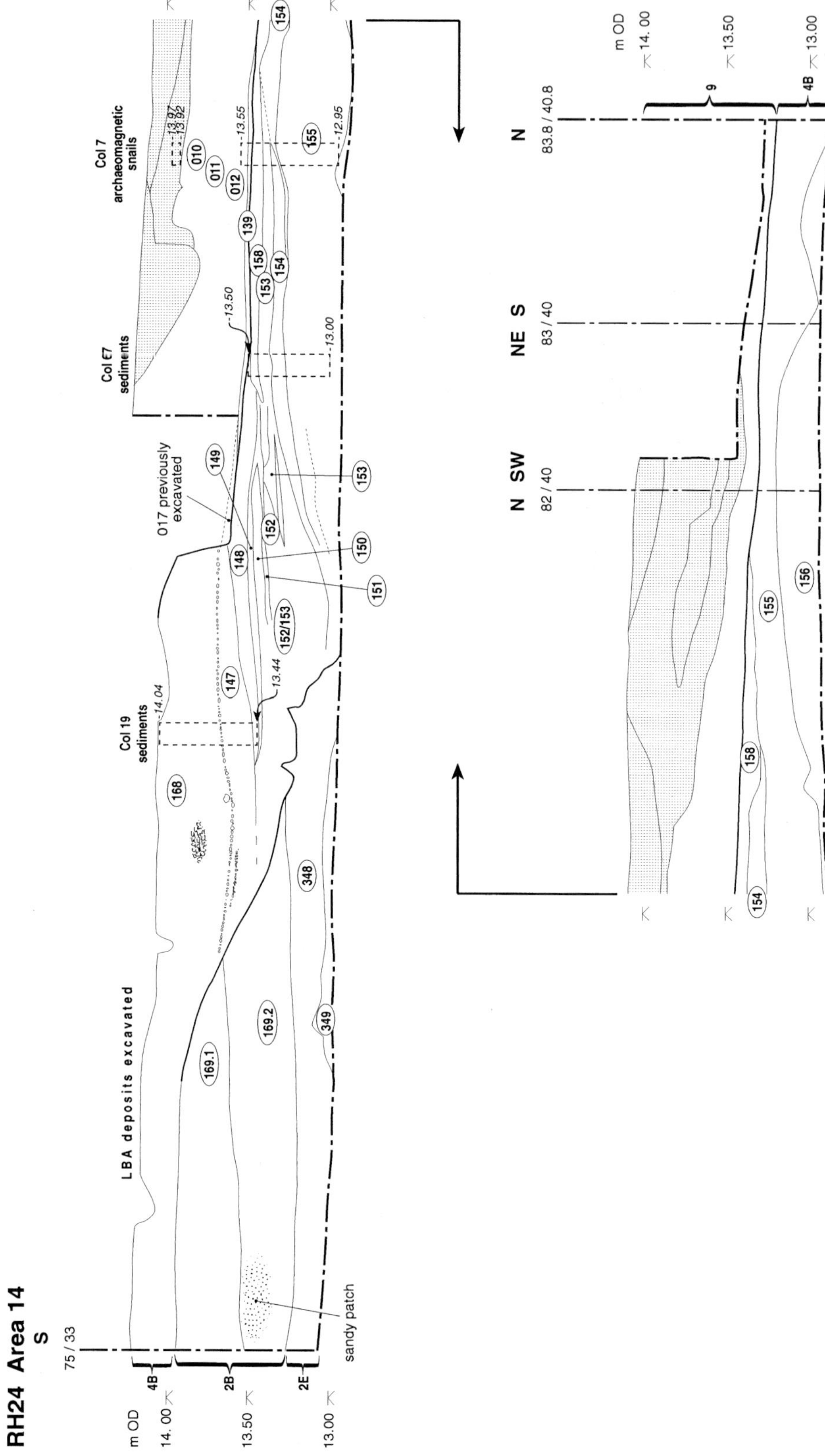

2.19 *Section RH24, Area 14 north extension, west wall.*

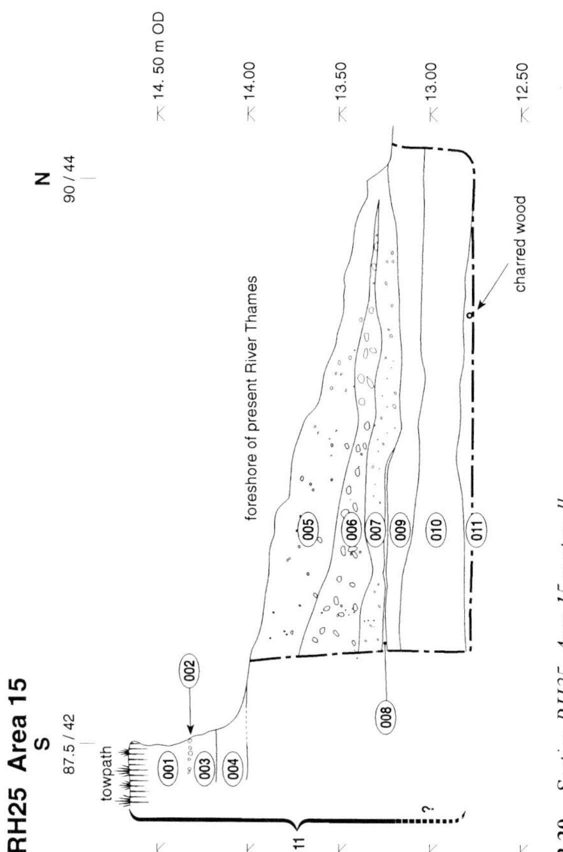

2.20   *Section RH25, Area 15 west wall.*

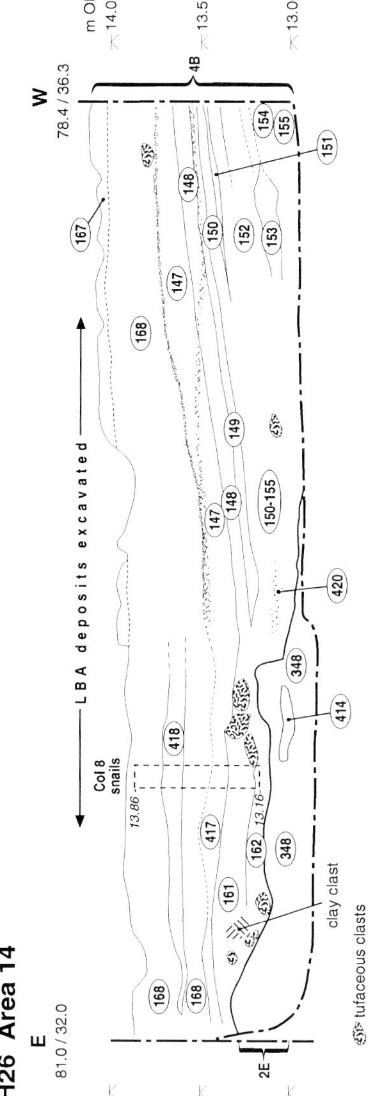

2.21   *Section RH26, Area 14 north corner, face of Post–Medieval cut (not vertical).*

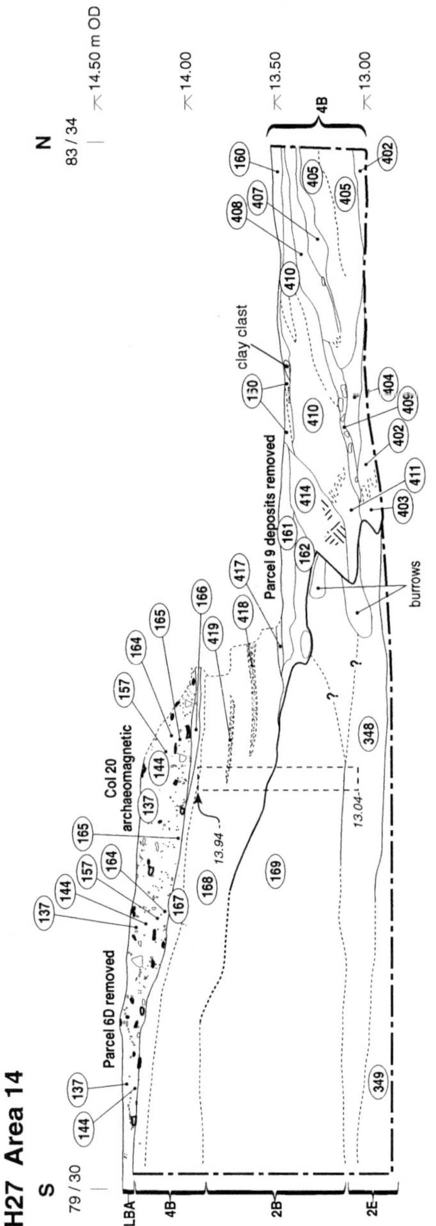

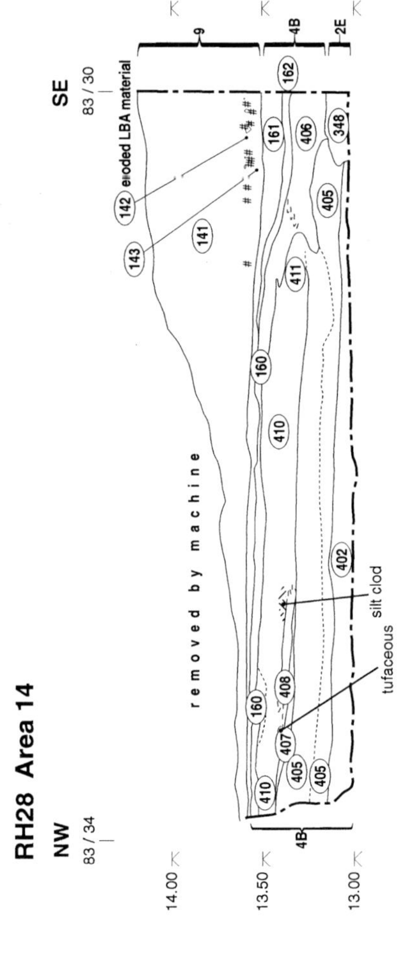

2.22 Sections RH27 and RH28, Area 14 north-east extension, west wall and north-east wall respectively.

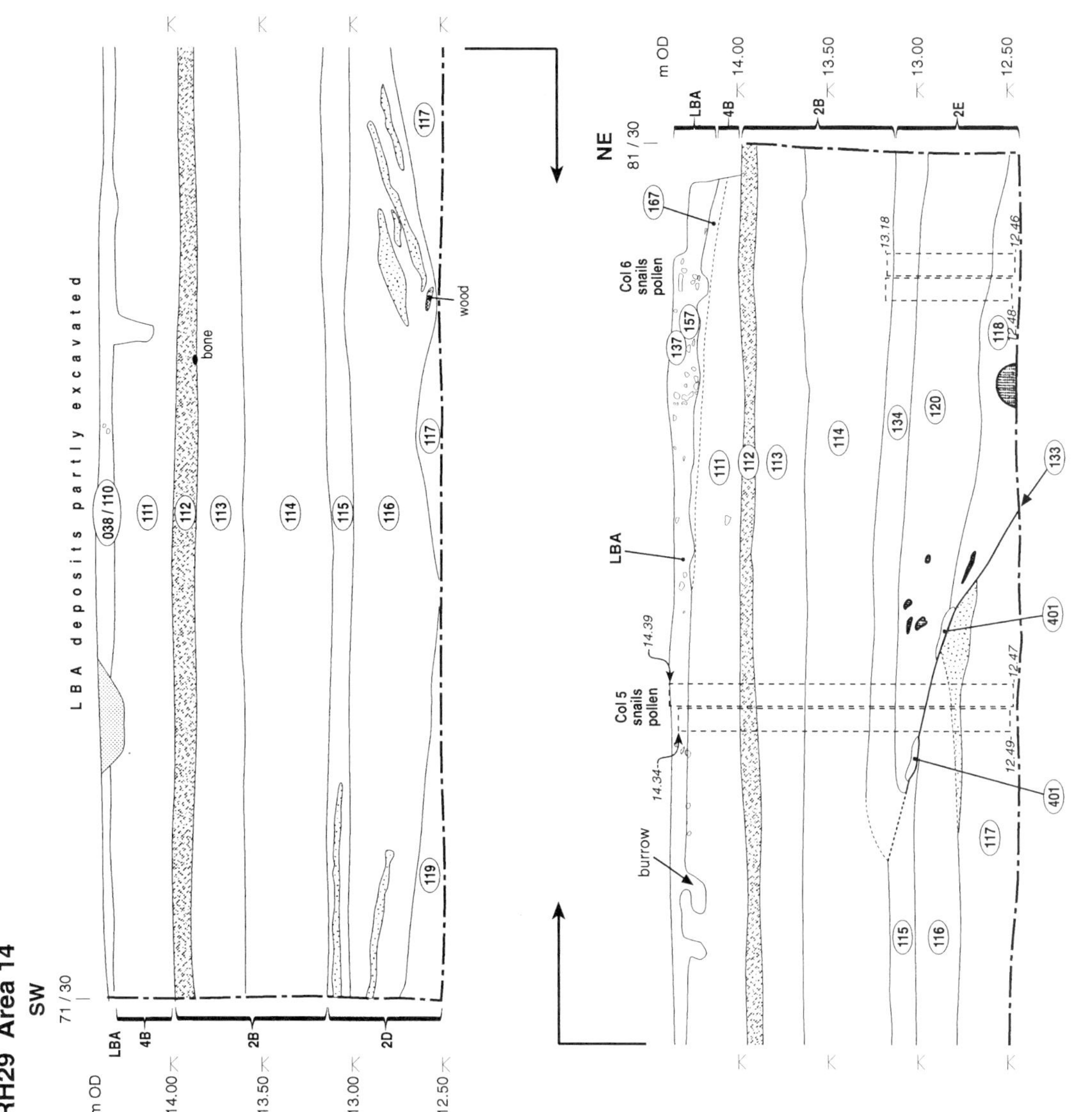

2.23   *Section RH29, Area 14 south-east deep cutting, north-west wall.*

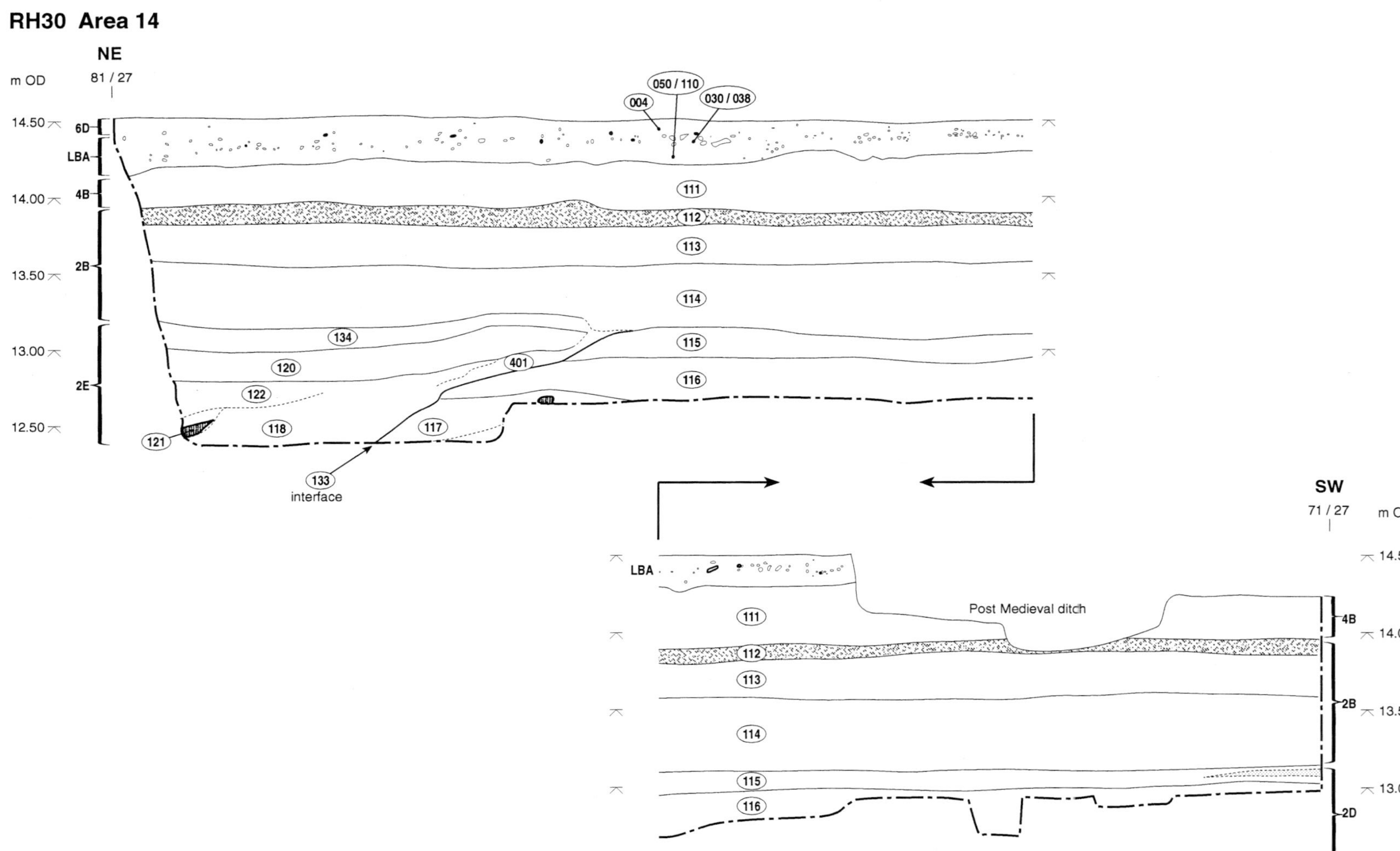

2.24 *Section RH30, Area 14 south-east deep cutting, south-east wall.*

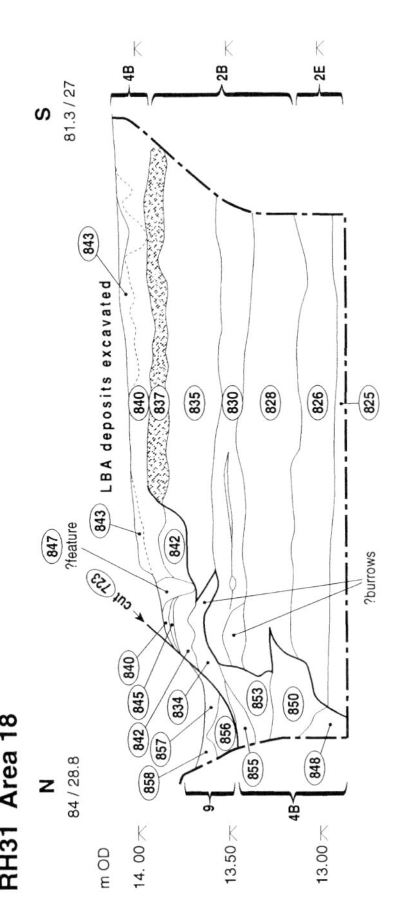

2.25 *Section RH31, Area 18 deep cutting, east wall.*

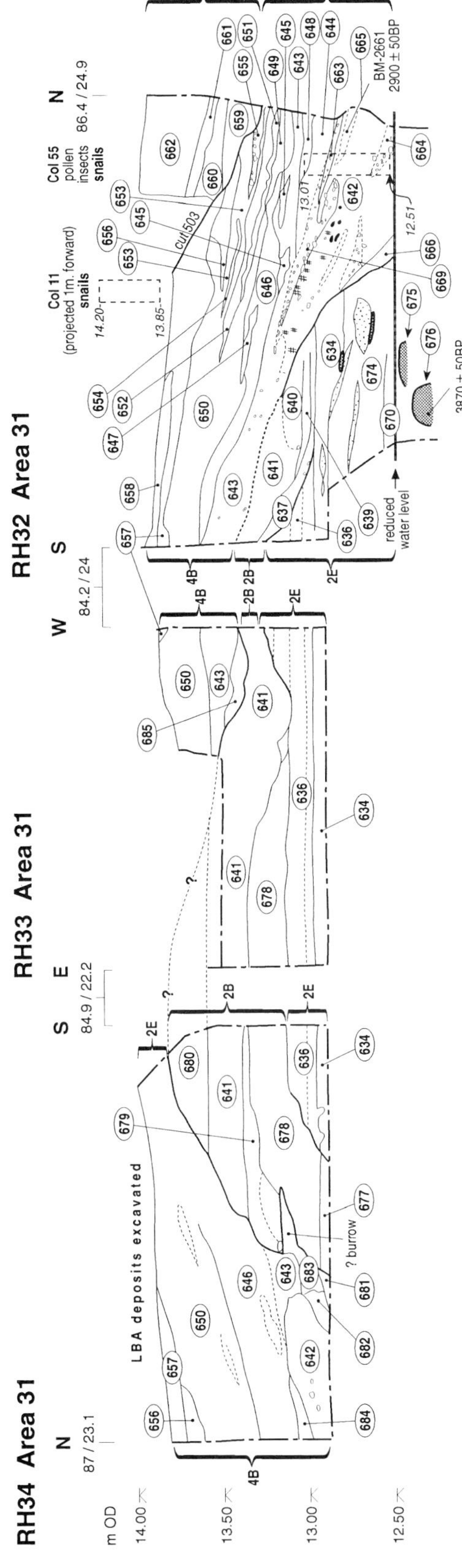

2.26 *Sections RH32, 33 and 34, Area 31 deep cutting, west, south and east walls respectively [RH33 and RH34 based on sketches with measured datums].*

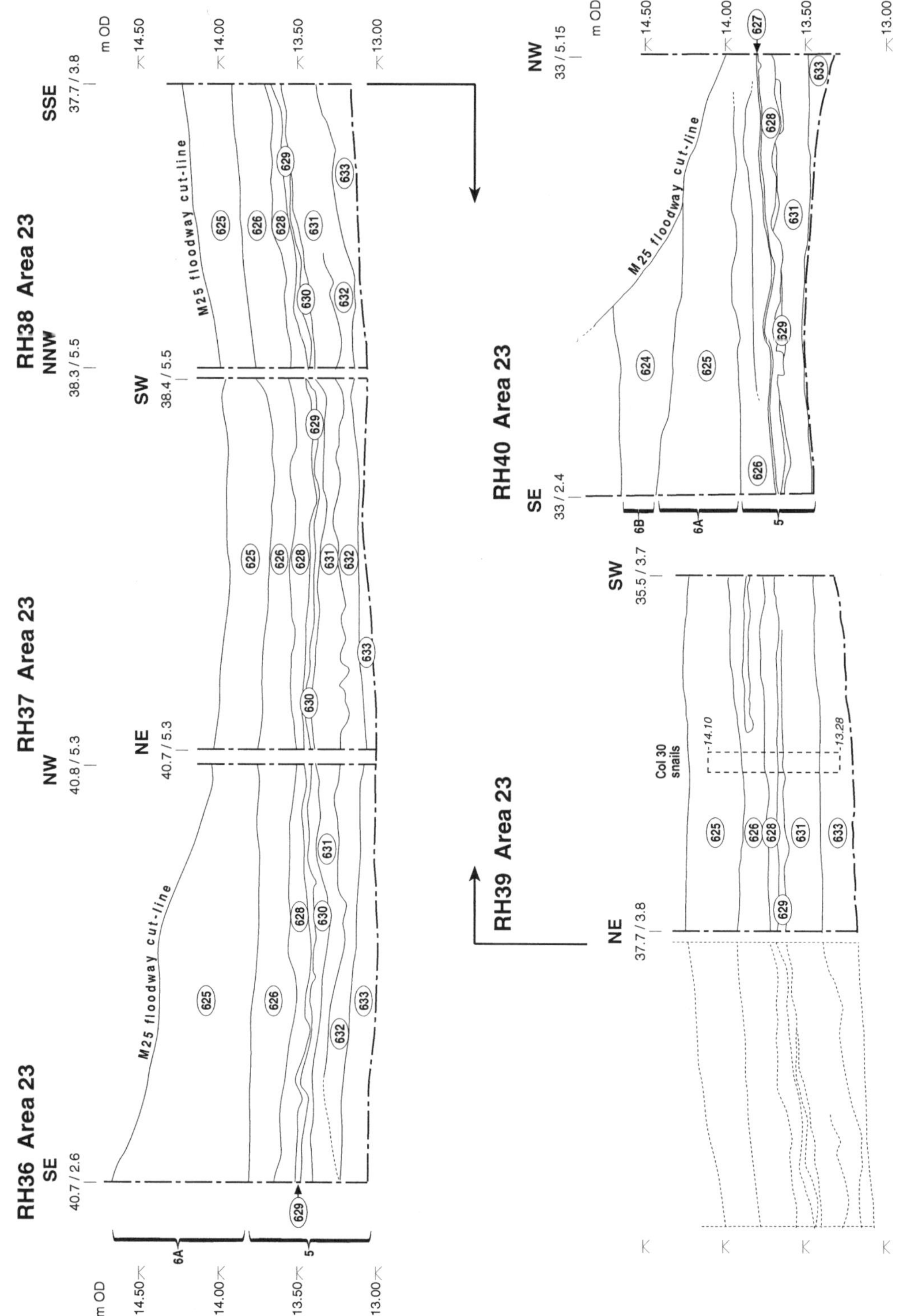

2.27 *Sections RH36–RH40, Area 23 north-east, segments of exposed profile.*

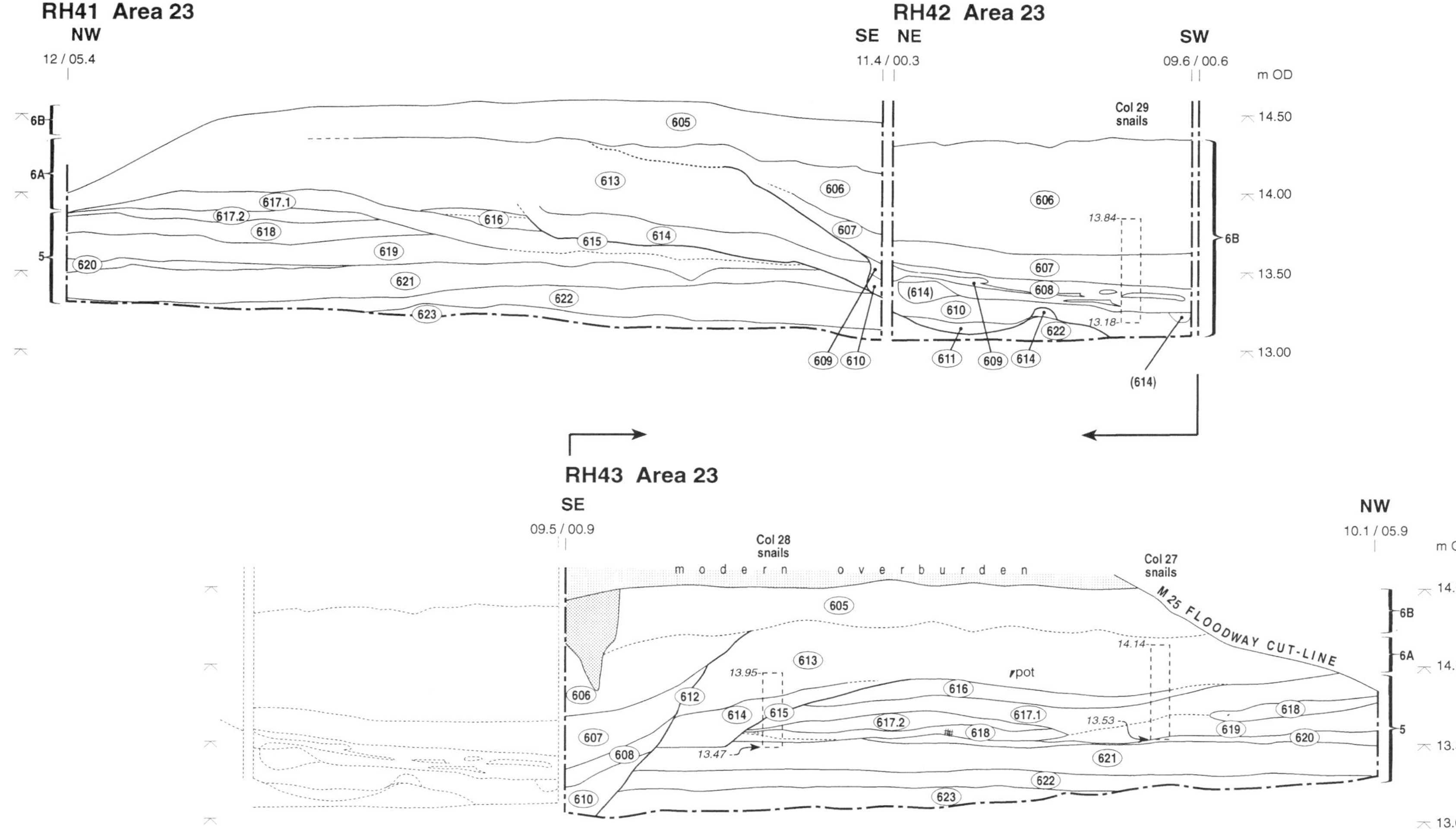

2.28 *Sections RH41–RH43, Area 23 south-west cutting.*

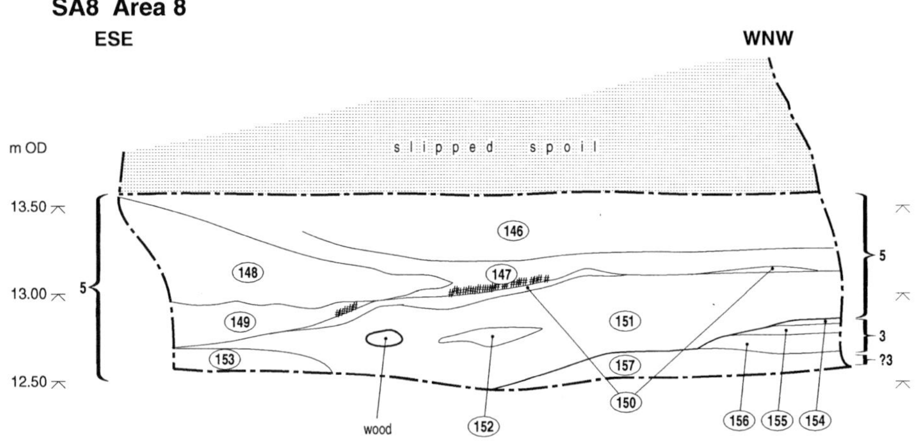

2.29 *Section SA8, Area 8 East.*

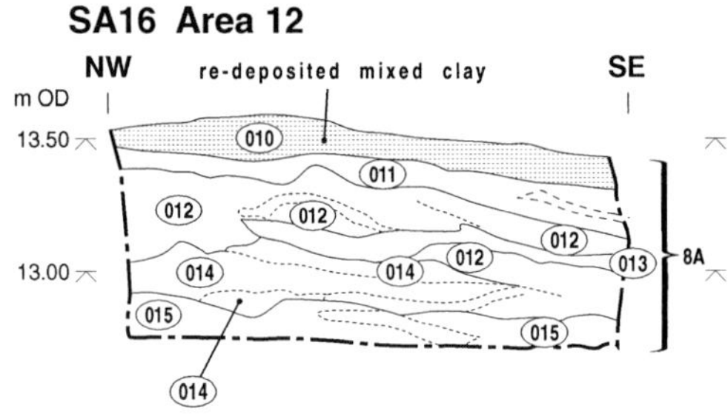

2.30 *Section SA16, Area 12 north-east wall.*

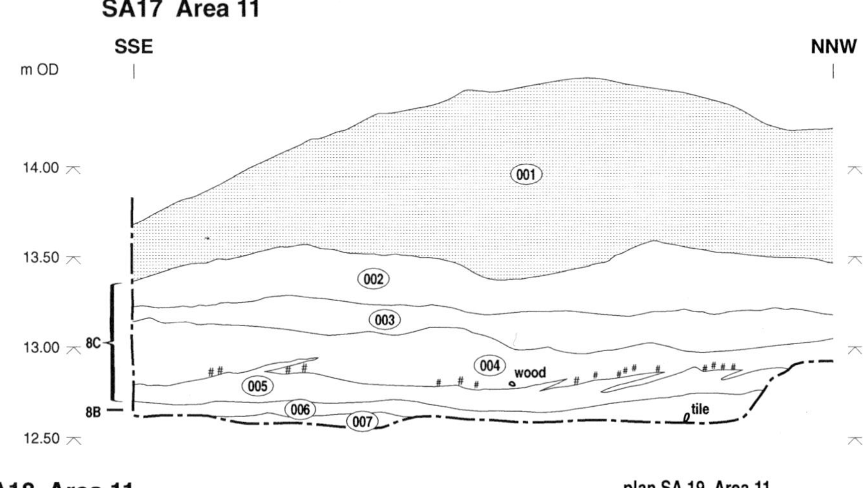

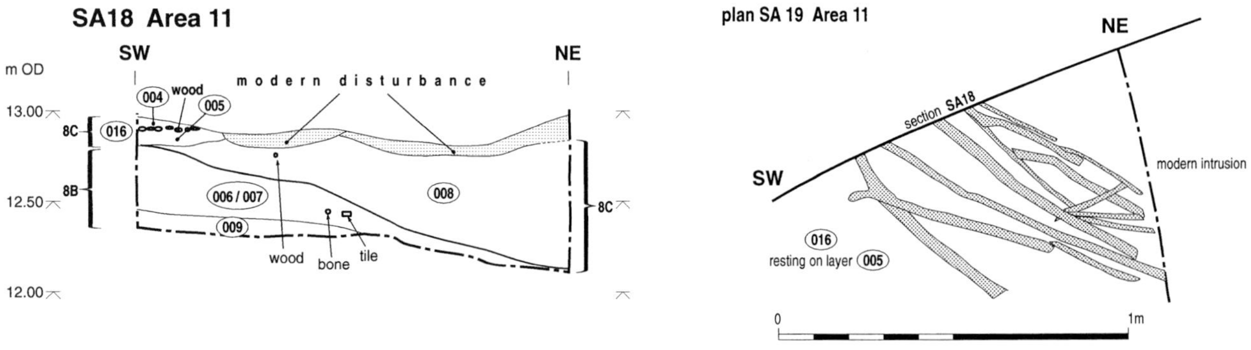

2.31 *Sections SA17 and SA18, Area 11 south-west wall and north-west wall (part) respectively; plan SA19 of laid poles in west corner of Area 11.*

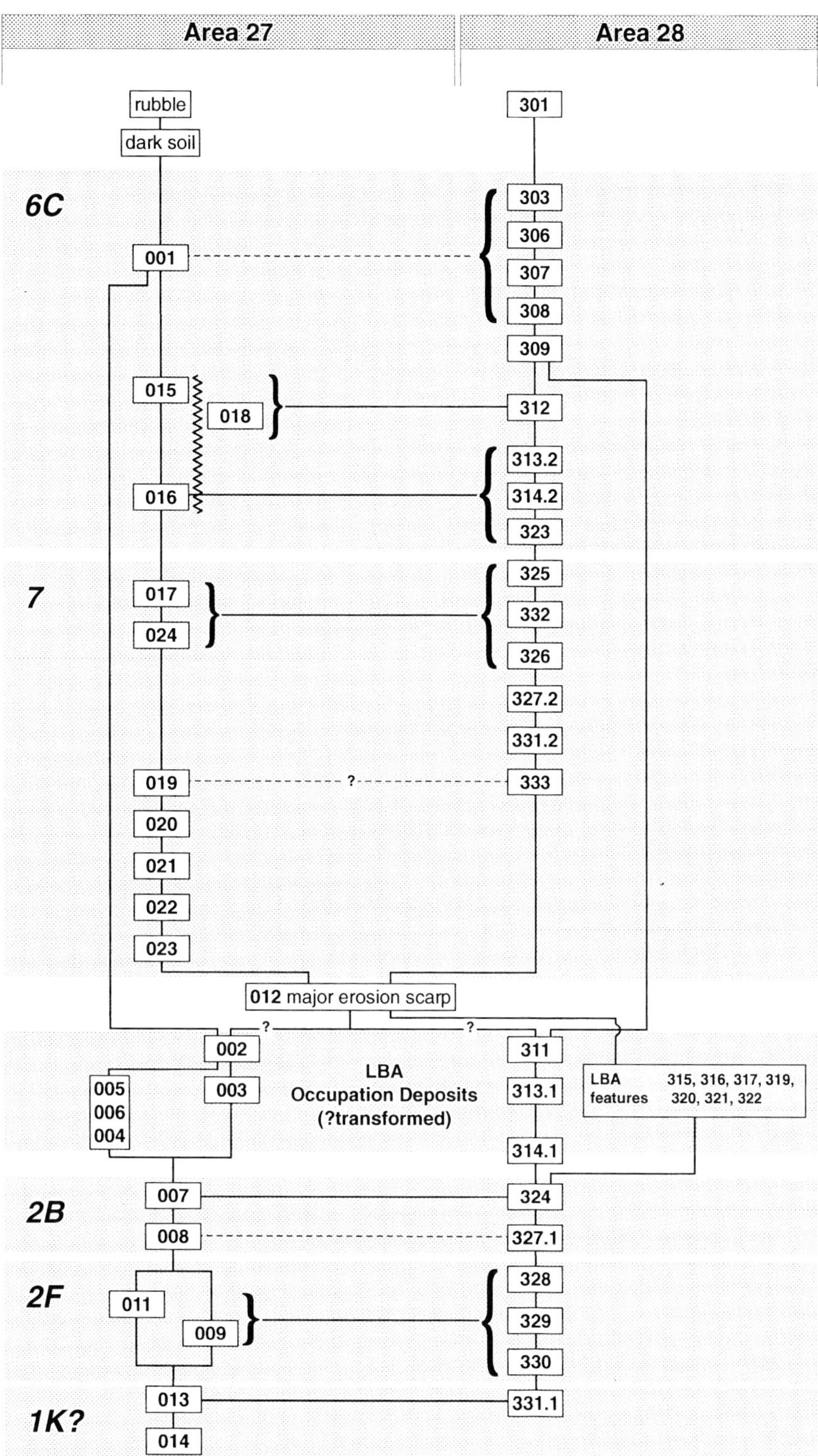

2.32 *Stratigraphic matrix for alluvial parcels in Areas 27 and 28 (full sequence).*

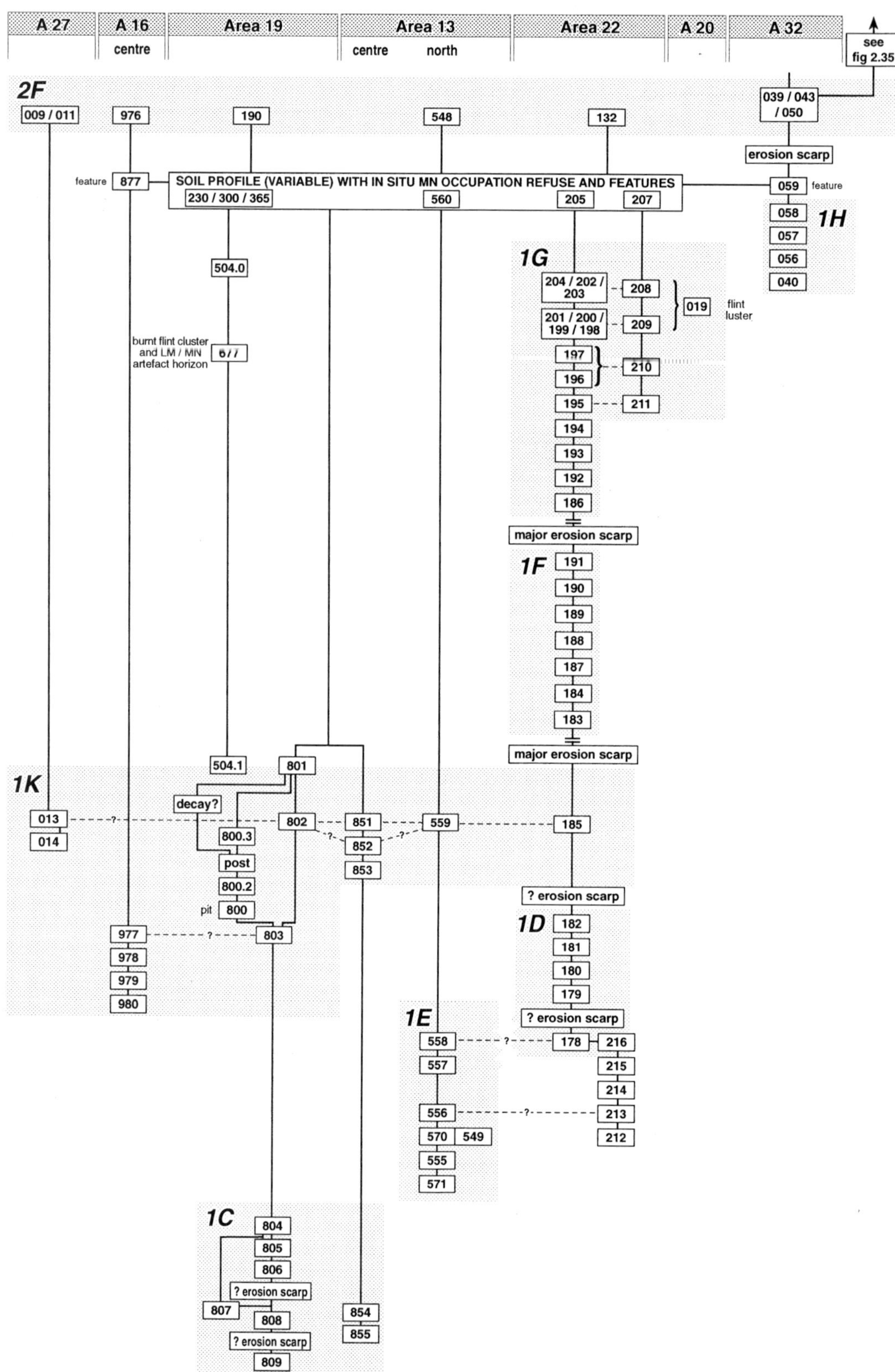

2.33 *Stratigraphic matrix for pre-Middle Neolithic, alluvial parcel 1; Area 27, interior zone and intervening zone.*

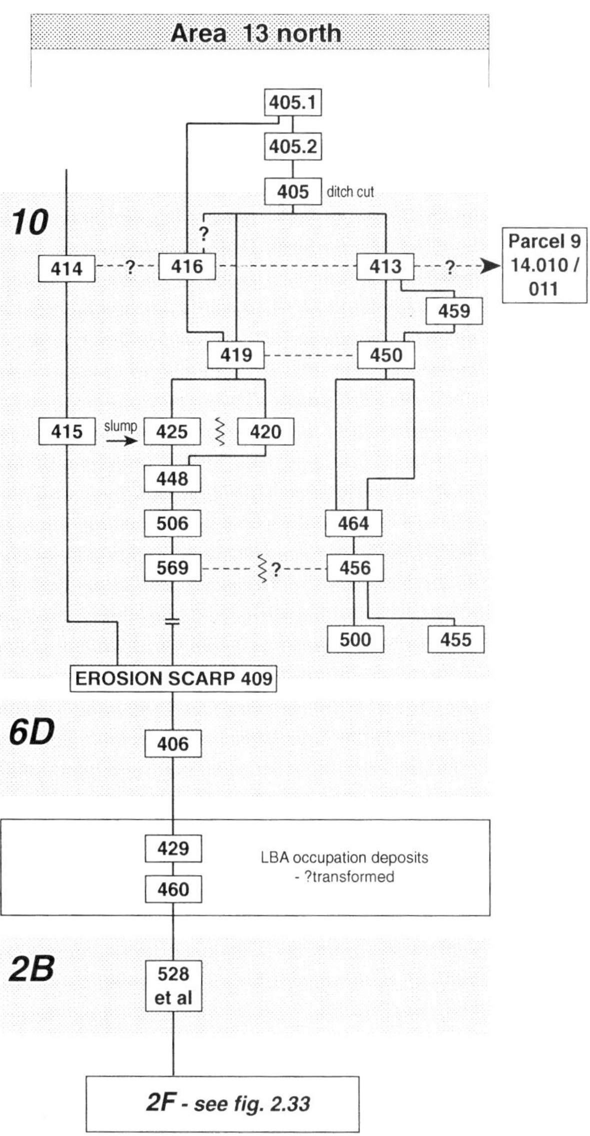

2.34 *Stratigraphic matrix for Bronze Age to modern alluvial parcels; Area 13 North.*

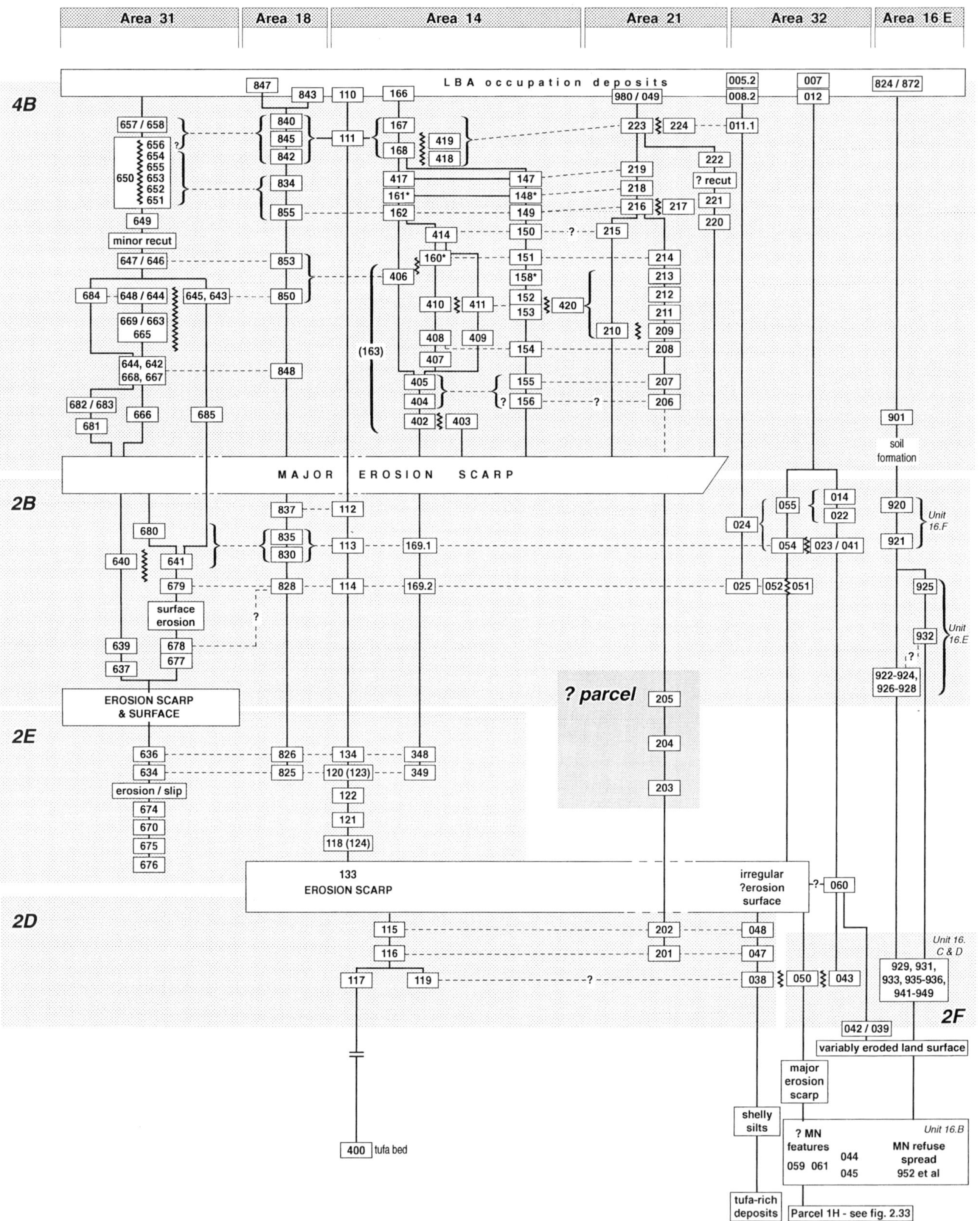

2.35 *Stratigraphic matrix for post–Middle Neolithic and pre–Late Bronze Age alluvial parcels; riverside and intervening zones. *Asterisked contexts in 4B were impinged upon by the Post-Medieval cut and thus at risk of intrusive finds (see Fig. 2.36).*

Area 21
Area 14
north-east
south-east
Area 18
Area 31
19th & 20th century AD dumped layers and features
9
808
010
011
parcel 10
13.413 / 416
141
858
662
012
142
710
661
503.1
014
143
857 / 728
660
015
017
139
139
856
659
140
723
503
Post-Medieval cut bank
14.018
148
158
160
161
6D
817
004
703
001
LBA occupation deposits
(see fig. 2.35)

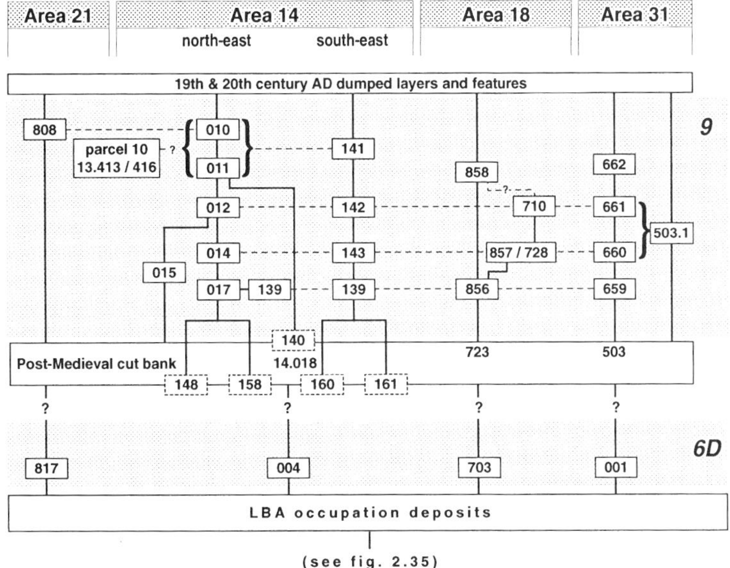

2.36 *Stratigraphic matrix for post-Late Bronze Age alluvial parcels; riverside zone.*

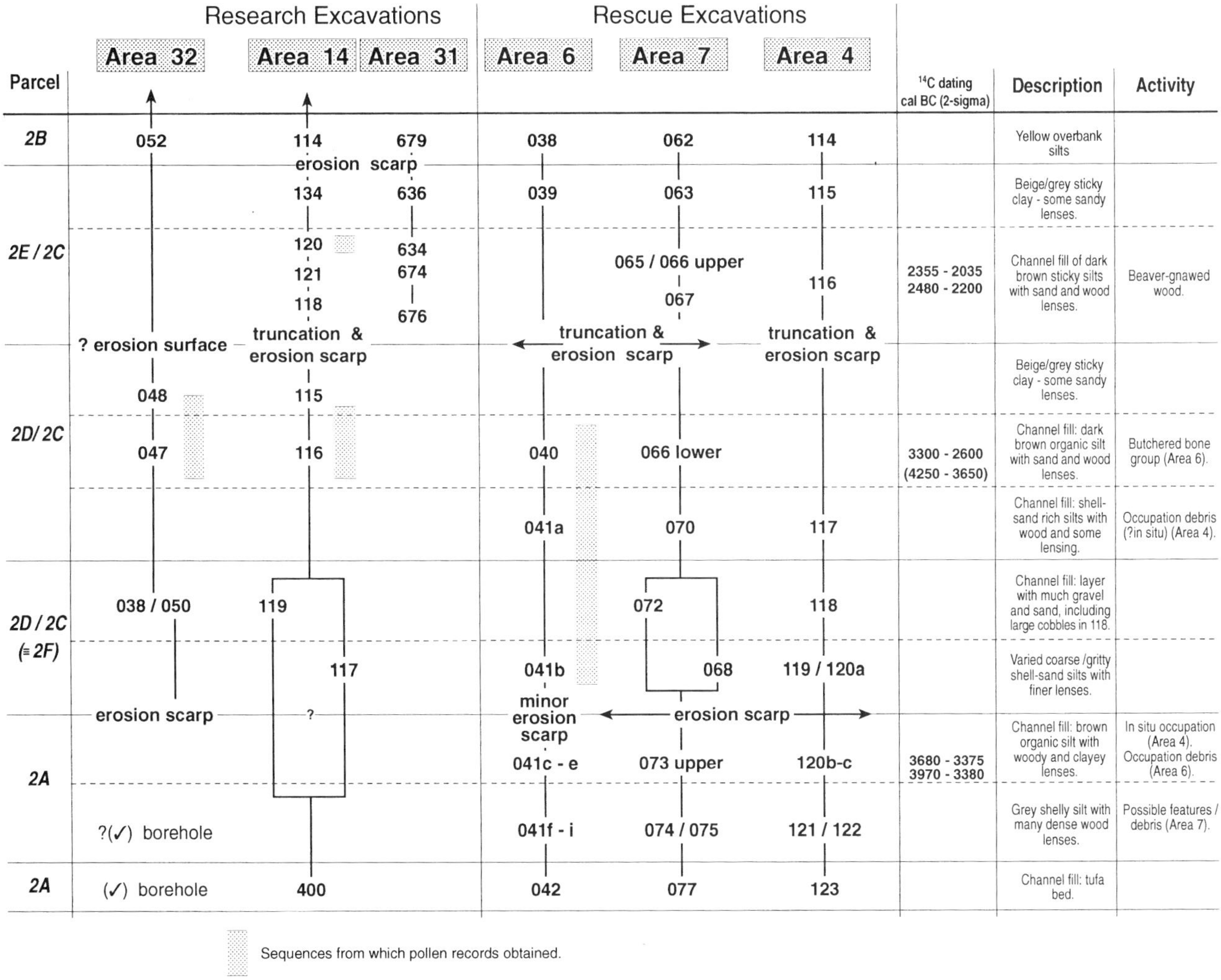

Research Excavations
Rescue Excavations
Area 32
Area 14
Area 31
Area 6
Area 7
Area 4
Parcel
^{14}C dating cal BC (2-sigma)
Description
Activity
2B
052
114
679
038
062
114
Yellow overbank silts
erosion scarp
134
636
039
063
115
Beige/grey sticky clay - some sandy lenses.
2E / 2C
120
121
118
634
674
676
065 / 066 upper
067
116
2355 - 2035
2480 - 2200
Channel fill of dark brown sticky silts with sand and wood lenses.
Beaver-gnawed wood.
? erosion surface
truncation & erosion scarp
truncation & erosion scarp
truncation & erosion scarp
048
115
Beige/grey sticky clay - some sandy lenses.
2D/ 2C
047
116
040
066 lower
3300 - 2600
(4250 - 3650)
Channel fill: dark brown organic silt with sand and wood lenses.
Butchered bone group (Area 6).
041a
070
117
Channel fill: shell-sand rich silts with wood and some lensing.
Occupation debris (?in situ) (Area 4).
2D / 2C (= 2F)
038 / 050
119
072
118
Channel fill: layer with much gravel and sand, including large cobbles in 118.
117
041b
068
119 / 120a
Varied coarse /gritty shell-sand silts with finer lenses.
erosion scarp
?
minor erosion scarp
erosion scarp
2A
041c - e
073 upper
120b-c
3680 - 3375
3970 - 3380
Channel fill: brown organic silt with woody and clayey lenses.
In situ occupation (Area 4). Occupation debris (Area 6).
?(✓) borehole
041f - i
074 / 075
121 / 122
Grey shelly silt with many dense wood lenses.
Possible features / debris (Area 7).
2A
(✓) borehole
400
042
077
123
Channel fill: tufa bed.
Sequences from which pollen records obtained.

2.37 *Suggested correlation of parcel 2 deposits between key excavated sequences across the site.*

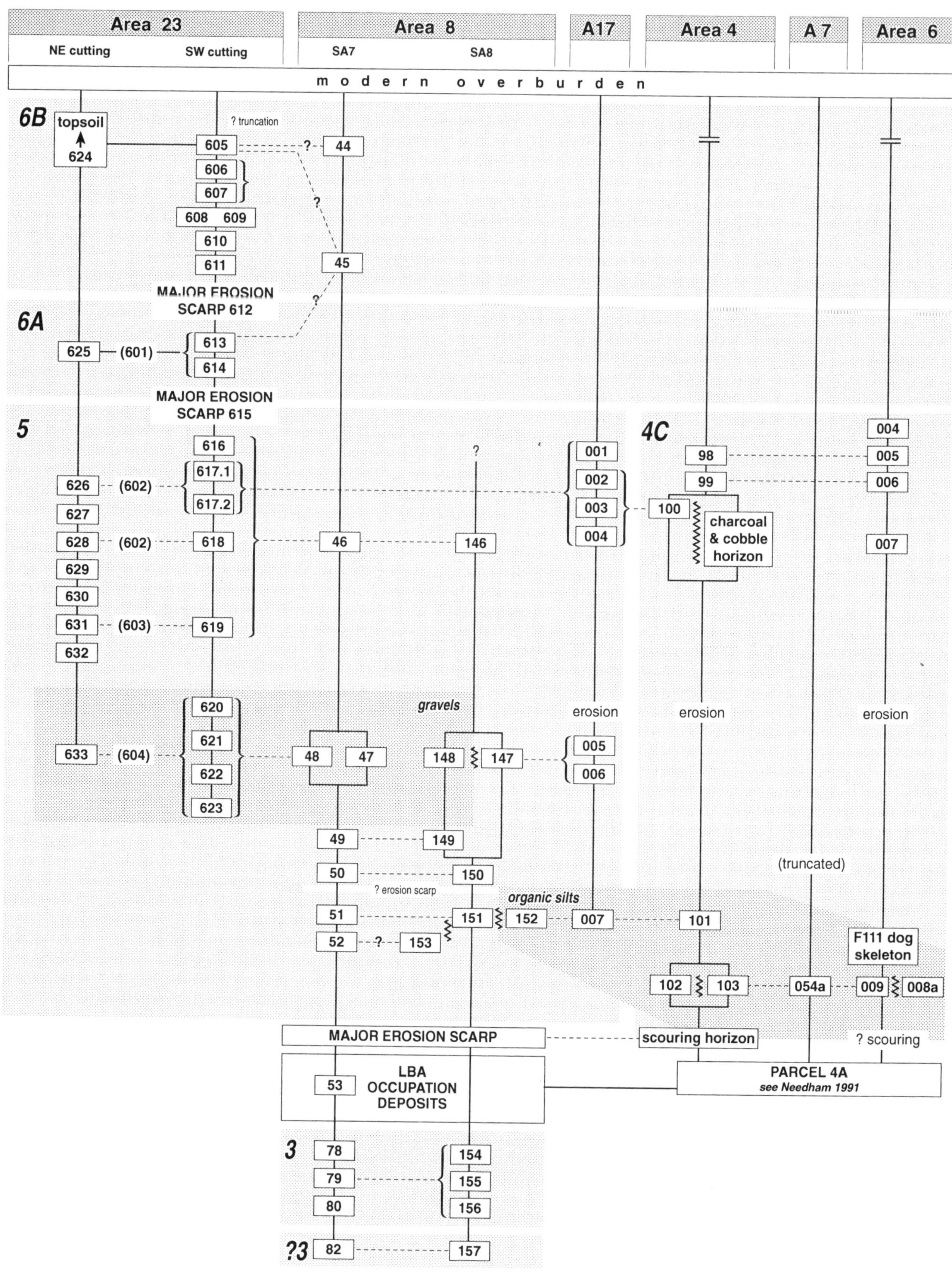

2.38 *Stratigraphic matrix for Late Bronze Age and later alluvial parcels at the east end of the site (Areas 4, 6, 7, 8, 17, 23).*

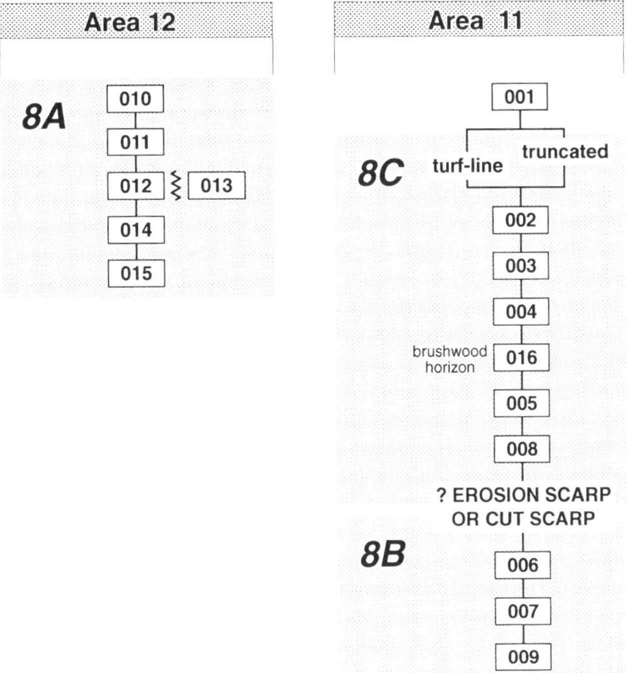

2.39   *Stratigraphic matrix for alluvial parcels in Areas 11 and 12.*

Table 2.2   *The stratigraphic sequences.*

Context	Description	Top (m OD)	Comments
	AREAS 27 and 28 sections RH1-RH4		**Figs. 2.3-2.5; columns 52 and 65; matrix, Figs. 2.32 and 2.33.**
	Parcel 1K?		
–	Base of section.	13.30	
27.014	Greyish yellow-brown silt - soft, sticky and slightly gritty; little Fe stain.	*c.* 13.60	Grades into 27.013 above.
27.013 28.331.1	Greyish yellow-brown clayey silt, slightly lighter and less sticky than 27.014; heavy Fe streaking.	13.90-13.92	Some charcoal towards top.
	Parcel 2F		
28.330 27.011	Medium to large gravel in grey-brown clay matrix.	13.88-14.00	
28.329 27.011	Lens of orangey loose gravel.	13.94-14.02	Presumably iron-rich.
28.328 27.011	Mixed gravel in grey-brown matrix.	13.95-14.05	
27.009	Predominantly small gravel (2-3 cm) with some larger, in gritty greyish-brown matrix.	13.98-14.07	Appears to grade laterally into 27.011 as it approaches the palaeobank.
	Parcel 2B		
27.008 28.327.1	Dark greyish-yellow silty clay with some manganese staining.	14.03-14.05	Intermittent layer; in Area 28 a string of cobbles lies at interface to 28.324.
27.007 28.324	Yellowish-grey silty clay with orange/brown mottling and rare pebbles.	14.23-14.42	Surface cut by LBA features.
	Late Bronze Age occupation deposits		
27.002 28.313.1 28.311	Light grey silt with cultural inclusions; cobble-rich horizon at top (28.311).	14.37-14.50	Uncertain to what extent transformed by natural agencies after formation; part of the sequence may be contemporary with post-LBA channel fill (next two tables).
	Parcel 7		Sequence in parallel with that below.
27.012	Major erosion scarp through parcels 2F, 2B and possibly part of cultural horizon.	(13.50)-*c.* 14.40	
–	Base of section.	13.22	Water table.
27.023	Somewhat mottled white/yellowish-grey/brown, loose coarse sand and grit with frequent small rounded gravel (<1 cm diameter).	13.40-13.47	Relationship to 27.012 not observed, but presumed to overlie it; extends below water table.
27.022	Variegated lenses of fine sand to clayey silt, compact and greyish- yellow with sparse iron staining.	13.48-13.52	Rises gently to SE.
27.021	Compact lensed deposit of pale yellow to yellowish-grey coarse shell-sand; varies in colour and coarseness.	13.62-13.68	One lens with probable charcoal fragments.
27.020	Similar character to 27.022.	13.68-13.72	
27.019	Yellowish-grey slightly sandy silt, a little cohesive; moderate to sparse iron staining, increasing upwards, especially at NW end.	13.81-14.02	Surface dips to SW and palaeobank; unclear whether cut through or interleaved with 27.020, 021, 022; merges into 27.017 where 27.024 not in between.

Context	Description	Top (m OD)	Comments
27.024	Dull yellow-orange shell-sand with moderate pea gravel.	13.82-14.10	Well separated from 27.019 beneath by a thin iron-rich band.
27.017	Dull yellow-orange sandy silt.	14.00-14.14	Occupies shallow trough between 27.024 and palaeobank.
	Parcel 6C		
27.016	Greyish-yellow clayey silt. Sparse to moderate manganese, some charcoal and possible fired clay fragments - giving variegated appearance.	14.10-14.28	Grades into 27.018 at SE end; boundary with 27.017 of variable clarity. Equivalent to 28.323 *et al.*
27.018	Rather mixed, intermediate deposit with ?manganese flecks concentrated in upper part (cf. 27.016) and yellowish-grey silt (cf. 27.007).	14.30-14.42	Grades down-slope quickly into 27.016.
27.015	Horizon of burnt flint and pebbles, approximately one cobble thick.	(*c.* 14.30)	Material eroded off adjacent bank, equivalent to 28.312.
27.001	Dull yellowish-brown clayey silt, friable.	14.73-14.93	Occasional artefacts. Boundary to overlying dark soil varies from clear to slightly diffuse.
–	Overlain by thick deposit of dark soil and cut by modern pipe and cable trenches.		
	Parcel 7		Sequence in parallel with that above.
–	Major erosion scarp through LBA surface and underlying deposits.	(13.75)-14.36	
28.333	Grey-brown silty clay with much Fe streaking.	13.80-13.86	At same level as 27.019 and may equate with upper part, where Fe increases.
28.331.2	Lens of greenish-grey clay.	13.82-13.85	Locally washed out or slumped 'tongue' of bank material 28.331.1, which it runs into.
28.327.2	Intermittent lenses of gravel in grey-blue sandy silt; variable pebble size up to 8 cm.	13.86-14.00	Slumped material adjacent to palaeobank and in part merging into source gravels under bank (28.328, 329, 330). Similar lenses not present close by in Area 27.
28.326	Yellow/buff silt with much shell- sand and some iron staining; very few pebbles.	13.96-14.08	
28.332	Thin band of grey silt down-slope of upper palaeobank locally (N section).	14.05-14.30	Washed-out material from LBA occupation deposits. Post-dates 28.327.2 and probably also 28.326.
28.325	Grey to orangey silty clay with dark flecks (?manganese and charcoal), burnt flint, bone, burnt clay fragments and a few pebbles.	14.10-14.24	Poorly defined upper boundary.
	Parcel 6C		
28.323 28.314.2 28.313.2	Greyish sandy clay with pottery, burnt flint, struck flint and pebbles.	14.22-14.35	Excavated as three spits, but all similar. Equates with 27.016.
28.312	Thin horizon of very light greyish- yellow clayey silt with moderate density of burnt flint cobbles and pebbles.	14.26-14.39	Equates with 27.015. Dips gently away from palaeobank; density of stone less than contiguous deposits on bank (28.311), which it is either pene-contemporaneous with, or represents its later erosion.

Context	Description	Top (m OD)	Comments
28.309-28.303	Buff/greyish-yellow firm clay-loam; cultural material and especially burnt flint increasing towards bottom.	*c.* 14.90	Excavated as five spits (28.303, 306, 307, 308, 309), equating with 27.001.
–	Cut by modern pipe trench.		
AREA 10 (also 9) **sections SA20 - SA22**			**Not illustrated.**
10.038	Orange/yellow clayey silt.	13.25-13.55	Established by augering.
-	Base of drawn section.	13.80-14.00	
Middle Neolithic occupation/Parcel 2F			
10.560	Neolithic sub-soil.	13.92-14.04	Few artefact finds.
10.566 10.501	Dense gravel.	14.05-14.25	Some cultural finds beneath.
10.548	Pocket of silt with sparser gravel.	14.09-14.25	Good assemblage of Neolithic material.
Parcel 2B			
10.499	Yellow-brown clay-loam.	14.35-14.51	
Late Bronze Age occupation soil			
10.018 *et al.*	Dark grey-brown loam.	14.59-14.72	Rich in cultural material and in situ structures.
Parcel 6C			
10.006 *et al.*	Yellowy clay-loam, grading into a soil profile on top.	(15.10)-15.25	Reworked artefacts increasing in density downwards.
AREA 19 West **section RH5**			**Fig. 2.6; columns 31, 32 and 63; matrix, Fig. 2.33.**
Parcel 1C			
-	Base of pollen monolith.	12.88	
19.809	White marl.		Lower extension of deposit seen in pollen monolith.
–	Base of drawn section.	13.13	Water table.
19.809	White marl.	13.24-13.30	Possible erosion scarp to north prior to 19.808.
19.808	White marl with some orange iron mottle and iron concretion band at base.	13.20-13.23	Possible erosion scarp to north prior to 19.806.
19.807	Thin band of iron concretion.	13.21-13.37	Probably not an original layer, but may be aggregation at a genuine interface. Overlies 19.808 and 809; uncertain relationship to 19.806.
19.806	White marl with iron mottle (more than 808).	13.20-13.33	
19.805	Marl with dense dark brown manganese mottles.	13.33-13.35	Likely post-depositional aggregation within deposit 19.806.
19.804	White marl containing large and small patches of dark brown manganese mottles.	13.49-13.80	Base may be genuine interface with various underlying deposits. Inclined top grades into 19.803 (*c.* 5 cm zone).
Parcel 1K			
19.803	Orange, soft silty clay, slightly crumbly.	13.91-14.15	Top undulates gently and grades into 19.802 (*c.* 15 cm zone).
19.802	Light grey-brown silty clay with some orange iron mottles.	–	Grades steadily into 19.801 without definable boundary.

Context	Description	Top (m OD)	Comments
19.800	Cut/driven feature with sharp 'V' profile.	13.26-14.00	Certainly post-dates 19.803; may also post-date 19.802, if 19.800.3 is slump.
19.800.2	Grey claggy clay.	13.57-13.78	Lower fill; graded upwards into 19.800.1.
19.800.3	Silvery-grey crumbly silt with orange iron mottles and 'clods' of pale marly silt.	*c.* 14.10/14.15	No boundary determined with 802 and may thus be 802 slumping into fill rapidly after cutting. Both sides butt up to 800.1.
19.800.1	Light grey silty clay with orange mottles.	–	Vertical boundaries with 19.800.3 suggest this is a post-pipe, *c.* 25 cm across. Top grades imperceptibly into 19.802 - thus decay perhaps contemporary with 19.801 formation. Contained charcoal fragments ^{14}C dated (OxA-3580).
19.801	Light brown silt or soil.	14.37-14.47	Probably a soil developed from upper part of 19.802, this being the lower profile not disturbed or infiltrated by MN occupation. Soil formation seems to post-date final fill of 19.800.
	Middle Neolithic occupation soil		
19.230 19.300 19.365	Brown silty soil with in situ MN occupation deposits.	14.45-14.53	
	Parcel 2F		
19.190 *et al.*	Gravel, predominantly 2-5 cm, in grey-brown loam matrix.	14.55-14.60	Contains quantities of MN artefacts partly through flood reworking, but at base probably also due to downward migration of gravel.
	Parcel 2B		
19.185 *et al.*	Pale yellow-brown silty loam, becoming clay-loam at base.	14.60-14.65	Contains small number of LBA finds infiltrating from above.
	Late Bronze Age occupation soil		
19.008 19.009	Grey-brown loam.	14.70-14.75	Includes cobble 'threshold' at NW end.
–	Overlain by parcel 6D		Machine excavated.
	AREA 19 North section RH6		**Fig. 2.6; column 13; matrix, Fig. 2.33.**
	Parcel 1K		
19.504.1	Grey-brown clay-loam.	*c.* 14.26	Lower part of a deeper soil profile (with 19.504.0) across the W corner of Area 19.
19.677	Artefact horizon.	*c.* 14.30	Diffuse horizon of artefacts, mainly thinly spread, but alongside this was a concentration of burnt flint (19.677), yielding TL date (Chapter 4).
19.504.0	Grey-brown clay-loam.	14.39-14.41	Upper part of soil profile, which should broadly equate with 19.801.
	Middle Neolithic occupation		
19.812 19.811 19.810 19.310	Neolithic cut features.	14.26-14.40	Features cutting soil 19.504/801. Bottom of 19.810 cuts into yellow to pale grey marly deposit, probably a peak of 19.804 (section RH18).
19.230 *et al.*	As described in previous section.	14.34-14.50	

Context	Description	Top (m OD)	Comments
	Parcels 2F and 2B		
19.190 19.185	As described in previous section.	14.45-14.55 14.52-14.59	
	Late Bronze Age occupation soil		
19.012 19.004	Grey-brown loam.	14.62-14.69	Includes yard surface cobbled with large river pebbles.
	Parcel 6D		
19.002	Light grey-brown silt.	(14.85)	
	AREA 16 Centre section RH7-RH8		**Fig. 2.7; column 9; matrix, Fig. 2.33.**
	Parcel 1K		
–	Base of drawn section.	13.12	Water table.
16.980	Grey clayey silt.	13.33-13.42	
16.979	Orangey-yellow silt.	13.35-13.45	
16.978	Grey clayey silt.	13.43-13.54	Undulating top.
16.977	Orangey-yellow silt.	13.94-14.05	No obvious soil profile along top.
	Middle Neolithic feature		
16.877	Yellowy silt fill.	14.05-14.07	Probably a pit; bottom lined with layer of stacked pot sherds forming pot NP94 (Fig. 3.3b).
	Parcel 2F		
16.976	Dense gravel, also incorporating pottery, animal bone and flints.	14.00-14.16	Undulating top at E seems likely due to reworking before 16.975.
	Parcel 2B		
16.975	Bluey-grey silt (E only); occasional pebbles.	14.07-14.12	Fills depressions in gravel.
16.982	Sandy bed at interface of 16.975 and 974.	14.09-14.14	
16.974	Yellowy-grey silt with sparse pebbles.	14.25-14.37	Possible indication of a division marked by sparse pebble-line, cf. Area 16 East (Volume 2).
	Late Bronze Age occupation soil		
16.981	Charcoal-rich lens with burnt flint.	14.30	Occupies slight hollow.
16.973	Lens of dirty yellow sandy silt.	14.31-14.35	Probably upcast from digging.
16.972	Dark grey cultural soil, rich in debris, especially bone, burnt flint, pottery and daub.	14.44-14.54	In situ LBA dark earth.
	Parcel 6D		
16.971	Light grey-brown silt, incorporating some cultural debris, especially burnt flint - density thinning upwards.	14.80(?+)	
–	Overlain by modern rubble layer.		
	AREA 16 East section RH9		**Fig. 2.8; columns 10, 12, 22-25; matrix, Fig. 2.35.**
	Sequence published in detail in Needham and Spence 1996; includes Middle Neolithic, parcels 2F, 2B, Late Bronze Age, parcel 6D.		

Context	Description	Top (m OD)	Comments
	AREA 13 Centre section RH10		**Not drawn; column 64.**
	Parcel 1C		
–	Base of recorded section.	13.16	Water table.
13.855	Marl with Fe mottling increasing downwards and strongest within lowest 15 cm.	13.71	
13.854	White marl.	13.81	
13.853	Intense band of manganese mottling.	13.83	
	Parcel 1K		
13.852	(Merging zone.)	14.01	
13.851	Greyish-brown mottled soil.	14.21	Limbrey defines as A/Bg horizon (Chapter 5).
	Middle Neolithic occupation soil.		All overlying deposits previously excavated.
	AREA 13 North sections RH11-RH13		**Figs. 2.9-2.10; columns 3, 4 and 17; matrix, Figs. 2.33 and 2.34.**
	Parcel 1E		
-	Base of drawn section.	12.57	Reduced water level, falling NW towards sump.
13.571	Pale grey coarse shell-sand; small fibrous and tufaceous components.	12.60-12.69	Dipping gently to NW.
13.555 13.549	Lensed deposit of sandy and fibrous organic silts, very dark brown with grey lenses.	12.83-12.86	Branch-wood and small trunk lying in top, including 13.549, ^{14}C dated (BM-2550).
13.570	Soft pale-coloured clay.	12.91-12.93	Seals branch-wood 'horizon'.
13.556	Grey sandy clay, firm; darker flecks, possibly organic.	13.11-13.17	Top dipping gently to SE. Contained one piece of wood.
13.557	Firm, dark orange/brown mottled sandy silt, rich in mineral (iron) precipitates.	13.17-13.20	Occupies depression or laps over slope in top of 13.556.
13.558	Very pale (whitish) shell-sand with slight orange (iron) mottling.	13.32-13.41	Top dipping to SE.
	Parcel 1K		
13.559	Silty clay, light greyish yellow-brown towards bottom, becoming more orange higher due to bands of mineral (iron) enrichment.	13.94	
	Middle Neolithic silt/soil		
13.560 *et al.*	Yellowish-grey silty clay.	14.09-14.21	Hand-excavated deposits; undulating surface due to Neolithic features and/or erosion.
	Parcel 2F		
13.548 *et al.*	Gravel, predominantly pea-gravel, in dark grey silt matrix.	14.15-14.32	Undulating bed of varying thickness.
	Parcel 2B		
13.528 *et al.*	Yellow-grey silty clay with shell and Fe mottles.	14.40-14.45	Small flecks of charcoal.
	Late Bronze Age soil: flood diffused?		
13.429 13.460	Grey-brown silt.	*c.*14.55-14.60	Some large cultural debris, but no certain in situ material.

Context	Description	Top (m OD)	Comments
	Parcel 6D		
13.406	Light yellow-brown silt.	(14.77)	Top truncated by machining.
	Parcel 10		
13.409	Steep erosion scarp truncating parcel 6D, LBA and underlying deposits.		
13.455	Dense gravel, white to dark grey; pebbles about 2 cm diameter; contained bone and probably wood. Tufa also noted.	(Below 12.24)	Observed during excavation of sump, but always beneath reduced water table.
–	Base of drawn section.	12.24-12.56	Water table (pump-reduced).
13.500	Black, slightly fibrous clay.	12.49-12.71	Stratigraphic relationship with 13.409 not established (below level reached), but assumed to be later. Dips gently towards palaeobank. Yielded clay pipe of *c.* AD 1680-1710.
13.456	Mid-grey to bluey grey coarse shell- sand, loose; contains quantities of wood.	12.49-12.75	Recorded in two separate exposures 3.2 m apart: to N it was fairly thick and probably sat directly on 13.455; to S it was a thin bed lying over 13.500. Although the same matrix, it is possible that they were not strictly contemporary.
13.569	Bright yellow-green coarse sand.	12.56-12.67	Laps up against palaeobank. Could be discoloured lateral continuation of 13.456.
13.464 13.506	Grey sticky clay, becoming bluer-grey (13.506) towards palaeobank, and also becoming darker with depth; localised mineral concretions (Fe).	12.88-13.17	Dips steadily to S, but at S end it merges into 13.448 above. Yielded clay pipe of mid-nineteenth century AD.
13.448	Pale to mid-grey firm clay; contained some wood.	12.95-13.48	Sloped deposit against palaeobank; merges into 13.464 below; likely same deposit differentiated only by wetness.
13.425	Light grey-brown loamy silt, friable. Includes derived artefactual material.	(14.37)	Sits against upper palaeobank dipping rapidly and merging into 13.420.
13.420 13.450	Light grey-brown clay, slightly sticky; N tail (under ditch) is mineral stained (Fe).	13.20-13.91	Dips strongly away from palaeobank and seems to taper out just beyond ditch 13.405 in west section.
13.419 13.413	Very pale brown clay, firm, becoming more silty at S end (13.419) with occasional cobbles and artefacts.	13.13-13.95	Almost certainly same deposit, but interrupted by ditch 13.405. Trapped Post-Medieval tiles at steep angle against interface with 13.425 in E section. Yielded clay pipes of *c.* AD 1700 and mid-nineteenth century AD.
13.459	Pale yellow-brown to pale grey- brown, soft sandy clay, rich in orange mineral precipitates.	13.33-13.47	Transition zone with post-depositional accumulation of precipitates.
13.416 13.414	Yellow-brown to yellow-grey silt, more firm and friable at S end (13.416) which also yielded very occasional cobbles.	13.60-14.38	Almost certainly same deposit, but interrupted by ditch 13.405. Rises in both directions forming gentle trough, possibly due to later scouring and ditch digging.
13.405	Cut of linear ditch aligned grid SSW-NNE.	–	Boundary ditch following edge of palaeochannel as seen on nineteenth- to twentieth-century maps.
13.405.2	Brownish-grey, plastic silty clay; contained occasional cobbles, plus artefacts including glass.	13.28-13.63	Lower fill of ditch.
13.405.1	Dark brown to black loam with occasional pebbles and much modern rubbish.	–	Upper fill of ditch.

Context	Description	Top (m OD)	Comments
–	Modern turf-line.		
	AREA 22 **sections RH14-RH16**		**Figs. 2.11-2.12; columns 39 and 40; matrix,** **Fig. 2.33.**
	Parcel 1D		
–	Base of pollen monolith.	12.70	
22.212	Highly humified black peat.	12.75	22.212-216 represented in pollen monolith sampled prior to rise of water table and drawing.
22.213	Pale brown to grey sand and silt with black organic fragments.	12.80	
22.214	Buff sand/silt with occasional black organic specks.	12.89	
22.215	Greyer silt with more organic material than 22.216.	12.92	
22.216	White to pale grey calcareous silt/marl with fine sand.	13.01	Probably same as 22.178.
–	Base of drawn section.	13.00	Water table.
22.178	Pale chocolate-coloured, soft slightly sandy silt.	13.07	Abrupt truncation at E end.
22.179	White marly silt with orange rootlet mottles (iron).	13.17-13.28	Seals truncation of 22.178, but is in turn steeply truncated at E end. Contained charcoal fragment, ^{14}C dated (OxA-3581).
22.180	Rich orange-coloured marl with localised iron concretions.	13.25-13.31	Top gently inclined to E; apparent stratigraphic relationship, overlying 22.187 and 188, probably false due to washing out and slumping of steep palaeobank.
22.181	Mottled pale yellow/orange marl, fairly granular from mineral precipitates.	13.43-13.46	Grades into 22.182 above; stepped erosion profile at E end.
22.182	Similar to 22.181, but pale yellow dominant and more granular.	13.60-13.62	Steeply truncated at E end.
	Parcel 1K		
22.185	Light grey crumbly soft silt with slight black mottling (?manganese).	13.83-13.88	Superficially similar to upper part of 22.191 alongside, but latter is firmer and a slump line between is likely, continuing the underlying palaeobank. Overlain by 22.201, top of parcel 1G.
–	Steep erosion scarp facing E truncates parcel 1D and probably 1K.		
	Parcel 1F		
–	Base of drawn section.	13.00	Water table.
22.183	Pale chocolate-coloured sticky silt with lenses of whitish fine sandy silt.	13.05-13.13	Butts up against 22.179 at the lower part of interpreted 1D/1F palaeobank; steeply truncated at E end.
22.184	Light grey sticky clay with patches of fine shell-sand.	13.00-13.20	Overlies 22.183 including its truncation; gently dipping E.
22.187	Yellow plastic clay with occasional shell pieces.	13.10-13.23	Possibly a merging zone between 22.184 and 22.188; truncated at E end.
22.188	Bright orange plastic clay.	13.15-13.27	Gently dipping E. Likely false stratigraphic relationship to 22.180 (see parcel 1D).

Context	Description	Top (m OD)	Comments
22.189	Deep orange/black mottled slightly sticky silt, very granular (iron pellets).	13.33-13.37	Boundary with 22.190 above partly poor. Steep scarp at E end and convolutions adjacent mark profile of palaeobank.
22.190	Firm, dry silt, light yellow-grey with deep orange mottles; less granular than 22.189 and perhaps more sandy.	13.47-13.56	Steeply truncated at E end, where convolutions may be due in part to slump of 22.191 above. Iron enrichment horizon in lower part continues throughout section eastward.
22.191	Slightly sandy grey silt, a little granular; similar to 22.190 but with mainly black pellets, especially dense towards base.	13.82	Whole of upper surface likely a gentle erosion scarp. Similarity with 22.185 (parcel 1D) may imply an element of bank slump. Overlain by 22.197 and 22.201, top of parcel 1G.
–	Erosion line facing E truncates parcel 1F.	13.00-13.82	
	Parcels 1F-1G		
–	Base of pollen monolith.	12.68	22.217-222 represented in pollen monolith, sampled prior to rise in water table and section drawing.
22.217	Grey organic fine silt.	12.75	22.217-219 are possibly the continuation below water table of 22.183.
22.218	Sand lens.	12.76	
22.219	Grey silt/clay; no visible plant structure.	12.83	
22.220	Transitional silt horizon.	12.88	
22.221	Dense shell lens.	12.90	
22.222	Dark grey organic silt; no plant macrostructure.	12.98	
	Parcel 1G		
–	Base of drawn section.	13.00	Water table.
22.186	Mottled grey/yellow silt with shell and marl pellets.	13.05-13.08	Overlaps 22.184, but also butts against the lowest part of interpreted 1F/1G palaeobank.
22.192	Yellow-beige soft sticky clay, a little granular with iron-rich and ?marl pellets; some orange mottling.	13.12-13.19	Undulations at W end likely due to scouring along palaeobank.
22.193	Beige to white shelliferous soft silt; some orange mottling.	13.16-13.27	Tapers gently away from palaeobank; W extremity occupies pot-hole in bank.
22.194	Similar to 22.193 but higher clay fraction and a little less shell-sand; some orange mottling.	13.27-13.34	Undulations in surface suggest phase of erosion prior to 22.195.
22.195	Pale yellow-grey clay with a little fine shell; iron-enrichment horizon clips W end.	13.38-13.42	Distinction between 22.195/196/197 not necessarily significant.
22.196	As 22.195 but slightly greyer; iron-enrichment horizon runs through lower part.	13.42-13.54	Upper surface appears to undulate, but poorly defined boundary to 22.197 above.
22.197	As 22.195, very slightly more crumbly; upper iron-enrichment horizon runs through top part.	13.54-13.59	
22.198	Light yellow-grey soft silt with much fine shell-sand; slight orange/black mottling.	13.68-13.75	Grades laterally and abruptly into 22.199 - significance not clear; steep slope at E end likely an erosion scarp.
22.199	Similar to 22.198, but a little less shell and less grey.	13.81-13.84	Conceivably a feature fill, cut into 22.198 and 22.201.

Context	Description	Top (m OD)	Comments
22.200	Essentially as 22.199 with more black mottles (?manganese).	13.75-13.86	Pockets within 22.199; perhaps areas of ancient disturbance by roots or burrows.
22.201	Quite hard pale yellow silt with slight orange mottling, drier than 22.200.	13.86-13.96	Grades laterally from 22.200.
22.203	Orangey-grey loam; some fine shell but less than 22.195.	13.69-13.81	Undulations and apparent cut in upper surface may in part be due to cultural (MN) activity.
22.204	Similar to 22.201 and 22.199, but slightly greener.	13.89-14.05	Undulations and cuts in upper surface likely due to cultural activity.
22.202	Rather stiff grey silt with dark orange-brown/black mottles and a little shell.	13.83-13.89	Complex lateral alternations with 22.204; area of disturbance or conceivably cut features, but not seen in plan.
	Middle Neolithic occupation soil		
22.205	Fairly dark orange-brown stiff soil, becoming paler downwards; becomes drier and slightly shellier in W part.	13.71-14.09	A soil profile, the upper part incorporating much cultural material and occasional stones; hand-excavated as various contexts.
	Parcel 2F		
22.132 et al.	Yellowish-grey to grey-brown silty clay matrix; dense but variable gravel including some large cobbles in lower parts; upper parts predominantly small to pea gravel.	13.93-14.16	Intermittent signs of lensing, especially two dipping lenses at 'shelf' halfway along. Incorporated reworked Neolithic artefacts; lower part of gravel probably infiltrates in situ cultural horizon.
	Parcel 2B		
22.118 et al.	Greyish-yellow silty clay with variable gravel component, mainly rather sparse; wedge at bottom of SW end is gravel-free (22.120).	13.98-14.33	
22.115 22.113	Greyish-yellow silty clay with sparse to moderate gravel (up to 5 cm) and some orange mottling.	c. 14.04-14.34	Dips consistently to NE; includes 'pebble-string' showing in the section.
22.101 et al.	Yellowish-brown silty clay with 'dirty' interface at top; lowest part at NE end stickier.	14.30-14.48	Wedge tapering towards SW.
	Late Bronze Age occupation soil		
22.020 22.029 et al.	Dark yellowish-grey silt.	14.50-14.57	Intrusions into top probably derive from roots.
	Parcel 6D		
22.001	Olive-brown silt.	(14.72)	Machine-truncated. Contained many reworked artefacts.
	AREA 32 sections RH17-RH19		**Figs. 2.13-2.14; columns 59-61; matrix, Figs. 2.33 and 2.35.**
	Parcel 1H		
32.062	Medium/coarse orange sand.	12.95	Recorded in pollen monolith.
–	Base of drawn section.	12.95	Minimum water table.
32.040	Dark brown/black organic silt with wood fragments; laminated with lenses of fine sand/shell; coarser lenses at E end.	13.00-13.05	Wood sample ^{14}C dated (BM-2657). Steep truncation at E end separates from superficially similar 32.046.
32.056	Pale grey-brown silty clay; some Fe streaks extending from above deposit.	13.09-13.15	Gleyed zone from fluctuating water table; superficially similar to 32.049 alongside.

Context	Description	Top (m OD)	Comments
32.057	Mid-grey-brown plastic silty clay; extensive Fe vertical streaks; shell-sand component.	13.24-13.46	Lens isolated from main body in N-S section.
32.058	As 32.057 but lacking noticeable shell-sand.	13.67	Whole of upper surface likely to be gentle erosion scarp at top of palaeobank which becomes steeper lower.
	Middle Neolithic features and soil		
32.059	Darker grey silty clay with occasional charcoal flecks.		Probable feature fill; top likely truncated by erosion scarp; directly overlain by 32.050.
32.061	Grey clay with bone, pot sherds, charcoal flecks and occasional stones.	13.70	Top planned, but not excavated (grid 51/39).
32.044 32.045	Grey-brown silty clay with scatter of pebbles and cultural debris.	13.63-13.74	Unexcavated except for bone group 32.045 (grid 47/39). In part at least thought to represent an 'A' horizon.
	Neolithic soil/ Parcel 2F		
32.039 32.042	Grey silty clay with variable density of gravel ranging from very sparse to fairly dense.	13.66-13.82	Contained large pot sherds and bone fragments, struck flint and charcoal. Probably a complex mix of gravel (2F) and underlying cultural soil infiltrated by gravel. Patch around 53/39, however, probably upper fill of later feature, 32.060.
	Parcel 2F		
-	Steep erosion scarp facing NE, truncates all silt beneath Neolithic land surface.		
32.043	Medium to coarse gravel, fairly dense in grey silty clay.	13.70-13.80	Occupies gently sloping ground at crest of palaeobank. In the section, appears to be cut by feature 32.060. May grade into, or be overlapped by 32.050. Little cultural debris.
32.060	Apparent feature: upper fill excavated - dense gravel and pea-grit.	13.88	Perhaps a natural scour feature refilled with gravel. Yielded few cultural finds.
32.050	Fine to medium gravel with occasional large cobbles in greyish gritty clay; occasional charcoal flecks and localised Fe concentrations.	(12.95)-13.76	Spreads whole way down palaeobank, becoming darker close to water table; probably links physically with 32.038.
	Parcel 2D		
32.038	Compact bed of mid-grey coarse sand and fine gravel.	12.94-12.99	Surface only revealed, but a borehole taken through to underlying sediments showed it to be a thin bed.
32.047	Dark grey-brown organic silt with localised coarse sand lenses; becomes more gritty with small pebbles towards W.	13.03-13.11	At W end merges into 32.050 - diffuse process of reworking of gravel.
32.048	Pale grey/beige silty clay with limited Fe streaks; includes some gravel at W end.	13.13-13.25	Top appears very uneven; probably merges into 32.051 and ?050.
	Parcel 2B		
32.052 32.025	Mid-brown clayey silt; lower part with occasional small pebbles and horizon of charcoal flecks.	13.65-13.80	
32.051	As 32.052 with sand, shell fragments and sparse small to large pebbles.	13.60-13.82	Can probably be regarded as part of 32.052, lapping bank and persistently reworking gravel exposed there.

Context	Description	Top (m OD)	Comments
32.054 32.024	Light yellow-brown silt with extensive dark brown/black (Mn) flecking and occasional orange (Fe) mottles.	13.94-13.99	At W end interleaves with 32.023.
32.023 32.041	Yellow-brown clayey silt with orange mottles and occasional charcoal flecks; moderately dense gravel, small to large, but unevenly distributed.	13.90-14.01	Interleaved within 32.054 at E end; later reworking of primary gravel of parcel 2F and its contained finds; hand-excavated westwards from 55E.
32.055 32.022 32.014	Light yellow-brown silt with some Fe and Mn flecking.	14.28-14.43	
	Late Bronze Age occupation soil		
32.007 *et al.*	Dark grey-brown silty loam rich in cultural debris.	14.50-14.64	In situ cultural earth with transition zone to 32.055 beneath.
	Parcel 6D		
32.028	Light brown silt; cultural finds thinning upwards.	(14.94+)	Only hand-excavated in grid square 46/39.
–	Overlain by modern rubble layer.		
	AREA 21 sections RH20, RH21		**Figs. 2.15 and 2.16; columns 36-38; matrix, Figs. 2.35 and 2.36.**
	Parcel 2D		
–	Base of drawn section.	12.99	Water table.
21.201	Dark grey clay with fine lenses of tufa and shell-sand; includes significant horizontal lens of coarse sand.	12.99-13.07	
21.202	Beige/grey clay with a few iron stains; includes discrete lens of pale yellow fine sand.	13.23-13.32	Dips lower at NE end, probably due to subsidence at palaeobank; somewhat diffuse boundary to 203.
	Parcel 2E/2B?		
21.203	Yellow-grey clay with Fe mottling.	13.35-13.45	Dips lower at NE end, but probably due to subsidence at palaeobank.
21.204	Similar to 21.203 but paler bluish-grey hue.	13.65-13.67	Subsidence at NE end.
21.205	Similar to 21.203.	(13.94)	Upper surfaces entirely product of truncation by palaeobank.
–	Truncated by N-facing palaeobank with associated mud stacks.		
	Parcel 4B		
–	Abuts palaeobank eroded through parcel 2.		
–	Excavated to maximum depth of	13.01	
21.206	Beige sandy matrix with dense pea gravel and a little medium-sized gravel.	13.11-13.34	Gently domed surface.
21.207	Medium gravels in beige matrix of dense coarse sand.	13.08-13.56	
21.208	Very sandy silt, grey with orange mottles; some clayey lenses within.	13.07-13.56	Thick deposit on SW flank of bar, thinner one to NE.
21.209	Bedded medium gravels in whitish dense shell-sand matrix.	13.28-13.70	Becomes 21.210 at SW limit.

Context	Description	Top (m OD)	Comments
21.210	As 21.209, but stained black/brown.	13.14-13.28	Continuation of 21.209; presumably local mineral precipitation due to within-bed water drainage.
21.211	Dense fine white sand, grading upwards to light grey clay with fine white sand lenses.	13.52-13.76	Only survives over top of bar; SW end of lens dips into likely erosion gully cut through 21.209.
21.212	Light yellow/grey coarse sand, encapsulating lens of light grey clay.	13.56-13.60	Fills rest of erosion gully cut into 21.209.
21.213	Beige sandy matrix with mainly pea gravel and some medium sized gravel (very similar to 21.206).	13.60-13.88	SW tail just overlaps 21.212, but this could be due to secondary slippage and sequence could thus be reversed.
21.214	Pale yellow fine sand/silt incorporating lenses of small gravel (NE), light grey clay with fine white sand (lower part), and mottled grey/white coarser sand with pea gravel.	13.32-13.76	SW tail merges into 21.216.
21.215	Mottled grey/white coarse sand and pea gravel.	13.28-13.33	Against palaeobank.
21.216	Light yellow/grey coarse sand enveloping band of light grey clay with fine white sand lenses.	13.34-13.78	
21.217	Fine grey silt.	13.36-13.46	Against palaeobank.
21.218	Mottled grey/white coarse sand with pea gravel.	13.40-13.66	Occupies slight hollow in top of 21.217 and 21.216.
21.219	Light grey clay with fine white sand lenses; becomes iron-stained adjacent to palaeobank; incorporates lenses of coarse light yellow-grey sand, coarse sand/pea gravel, and coarse black-stained sands.	13.51-13.78	NE end feathers into lowest part of 21.223, which otherwise overlies.
21.220	Light grey clayey silt with patches of fine white sand.	13.52-13.68	Occupies basal fill behind mud stack on palaeobank, including fill of likely burrow.
21.221	Light yellow sandy silt with fairly strong iron staining.	13.73-13.87	Overlies 21.220 and also occupies hollow in front of main mud stack.
21.222	Pale yellow silt, dense shell-sand.	13.70-13.83	Possibly fill of recut through 21.221 behind mud stack.
21.223 21.980	Pale yellow fine sand/silt grading up-slope in both directions into slightly sandy silt; lenses of coarser sands, light yellow/grey or black-stained; upper part of deposit in trough is 'dirtier' and more iron-stained (excavated context 21.980).	13.81-14.27	Above bar it incorporates thin lenses of small gravel, 21.224; overlies 21.219, 220, 221 and earlier deposits where exposed.
21.224	Two main and other minor lenses of small gravel.	13.85-13.97 and 13.96-14.06	Entrapped within 21.223 above bar crest.
	Late Bronze Age occupation soil		
21.942 *et al.*	Blackish soil, rich in burnt debris and cultural artefacts; sandier towards bottom. Becomes paler as it rides up onto levee.	14.10-14.31	Comprises stratigraphic units 33.A, 33.C and 33.D.
21.830 *et al.*	Brown loamy soil.	14.20-14.26	Still in situ material, confined to trough. Stratigraphic unit 33.E.
21.826	Grey-brown loam with very dense burnt flint.	14.26-14.36	Conflation horizon at interface with flood-reworked deposits. Stratigraphic unit 33.I.

Context	Description	Top (m OD)	Comments
21.815 21.227 *et al.*	**Parcel 6D**		For description see Area 14.
	AREA 14 South sections RH26-RH30		**Figs. 2.21-2.24; columns 5, 6, 20 and 31; matrix, Figs. 2.35 and 2.36**
	Parcel 2?		
14.400	Tufa bed.	12.30	Not in drawn sections, but seen in sump at *c.*12.20-12.30m OD.
	Parcel 2D		
–	Base of drawn section.	12.42	Pump-reduced water level.
14.117	Grey lensed silt, gritty and sandy with lenses of small gravel and tufaceous pellets.	12.52-12.79	Internal lenses undulate, as does upper surface - uncertain whether erosion. Penetrated by some rootlets.
14.119	Similar to 117, but with more gravel and tufa.	12.52-12.76	Uncertain stratigraphic relationship to 14.117. Bone pieces noted (not retrieved).
14.116	Predominantly brown silt but incorporating some dipping sand lenses.	12.91-13.06	Overlies both 14.117 and 119; most of sand lenses dip in parallel with underlying interface. Penetrated by rootlets; ^{14}C date on animal bones 5100 ± 100 BP (BM-3039).
–		12.95-13.10	Water table.
14.115	Beige or creamy clay, sticky and plastic; includes sandier lenses, notably one at SW end.	13.07-13.16	Apparently merged into 14.134 (parcel 2E) in northern section, but clearly truncated by erosion scarp 14.133 and parcel 2E in southern.
14.133	Truncated on NE side by erosion scarp.		
	Parcel 2E		
–	Base of drawn section.	12.42	Pump-reduced water level.
14.401	Lenses of blue clay.	12.84-13.14	Lying on upper slope of erosion scarp 14.133; continues as lenses and clods within 14.118.
14.118	Sticky brown lensed silt, some sand, plus blue clay inclusions as lenses or clods; also large wood piece.	12.48-12.95	Blue clay components, especially against scarp, are continuation of 14.401 on upper slope. Surface dips to NE.
14.121	Tight bundle of pieces of roundwood within matrix similar to 14.118, in SE corner of trench.	12.55-12.66	Many fragments with beaver gnawing; ^{14}C dated 3760 ± 50 BP (BM-2435 and 2436).
14.122	Greyer sandy layer in SE corner of trench.	12.79(-?)	Not clearly differentiated from 118 at W end.
14.120 14.349	Sticky brown silt, relatively homogenous.	12.93-13.17	Surface dips gently to E.
14.134 14.348	Beige/grey sticky clay with sandier lenses.	13.10-13.30	Rises at edge of channel fill to SW; also rises gently to N before truncation by erosion scarp with mini-stacks.
	Parcel 2B		
14.114 14.169.2	Light yellow-brown sticky clay: some orange Fe mottling.	13.54-13.64	
14.113 14.169.1	Light yellow-brown clay with orange Fe mottling giving gritty texture.	13.79-13.95	
14.112	Greyish yellow silt with intense purple (Mn) staining and Fe-rich pockets.	13.89-14.09	

Context	Description	Top (m OD)	Comments
–	Truncated by major erosion scarp facing N.		
	Parcel 4B (in parallel with sequence below)		
–			NB. Layers 14.402-14.411 and 14.414 all excavated as single context, 14.163.
–	Base of drawn section.	12.95-13.10	Water table.
14.402	Large to fine gravel with tufaceous clasts.	13.01-13.15	Possibly the bed gravel of an active channel.
14.403	[NW section only] Small gravel and silt.	13.04	Grades from 14.402 very close to palaeobank.
14.404	[NW only] Gravel with heavy iron staining.	13.03-13.14	Probably scoured on NE edge prior to 14.405.
14.405	Small to medium gravel, lower part with significant coarse shell-sand matrix, yellow.	13.13-13.42	Top dips towards SW and palaeobank.
14.406	[E only] Yellow-grey silt.	13.32-13.35	Top grades into 14.160; part seems to underlie 14.410.
14.407	Grey silty sand.	13.14-13.43	Grades into 14.408 above; dips to SW.
14.408	Coarse to medium shell-sand with lensing; incorporates tufaceous lump.	13.24-13.45	Bottom end grades into gravel deposits; dips to SW.
14.409	[NW only] Heavy gravel lens with large cobbles (up to 15 cm).	13.04-13.23	Overlies 14.404 and 405; dips to SW.
14.410	Partially banded fine to medium gravels with moderate shell-sand component; incorporates silt clod [E].	13.33-13.52	Evidently truncated by steep scarp along SW edge, prior to 14.414.
14.411	Coarse grey sands and silts.	13.34	Grades from 14.410 along its SW margin.
14.160	Thin variegated sand layer, yellow to grey, fine to coarse, with some gravel; incorporates clod of red clay in NW.	13.36-13.56	Much exposed after removal of Post-Medieval disturbed gravel 14.139, but SW margin dipping under 14.162; at lowest point to S it grades into top of 14.406.
14.414	[NW only] Loose gravel with a little sand and clay component; incorporates clay clods.	(13.26)-13.41	Deposited against steep scarp through 14.410/160, possible erosion gully along foot of palaeobank; also noteworthy that it leads down to a possible burrow in the bank (which has a silty fill).
14.162	Lens of mainly coarse shell-sand, yellow-grey, partly iron stained.	13.36-13.49	Overlies 14.160 and 414; grades into 14.161 above; confined to gully along foot of palaeobank.
14.161	Band of medium gravel (2-5 cm) in yellow/grey clay matrix.		Sits in shallow hollow above 14.162; merges almost imperceptibly into Post-Medieval disturbed gravel 14.139.
14.417	[NW only] Pale-coloured sandy clay with fine clayey banding.	13.46-13.49	Only tail of deposit survives in this quadrant, but present in sections to NW.
14.168 14.111	[NW only] Yellow-grey silt with strong component of fine sand; variable iron mottling.	13.89-14.24	Encapsulates very coarse lenses, 14.418 and 419.
14.418	[NW only] Lens of coarse shell-sand.	13.63-13.66	Within 14.168.
14.419	[NW only] Lens of coarse shell-sand.	13.75-13.78	Within 14.168; does not quite extend to Post-Medieval cut and thus does not appear in conjoining section.
14.167 14.111	[NW only] Yellow-grey silty loam.	13.92-14.24	
14.166	Pale shell-sand lens.	13.94-13.99	

Context	Description	Top (m OD)	Comments
	Late Bronze Age occupation soil		
14.110 14.050.2	Dirty yellow-grey silty loam.	-	Subsoil infiltrated by some cultural material?
14.038 14.030 *et al.*	Light grey-brown silt, becoming darker to N as it dips into trough.	14.29-14.42	In situ occupation deposits, rich in refuse.
	Parcel 6D		As described in next section.
	AREA 14 North-east sections RH22-RH24 and RH 26		**Figs. 2.17-2.19 and 2.21; columns 7, 18, 19, 66, 67; matrix, Figs. 2.35 and 2.36.**
	Parcel 4B (in parallel with sequence above)		
–	Abuts palaeobank eroded through parcel 2B (see table above).		
–	Base of drawn section.	*c.*12.90	Water table.
14.156	Clean gravel in lenses.	12.90-13.23	Mounded profile forming bar.
14.155	Lenses of coarse sands and gravels.	13.01-13.38	Surface dips towards palaeobank.
14.154	Relatively homogeneous layer of shell-sandy silt; lens of silt trapped beneath at one point.	13.02-13.45	Covers most of 14.155, except at NE end of extension trench, where probably truncated by Post-Medieval surface.
14.153	Lenses of mixed gravels with coarse sands.	(?13.10)-13.46	Lying over SW flank of developing bar.
14.152	Coarse sand in clay with gravel component (*c.* 35%).	(13.33)-13.42	This and overlying layers to 14.147 only encountered on SW side of bar.
14.420	Grey sand lens.	13.12-13.15	Feathers into 14.152 and 153 alongside palaeobank.
14.158	Sandy gravel (*c.* 70%, up to 5 mm).	13.39-(13.52)	SW end sealed by 14.148, but most of layer overlain by, and poorly differentiated from, basal deposits of parcel 9. No burnt flint or artefacts.
14.151	Thin lens of grey sandy clay with a little gravel (*c.* 20%).	13.36-13.46	Grades out along SW edge; probably graded into 14.149 at top (NE) edge.
14.150	Grey lens of coarse shell-sand and small gravel (c. 50%).	13.20-13.53	Grades into finer material at palaeobank.
14.149	Lens of grey coarse sands and silts; at highest it is enriched with iron precipitates.	13.25-13.57	SW section shows likely linkage with 14.162, of similar character.
14.148	Coarse yellow/grey sands with small to medium gravel (2-5 cm; 30-50%).	13.31-13.66	SW section demonstrates linkage to 14.161 against palaeobank, where there is a concentration of eroded clasts of tufaceous and clayey deposits. At highest comes into contact with 14.139 with no clear boundary.
14.147	Finely laminated pale grey deposit of muds/silts/fine sands/shelly sands; the upper surface marked by a sparse string of small pebbles; also contained an isolated pebble lens.	13.45-13.76	SW section demonstrates linkage to 14.417.
14.168	Description as in previous section.	13.86-14.05	
	Late Bronze Age occupation soil		
14.315 14.305 *et al.*	Dark grey silty loam.	14.21-14.33	In situ occupation deposits sitting in trough.

Context	Description	Top (m OD)	Comments
14.265 *et al.*	Grey silty loam with dense burnt flint cobbles.	14.23-14.38	Conflation horizon at interface with flood-reworked silt.
	Parcel 6D		
14.004 14.001 *et al.*	Pale yellow-grey clay-loam.	14.70	Contains quantities of reworked LBA artefacts, thinning upwards. On N edge, possible diffuse boundary observed, representing upper continuation of 14.018 (section RH22).
	Parcel 9		
14.018	Cut-line of landscaped steep bank with almost level foreshore.	13.22-14.55	Truncates upper part of parcel 4B including LBA on top; may also truncate parcel 6.
14.140	[NE] Grey sandy silt.	13.41-13.43	Overlies 14.158 of parcel 4B; uncertain whether cut-line 14.018 runs along base or top of this layer; this might thus alternatively be in situ or minimally disturbed parcel 4B, for example an extension of 21.211 dipping away to the NE. Excavated part yielded only four bone fragments. Overlain by 14.011.
14.139 14.017	14.017 to SW (1984); surface layer of large gravel 'cobbles', extending at least 5 m out from foot of steep bank.	?13.41/13.49-13.68	Whenever in direct contact with underlying gravel deposits (14.148, 158 and 161), boundary difficult to discern; however, in contrast to those, 14.139 incorporated much burnt flint and LBA pottery. It constitutes a redistribution and resorting of exposed gravel surfaces plus material eroded from the cut bank.
14.015	Timber post/beam socket cut further back than 14.018.	base 13.94	Contemporary with or slightly later than 14.018 and sealed by 14.012.
14.014 14.143	Sticky yellow-brown clay.	13.53-13.82	Early erosion deposit caught in angle at foot of steep bank; overlies 14.139/017.
14.012 14.142	Lensed deposit of grey-brown clays and shell-sands; some lenses rich in charcoal.	13.56-14.15	Dipping strongly away from cut bank, and petering out rapidly.
14.011 14.141	Light yellow-brown clay-loam.	c.13.72-14.01 (+?)	Boundary with 14.010 above often diffuse.
14.010 14.141	Yellow-brown clay-loam.	13.40-14.50(+)	Surface dipping to NE, perhaps truncated prior to nineteenth- and twentieth-century AD dumping.
–	Directly overlain by modern dump sequence (sections RH10 and RH12).	14.90-15.15	Part at least thought to be associated with early nineteenth-century lock construction (Chapter 10).
	AREA 15 **section RH25**		**Fig. 2.20**
	Parcel 11		
–	River level at	12.80	
15.011	Pebbles with brown mud lenses.	12.80-12.87	Piece of charred wood present.
15.010	Orange sand.	13.05-13.18	
15.009	Grey sand with pebbles.	13.18-13.26	
15.008	Pale yellow sandy lens.	13.21-13.27	Possibly truncated at NE end by small bluff also cutting down into 15.009.
15.007	Brown sand with some small gravel.	13.28-13.40	
15.006	Large gravel.	13.31-13.73	Surface dipping strongly to NE probably as result of comparatively recent river erosion.

Context	Description	Top (m OD)	Comments
15.005	Sandy matrix with small gravel.	13.37-13.99	Sloping upper surface is current foreshore of Thames.
15.004	Pebbly silt (only in SW end section - river bank).	14.17	
15.003	Silt (only SW end).	14.30	
15.002	Very thin gravel lens (only SW end).	14.32	
15.001	Silt grading to modern top soil (only SW end).	14.64	
	AREA 18 **section RH31**		**Fig. 2.25; matrix, Figs. 2.35 and 2.36.**
	Parcel 2E		
–	Base of drawn section	12.87	Water table (no pumping out).
18.825	Sticky brown silt, slightly organic.	12.92-12.98	Continues layer 14.120 (baulk of >1 m between); merges into 18.826.
18.826	Beige/grey sticky silt with a little shell.	13.07-13.18	Top slightly undulating.
	Parcel 2B		
18.828	Pale yellow silt with slight shell.	13.36-13.47	Recorded in the field as three very slightly differing layers (18.827, 828, 829) of dubious significance. Top undulating.
18.830	Pale greyish-yellow silt with frequent Mn flecks and increasing Fe staining; more shelly towards NNE end. Contains lens of Mn stain.	13.54-13.60	Slightly undulating top. Possible burrow from palaeobank at NNE end.
18.835	As 18.830, but slight Mn flecking and more intense Fe mottling, and firmer texture.	13.83-13.89	Undulating top. Recorded in the field as two slightly differing deposits.
18.837	Pale yellow silt with a little shell, dense Mn and a little Fe staining.	13.90-13.96	Recorded in the field as two slightly differing contexts (18.837, 838); becomes softer towards NNE.
–	Truncated by steep erosion scarp at NNE end, abutted by:		
	Parcel 4B		
–	Base of drawn section.	12.87	Water table.
18.848	Mottled green/orange/grey sticky silt, with a little shell.	13.14	Recorded in field as two variants, here combined.
18.850	Beige sticky silt with a little shell; whiter patch within.	13.28-13.29	Recorded in field as three variants, here combined; 'false' interleaving with parcel 2B deposits at palaeobank due to undercutting or burrow?
18.853	Beige silt similar to 18.850 but with more shell, increasing upwards.	13.37-13.57	Recorded in field as two variants, here combined.
18.855	Banded yellow-orange coarse shell-sand.	13.46-13.53	Surface possibly slightly truncated by Post-Medieval cut to parcel 9.
18.834	Very pale yellow-grey silt.	13.69-13.73	SW end appears to interleave with parcel 2B deposits - likely burrow fill.
18.842	Pale yellow-grey silt with moderate amount of shell and some Fe staining.	13.78-13.89	
18.845	Very shelly, pale yellow-grey lens.	13.79-13.84	

Context	Description	Top (m OD)	Comments
18.840	Pale yellow shelly silt with Mn and Fe flecking; shell and Mn tend to increase towards NE.	13.83-14.13	Recorded in field as three variants, here combined, slope at NE end may have been increased by LBA activity.
18.843	Yellow/grey mottled silt.	14.04	Remnant of occupation - dirtied soil.
18.847	Very pale yellow shelly silt.	13.90	Likely LBA feature fill.
	Late Bronze Age occupation soil		See descriptions for Area 14.
	Parcel 6D		See descriptions for Area 14.
	Parcel 9		
18.723	Steeply cut scarp through upper parcel 4B, LBA deposits and ?parcel 6D		
18.856	Dense gravel (up to 2 cm diameter).	13.58-13.63	Possibly scoured by gully along foot of bank prior to 18.857.
18.857 18.728	Light brown loamy silt.	(13.65+)	Incorporated a few eroded LBA artefacts.
18.710	Light brown silt.	–	Horizon of significant redeposition of LBA material.
18.858	Orangey-brown clay.	(13.69+)	Overlies 18.856 and probably later than 18.857 and 18.710.
–	Truncated.		
	AREA 31 sections RH32-RH34		**Figs. 2.26; column 55; matrix, Figs. 2.35 and 2.36.**
	Parcel 2E?		
31.676	Sizeable tree trunk; enveloping sediments not visible.	12.40	^{14}C dated: 3870 ± 50 BP (BM-2662); aligned grid WNW-ESE.
31.675	Portion of tree trunk; enveloping sediments not visible.	12.48	
	Parcel 2E		
–	Base of drawn section.	12.50	Reduced water level.
–		12.88	Water table.
31.670	Light grey medium-sand silt.	12.54-12.56	
31.674	Darker brown silt with variable fine shell-sand, incorporating lenses of lighter grey medium-sand silt and wood fragments.	12.78-12.89	Step in upper profile at a dipping lens.
31.634	Soft grey-brown silt with variable fine shell-sand; incorporates a beige lens of shell-sand silt, as well as wood and charcoal fragments.	12.94-12.97	
31.636	Very pale beige sticky clay, banded with varied shell-sand component, generally slight; some orange mottles.	13.19	Clearly suffered truncation, both along top (SW section) and obliquely along NE edge.
	Parcel 2B		
31.677	Fairly dense shell-sand silt, brown.	c. 12.97	Seems to be a continuation of 678, differentiated only by colour due to retention of organic component.

Context	Description	Top (m OD)	Comments
31.678	Pale yellow, fairly dense shell-sand silt.	*c.* 13.35	Dipping of layer towards 677 in SE section probably original rather than slump, since it overlies pre-existing steep truncation of 634 and 636; apparently eroded away by gully in N of cutting (SW section).
31.679	Pale yellow/beige plastic clay with very slight shell; lower 'clod' slightly softer.	*c.* 13.40	May be subsumed within 641 in opposite (NW) section.
31.637	Light grey silt with shell-sand; stiffer with less shell in bottom stretch.	12.98-13.28	Occupies erosion gully; possibly derived from 636, but has more shell-sand.
31.639	Stiff orange silt, rather plastic.	13.02-13.07	Merges into 640 and 637 at NE end.
31.640	Very pale beige, almost white, soft sticky silt with a little orange mottling and moderate quantity of fine shell- sand.	13.15	Rapidly grades into 641.
31.641	Firm beige clay-silt with a little fine shell-sand and pale orange mottling.	*c.* 13.55	Horizontally bedded in SE section but fills ?gully running across W corner of cutting.
31.680	As 641, but slightly greyer and with more orange mottling.	*c.* 13.75	Only recorded in SE section where steeply truncated.
–	Truncated by erosion scarp facing N.		
	Parcel 4B		
–	Base of recorded sections.	12.50	Reduced water level.
31.685	[S section] Lens of fairly dense gravel, shell and shingle.	*c.* 13.40-13.50	Small deposit sitting in hollow part way up palaeobank; overlain by 31.643.
31.681	[E] Off-white sandy silt.	*c.* 12.90-13.00	Overlain by 31.682 and 31.683; dipping N away from palaeobank.
31.682	[E] Bluey clay with slight shell.	*c.* 12.90-13.10	Dips N; butts 31.683; overlain by 31.642.
31.683	[E] Pale yellow-beige plastic clay with very slight shell.	*c.* 13.10-13.15	Overlain by 31.643.
31.666	[W] Very soft shelly silt, brown with blue tinge.	12.50-12.80	Steeply inclined against palaeobank - ?slump; overlain by 31.642.
31.642	[W, E] Organic silt with soft, loamy texture and slight odour; brown with bluey-grey mottles; varied inclusions of sand, shingle, gravel, small tufa lumps and small wood/charcoal fragments.	12.70-13.15	Dips N; contains lenses 31.664, 667, 668; partly interleaves with overlying 31.665 and 669; butts or grades up-slope into 31.643, the equivalent above water table.
31.664	[W] Brown sand and shingle lens with wood fragments.	12.53-12.63	Lens within 31.642.
31.667 31.668	[W] Two lenses of light yellow-grey sandy silt/shingle/gravel.	12.68-12.80 12.76-12.93	Lenses within 31.642.
31.665	[W] Lens comprising dense mat of twigs.	12.76-12.83	Lens in top of 31.642; much of wood beaver-gnawed; sample ^{14}C dated (BM-2661).
31.669	[W] Mixed yellow-brown silt, fairly shelly and moderate gravel.	12.90-13.16	Overlies and partly interleaves with 31.642; interleaves with 31.663.
31.663	[W] Lens of light yellowy-grey sand and shingle, with wood fragments.	12.82-12.94	Interleaves into 31.669.
31.644	[W] Organic silt similar to 31.642, but more homogeneous matrix, also lacking stones and coarser sand.	12.96-13.06	

Context	Description	Top (m OD)	Comments
31.648	[W] Organic silt with strong component of yellow sand.	13.02-13.04	Likely equates with 31.684 in E section - but no physical linkage via N section.
31.684	[E] Sand lens.	*c.* 13.10-13.15	Overlies 31.642; overlain by 31.646.
31.643	[W, S, E] Beige plastic clay with pale orange mottles; a little more shell- sand than 31.641, slightly greyer colour and occasional pebbles; charcoal concentrations towards lower (N) end.	13.14-13.62	Up-slope continuation of 31.642, 644 *et al.* above organic preservation level. Contains lenses 31.645.
31.645	[W] Lenses of yellow-ochre medium sand.	-	Within lower part of 31.643.
31.646	[W, E] Pale beige loamy silt; paler than 31.643 with more very fine shell-sand, some as slight lenses.	13.23-13.61	Contains prominent lens 31.647; Lower (N) end perhaps erosion truncated, before 31.649.
31.647	[W] Lens of pale yellow fine to medium shell-sand.	13.28-13.38	Interleaves into 31.646.
31.649	[W] Mixed yellow/brown medium-sand silt.	13.15-13.24	
31.650	[W, S, E] Light yellow-grey, soft, fine to medium sand silts with fine lensing.	13.53-13.88 [- *c.* 14.00]	Down-slope end interleaves with 31.651-.656.
31.651	[W] Lens of plastic beige clay with some fine shell.	13.17-13.37	Overlies 31.649 and lowest part of 31.650.
31.652	[W] Whitish grey lens of dense medium to coarse shell-sand.	13.21-13.50	Interleaves into 31.650.
31.653	[W] Plastic beige clay with some fine shell (as 31.651).	13.37-13.48	Contains lenses 31.654 and 655; up-slope end grades into 31.650.
31.654	[W] Thin lens of plastic clay with extremely fine light yellow sand.	13.39-13.50	Within 31.650 and 653.
31.655	[W] Whitish grey dense medium to coarse sand (as 31.652) with addition of string of stones - medium size to shingle.	13.30-13.33	Within 31.653.
31.656	[W, E] Light yellow-grey soft silt similar to 31.650, but with moderate quantity of medium to coarse shell.	13.49-13.55 [- *c.* 13.70]	
31.657	[W, S, E] Extremely dense and coarse shell-sand; very friable.	13.77-13.88	
31.658	[W] Light yellow-grey, soft, fine to medium sand silt, as 31.650.	13.81-13.93	Top surface slopes to N, representing one side of broad trough in edge of palaeochannel.
	Late Bronze Age occupation deposits		For description see Area 14.
	Parcel 6D		For description see Area 14.
	Parcel 9		
31.503	Steeply cut scarp through upper deposits of parcel 4B, LBA deposits and parcel 6D.	13.25-13.77(+)	
31.659	Dense medium to large gravel, including some burnt flint.	13.44-13.48	Incorporating redeposited LBA material.
31.660	Stiff brown silt.	13.51-13.65	
31.661	Grey soil rich in LBA finds.	13.55-13.69	Redeposited LBA material, probably eroding from steep bank above.
31.662	Light brown silt.	(13.95+)	
-	Truncated by hand excavation.		

Context	Description	Top (m OD)	Comments
	AREA 17 **section RH35**		**Not drawn.**
	Parcel 5		
–	Base of inspected section.	12.72	Water table (temporarily lowered).
17.007	Light brown fibrous silt with many wood fragments, wood pebbles, other organic detritus and a little shell. Yellow sand also noted - ?below.	12.99	
17.006	Beige silt with little shell.	13.14	Probably a gleyed horizon.
17.005	Beige silt with bluey hues; occasional charcoal and some Fe mottles.	13.22	Probably followed by erosion (see 17.004).
17.004	Pale yellow shell-rich silt containing clasts of eroded 17.005/17.006.	13.28	
17.003	Pale yellow shell-sandy silt with occasional charcoal.	13.45	
17.002	Lens of less consolidated silt similar to 17.003.	13.48	
17.001	Pale yellow-grey silt with a little shell.	(13.78)	
–	Exposure of higher deposits prevented by thickness of overburden.		
	AREA 23 East **section RH36-RH40**		**Fig. 2.27; column 30; matrix, Fig. 2.38.**
	Parcel 5		
–	Base of sections.	13.00-13.51	
23.633	Small gravel in beige coarse shell-sand matrix; some Fe staining.	13.06-13.53	Surface dips to NW.
23.632	Dark brown mottled or, locally, banded sand with charcoal inclusions.	13.20-13.35	Occupies low-lying pocket and grades rapidly into 23.631 above; distinction may de due solely to position relative to water table.
23.631	Banded light grey/beige coarse shell-sand with grey clay lenses.	13.30-13.69	
23.630	Light grey sandy silt with a few charcoal flecks.	13.37-13.56	Occupies low-lying pocket at NE end.
23.629	Thin layer of beige-grey sandy silt with some charcoal flecks.	13.38-13.71	Becomes intermittent at SW end.
23.628	Light yellow shell-sand with Fe staining, particularly at base.	13.51-13.80	
23.627	Very thin lens of light grey clay (SW end only).	13.70-13.80	
23.626	Light yellow sandy silt with some beige banding and coarse shelly lenses.	13.66-14.00	Gently undulating top.
	Parcel 6A		
23.625	Light grey-brown to yellow-brown (higher) silt with Fe and Mn staining.	14.28-14.48	
	Parcel 6B		
23.624	Brown clay-loam topped by dark brown soil profile.	14.69-14.73	

Context	Description	Top (m OD)	Comments
–	Overlain by modern dumped material.		
	AREA 23 West **section RH41-RH43**		**Fig. 2.28; columns 27-29; matrix, Fig. 2.38.**
	Parcel 5		
–	Base of sections.	12.99-13.30	
23.623	Dense large to medium gravel with brownish sand.	13.12-13.28	
23.622	Medium to pea gravel with shell inclusions.	13.26-13.43	
23.621	Loose large to medium gravel in coarse sand matrix.	13.45-13.55	
23.620	Thin intermittent band of small gravel.	13.50-13.59	
23.619	Banded beige/pale yellow sandy silt with shell-rich lenses.	13.65-13.76	Dipping surface in SW section suggests oblique truncation prior to 23.618.
23.618	Whitish-yellow, shell-rich sandy silt.	13.54-13.86	
23.617.2	Pale yellow/beige sandy silt.	(13.56)-13.89	Thin and underlying deposits scoured by erosion gully running approximately SW-NE.
23.617.1	Banded light yellow/beige sandy silt.	13.71-14.01	
23.616	Light grey/pale yellow sandy silt.	13.79-13.90	Truncated towards SE end of cutting by erosion scarp 23.615.
	Parcel 6A		
23.615	Erosion scarp facing SE; fairly steep at top but rapidly shelving lower down.	13.38-13.90	
23.614	Light grey-brown sandy silt.	13.55-13.90	Dipping gradually south-eastwards in conformation with erosion slope.
23.613	Yellow-brown sandy silt with orange mottling (Fe) and Mn staining in upper zone.	14.15-14.34	Truncated towards SE end by erosion scarp 23.612. Layer yielded one sherd of pottery.
	Parcel 6B		
23.612	Steep erosion scarp facing SE truncating all observed earlier deposits.	13.00-14.30	
23.611	Gravel in sticky grey silt with a little shell (SE section only).	13.15-13.20	Occupies a small hollow in the palaeobank; probably material partly derived from flanking gravel deposits 23.620-623.
23.610	Mixed yellow/pale grey sandy silt with a few stones; incorporates clods of 23.614 eroded from bank.	13.22-13.46	
23.608 23.609	Banded grey sandy silt (608), interleaved with coarse yellow shell-sand with Fe staining.	13.30-13.63	Steeply inclined against palaeobank.
23.607	Light grey sticky silt.	13.59-14.00	
23.606	Mid-brown stiffish clay-loam with Fe mottles and Mn staining.	14.05-14.30	
23.605	Light brown silty clay with Mn staining.	14.30-14.58	
–	Overlain by modern dumped material, over 1 m thick.		

Context	Description	Top (m OD)	Comments
	AREA 8 East **section SA8**		**Fig. 2.29; matrix, Fig. 2.38**
	Parcel 3?		
–	Base of recorded section.	*c.* 12.40-12.55	(Water level reduced.)
8.157	Orange gravel.	*c.* 12.65	Approximately horizontally bedded. Probably equates with 8.082 in section SA7; may be a point-bar within the parcel 3 channel.
	Parcel 3		
8.156	Orange shell-sand.	*c.* 12.75	Approximately horizontally bedded.
8.155	Organic-rich deposit.	*c.* 12.80	Approximately horizontally bedded.
8.154	Shell-sand.	*c.* 12.83	Approximately horizontally bedded.
–	Erosion line obliquely truncates 8.154 and underlying deposits.	*c.* 12.40-12.83	Dips to SE.
	Parcel 5		
8.153	Orange shell-sand.	*c.* 12.50-12.65	Overlain by part of 8.151, but might have been interleaved.
8.151	Black/brown organic deposit including largish timber and lens 8.152.	*c.* (12.65)-13.10	Undulating surface dips to SE - perhaps an erosion surface.
8.152	Lens of shell-sand and gravel.	*c.* 12.78	Within 8.151.
8.150	Thin, intermittent band of orange sand with charcoal horizon along sloped part of top.	*c.* 12.75-13.13	
8.149	Pale sand.	*c.* 12.90	
8.148	Gravel in shell-sand matrix.	*c.* 13.03-13.50	Apparently edge of thick gravel bar tapering out to NW and interleaving with 8.147.
8.147	Light brown clay with some shell-sand.	*c.* 13.13-13.30	Bifurcates around tail of gravel 8.148 and rises up over it. Charcoal at interface with 8.150.
8.146	Pale yellow sand.	(13.50+)	
–	Exposure of upper section prevented by instability of overburden.		
	AREA 12 **section SA16**		**Fig. 2.30; matrix, Fig. 2.39**
	Parcel 8A		
–	Base of drawn section.	*c.* 12.70	Reduced water level.
12.015	Dark sand with high organic content including wood pieces; contains lens of dark clay.	*c.* 12.80-12.92	Undulating top, gently dipping to SE.
12.014	Lenses of sand, sandy clay and clay; shell content, wood fragments and charcoal.	*c.* 12.93-13.12	Gently dips to SE.
12.012	Bed of (?shell-) sand, some lensing including major lens, 12.013; a little wood.	*c.* 13.11-13.40	Dips to SE.
12.013	Lens of sandy clay rich in charcoal fragments.	*c.* 13.05-13.20	Low within 12.012.
12.011	Sandy clay/sand lenses.	*c.* 13.29-13.46	Gently dips to SE.
12.010	Mixed clay.	*c.* 13.41-13.59	Considered to have been redeposited material at base of new M25 floodway.

Context	Description	Top (m OD)	Comments
	AREA 11 **sections SA17 and 18, plan SA19**		**Fig. 2.31; matrix, Fig. 2.39.**
	Parcel 8B		
–	Base of drawn sections.	12.13	Pump-reduced water level.
11.009	Brown clay with a little gravel (NW-SE section only).	12.33-12.46	
11.007	Dark gravel in shelly silt.	12.72-12.74	Boundary with 11.006 above not definable in SW-NE section.
11.006	Gravel in mid-grey shelly silt.	12.64-12.81	Top undulates, but overall gentle dip SE. Included pieces of tile and a bone.
–	Possible erosion scarp.	12.15-12.80	Interpolation suggests approximately W-E alignment, dipping to N. If not erosional, this might be natural slope of flank of gravel bar.
	Parcel 8C		
11.008	Brown clayey silt with green/black mottling grading into deep greeny-brown to SW (SW-NE section only).	12.80-12.87	
11.005	Highly shelliferous silt.	12.77-12.91	Undulating surface, interleaving in places with 11.004 above, possibly due to disturbance associated with structure 11.016.
11.016	Group of wooden rods and branches mainly co-aligned W-E.	12.92	Likely represents deliberately laid timber on surface of 11.005 (Fig. 2.31).
11.004	Greeny-browny-grey clay with orange mottles and some shell; horizon of charcoal flecks trapped towards base.	12.97-13.16	Top undulates.
11.003	Pale grey/beige clay.	13.18-13.29	Observed to thicken to 0.50 m to the north-east of the drawn section.
11.002	Light yellow-brown clayey silt.	(13.37-13.59)	Top rather irregular and likely truncated by ?modern floodway construction.
11.001	Topsoil/overburden.	(-14.50)	Modern.

CHAPTER 3

Cultural material, charcoal and wood remains

The two main horizons of human activity on the site – Middle Neolithic and Late Bronze Age – are dealt with in detail in other published and planned volumes. A summary of their chronological implications for the whole alluvial sequence is, however, given in the appropriate parts of Chapter 10. The opportunity is taken in this volume to publish material from the alluvial 'non-cultural' contexts, but also finds which arrived on the site through desultory or low-scale human visitation. Not only do these have a bearing on the chronology of the sequence (although often as *termini post quem*), but they can also relate to the history of use of the site and its immediate hinterland.

In this chapter the essential details of the relevant groups of material are simply presented in tabular form. Discussion of their significance for the contexts is reserved for synthesis in Chapter 10.

Mesolithic flints

The crest of the silt island established during the Atlantic period was high enough not to suffer serious flooding very often. The mature soil profile that developed there was recipient not only to the dense refuse of the Middle Neolithic phase, but also to sparser traces of Early Neolithic and Late Mesolithic activity. Cut features and burnt deposits suggest that on occasion this was more than fleeting, and it is more likely that there was periodic, perhaps seasonal, use of the island.

Twenty-five microliths have been recognised, coming from widespread locations in the interior zone (Table 3.1; Figs 3.1 and 3.2). In some cases they came from positions low in the soil profile, but they also occurred within the main Neolithic contexts, both feature fills and surface deposits. Only in the western corner of Area 19 did there seem to be some separation between the refuse-rich horizon and an underlying thin spread of finds including microlith FL34, although both still occurred within the same, deep soil profile. Also here were microliths FL39–42 associated with the TL-dated burnt flint cluster (Chapter 4). Four microliths come from the fill of the early ditch running through Area 20 and into Area 13 which has yielded a Late Mesolithic radiocarbon date (microliths FL31, 52–54). A cluster of eight straight-backed bladelets recovered from

Area 20 (FL44–51) is important in indicating a primary context since it seems probable that they were deposited as a hafted set. No tip-piece is among them and they would appear, therefore, all to have been side-hafted. From another context there is an unfinished example (FL30), indicating on-site manufacture. All are very fresh with little sign of post-depositional damage and must represent losses or discards on the site. Also noteworthy are five examples of the 'needle-point' type (Fig. 3.1).

Roger Jacobi has inspected the group and would date virtually all to the later Mesolithic, after about 7500 BC. One piece, however, from 13.803.1 (FL32), is more likely to be earlier, and may therefore be in a redeposited context.

Earlier Neolithic pottery in secondary contexts

Inevitably some Neolithic finds (i.e., mainly Middle Neolithic) were reworked into later contexts. This is seen most prominently in the concentrations found throughout the 2F gravel. It also occurs in rather restricted quantity in the Late Bronze Age levels, almost exclusively in the interior zone, where the gravel was only capped by a thin silt layer which was consequently penetrated by many cut features. More significant for the topographic development of the site, however, is the occurrence of earlier Neolithic pottery in the gravels low in parcel 4B, dated unequivocally to the late second millennium BC. This pottery is detailed in Table 3.2 and eight featured sherds are illustrated (Fig. 3.3). In these same contexts were over ninety pieces of bone and twenty-five struck flints.

Late Neolithic to Early Bronze Age finds

Prior to the research excavations, the only evidence for activity during this period was a cluster of butchered bones lying in channel sediments (A6 F125). To add now to those are six transverse arrowheads, three barbed-and-tanged arrowheads, at least thirty diagnostic sherds of Beaker pottery and ten featured sherds of Early Bronze Age wares (Figs 3.4 and 3.5; Tables 3.3–3.5). There are certainly body sherds to go with the last and it is probable that the full tally of Beaker/EBA pottery will only emerge after all from the site have been thoroughly studied. These finds, which are scattered across the interior zone, collectively suggest episodic activity on or around the site, the arrowheads plausibly reflecting hunting. This group of finds is predominantly from the gravel parcel 2F and silt parcel 2B, and thus in the correct stratigraphic position relative to the two main cultural horizons. However, in detail their stratigraphic positions are not entirely consistent with supposed age; in particular, the nominally Late Neolithic arrowheads are mostly in higher stratigraphic positions than the barbed-and-tanged ones. The Beaker/Early Bronze Age pottery mirrors the positions of the latter. Possible explanations are discussed in Chapter 10.

Later Iron Age and Roman finds

While there is some uncertainty whether the span of early first-millennium BC occupation and activity on the site continued beyond the eighth century, there is certainly no major group of pottery post-dating the sixth century. However, there are two metal finds of later first-millennium BC date (Fig. 3.6), while thirty sherds significantly later than the post-Deverel-Rimbury assemblage have been isolated during study to date (Tables 3.6 and 3.7); it is not thought that this quantity will grow appreciably.

Up to eleven sherds are of Mid- to Late Iron Age, dating from the third century BC onwards, while a further nineteen are Roman. All are small and generally abraded sherds which

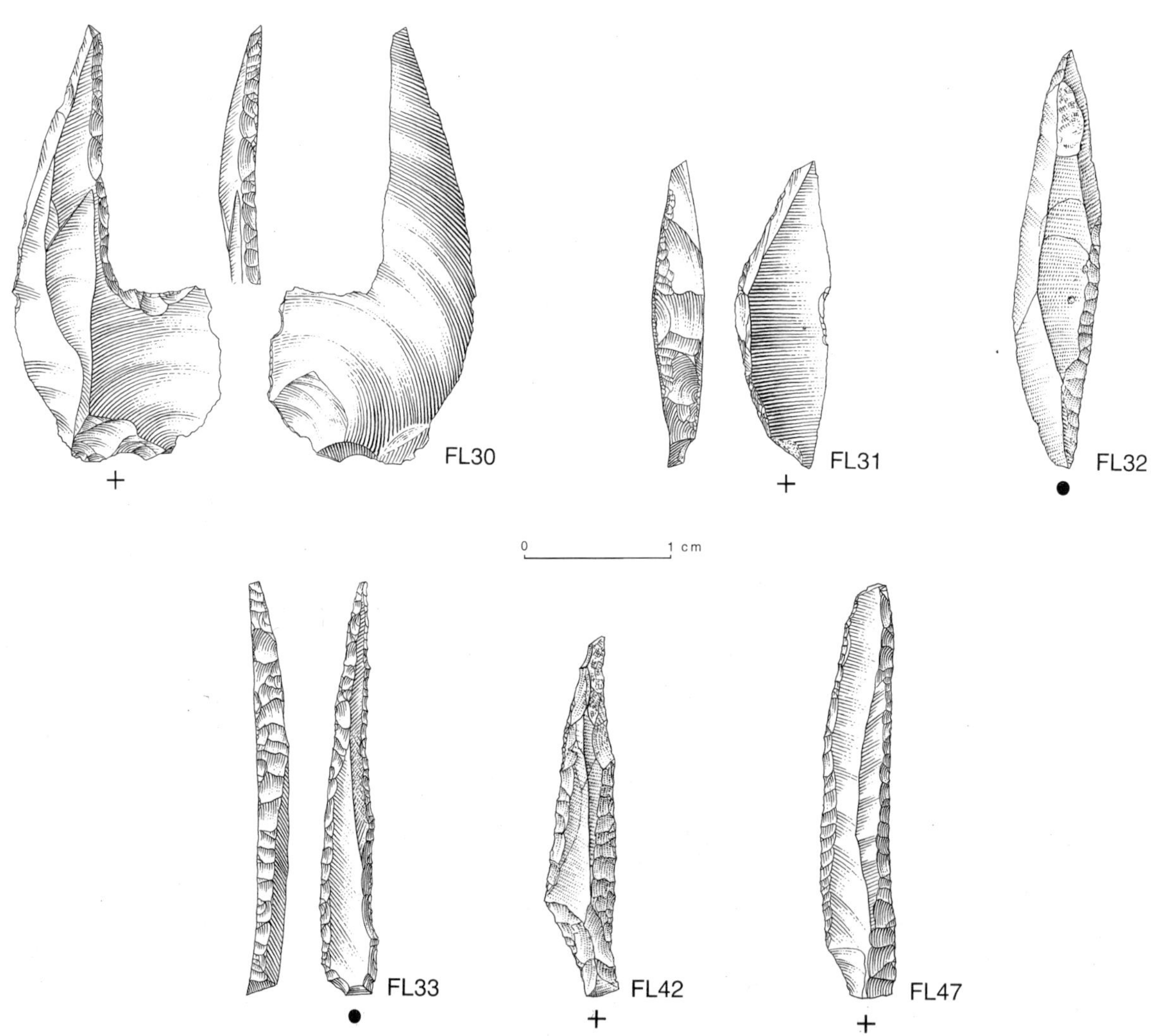

3.1 *Selection of flint microliths from the interior zone (scale 2/1).*

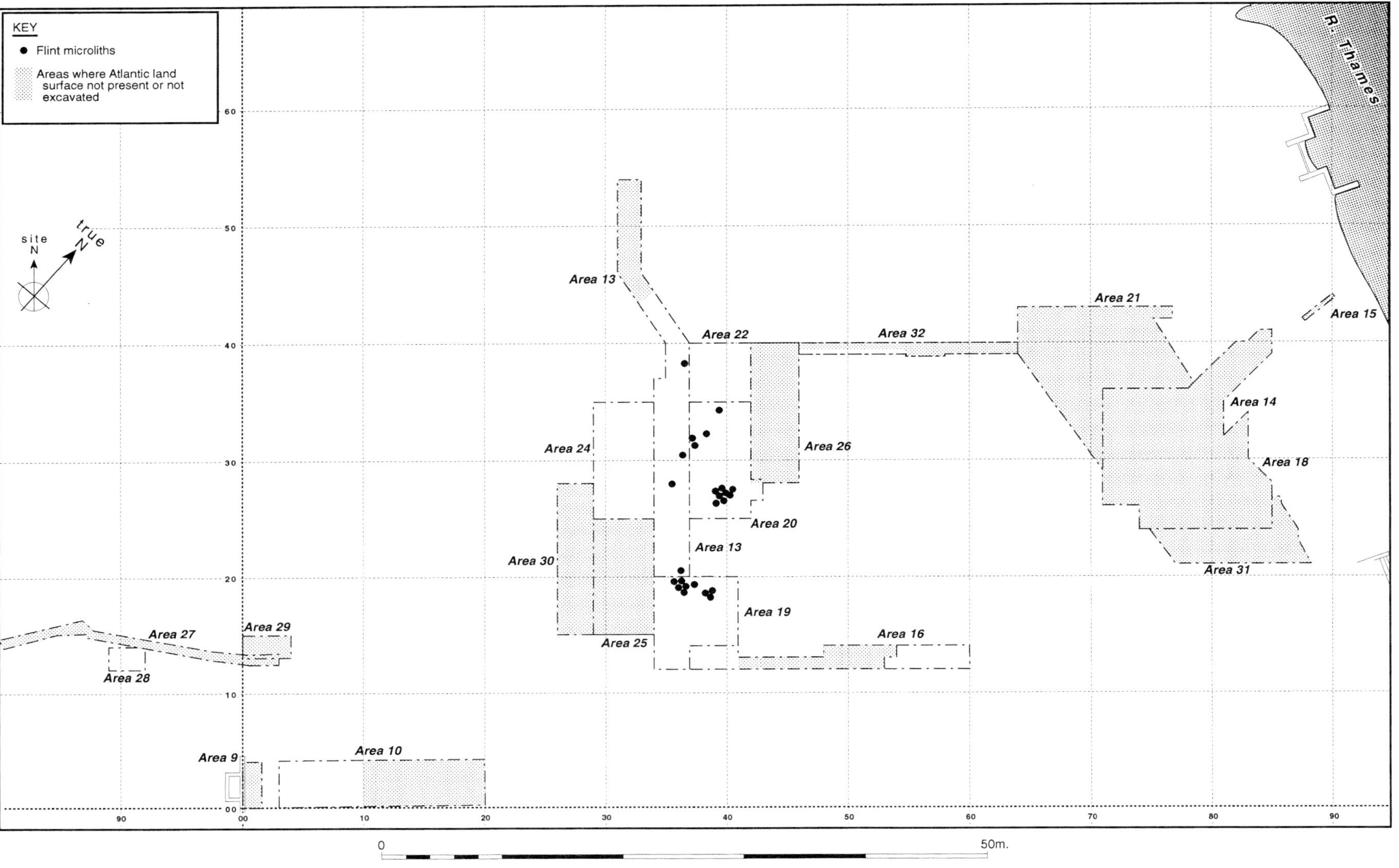

3.2   *Distribution of microliths.*

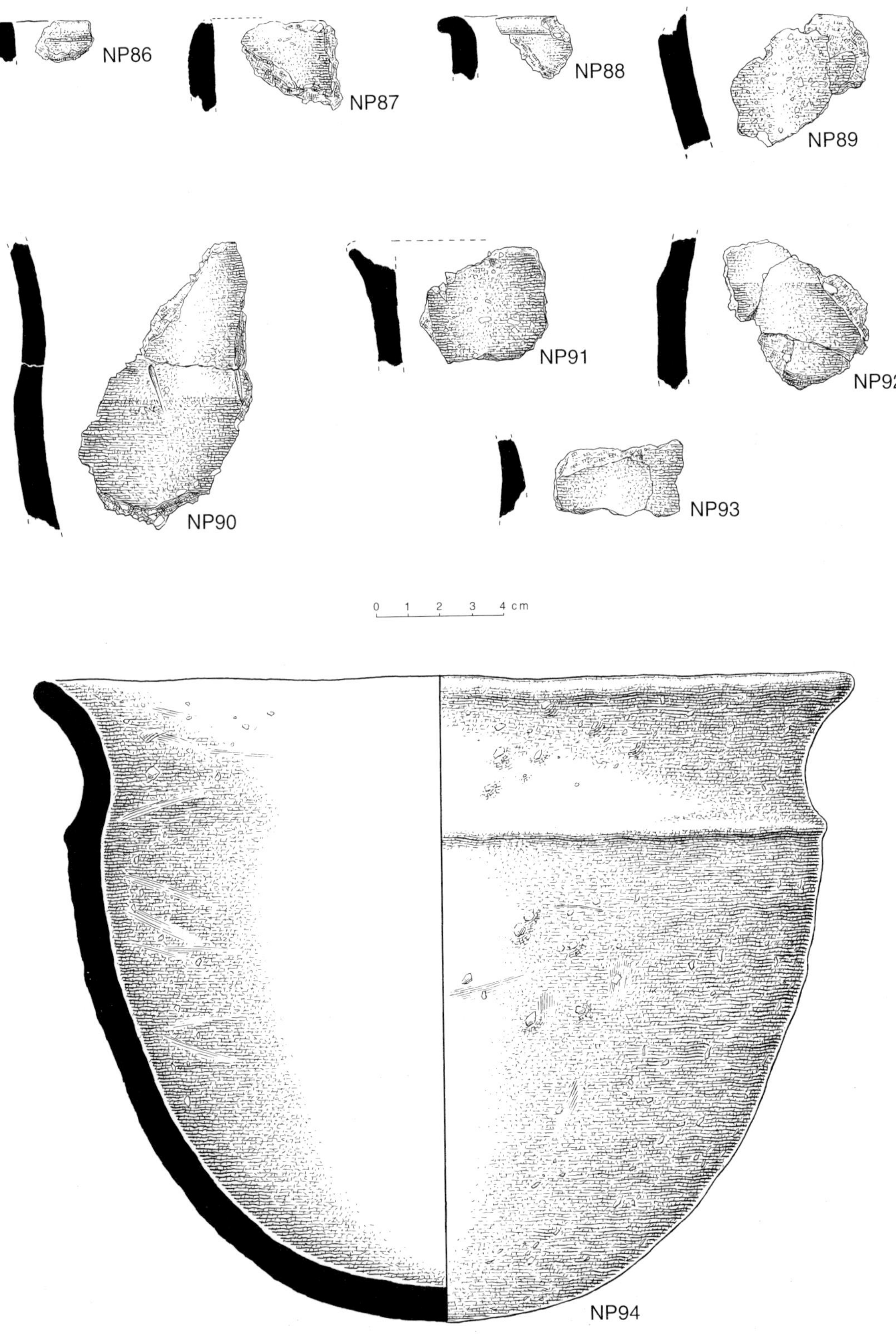

3.3   *Neolithic pottery: a) feature sherds from parcel 4B; b) restored pot from sherds in Area 16 feature (scale 1/2).*

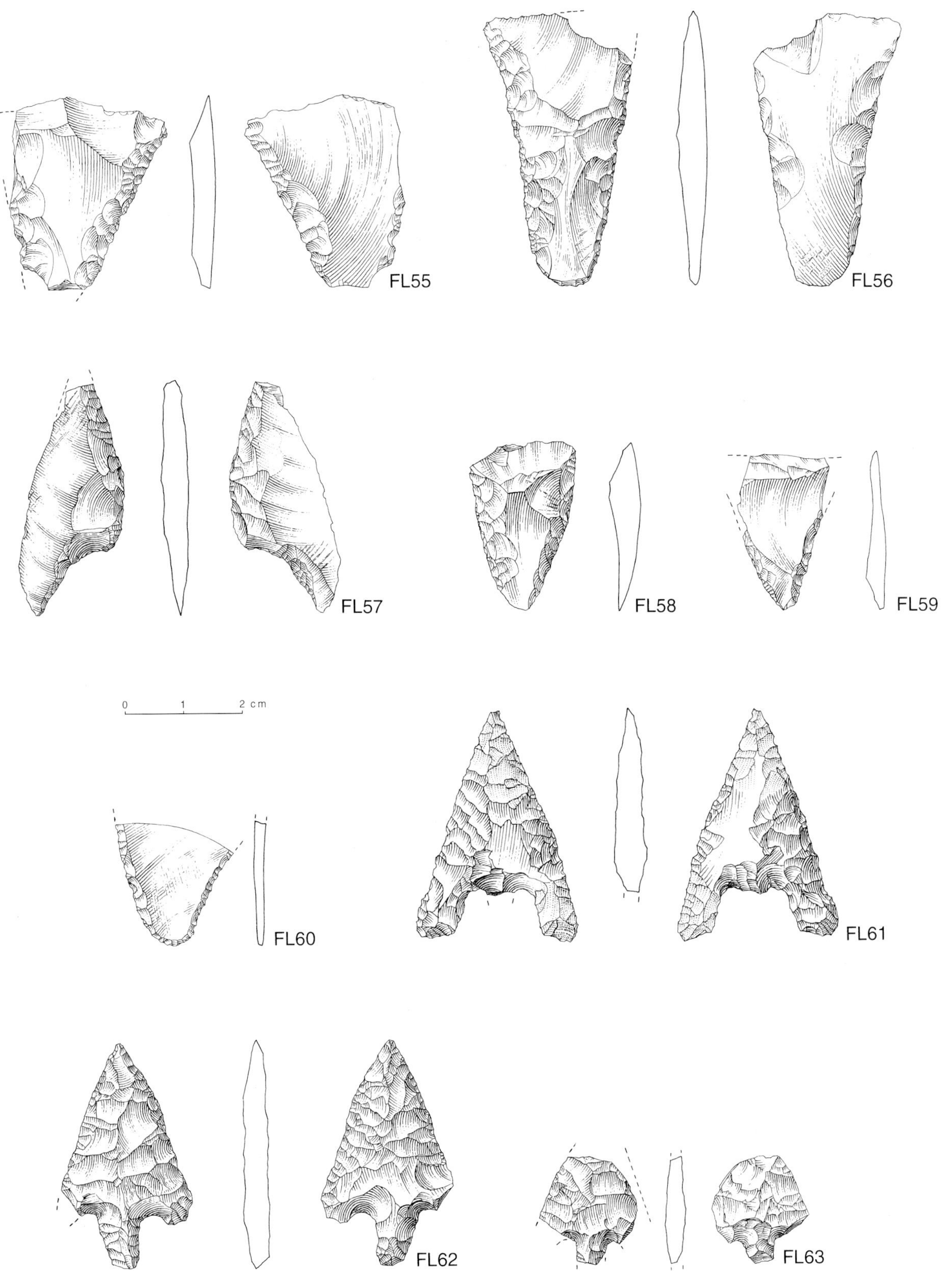

3.4 *Later Neolithic and Early Bronze Age flint arrowheads (scale 1/1).*

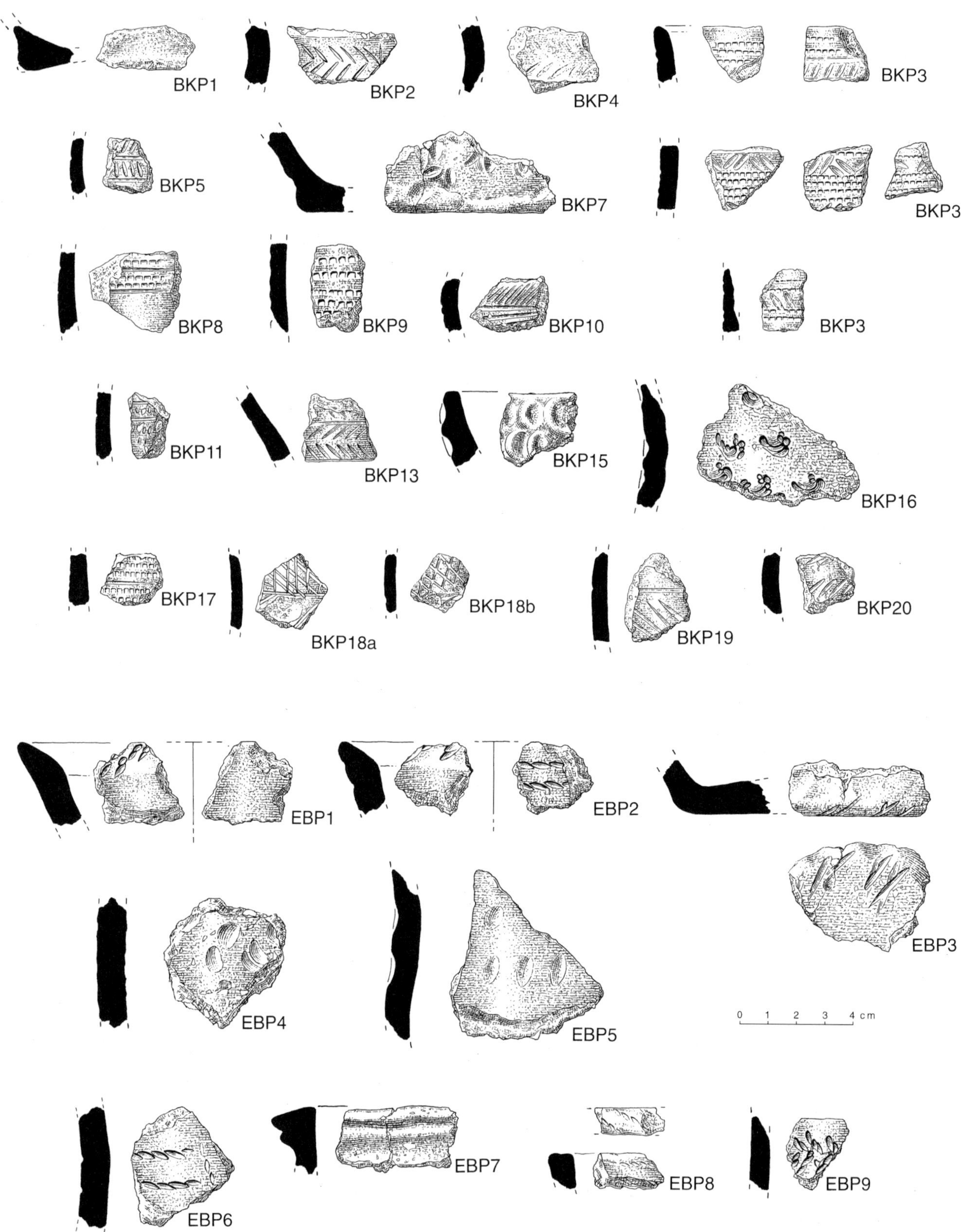

3.5   *Beaker and Early Bronze Age pottery from the interior zone (scale 1/2).*

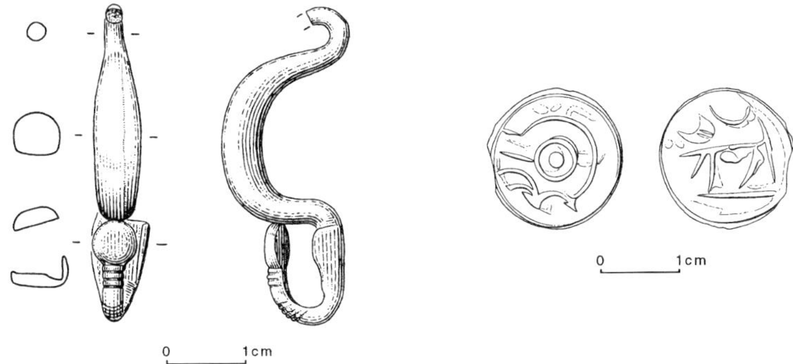

3.6   *La Tène brooch and Potin coin (scale 1/1).*

have suffered much attrition. They are widely scattered across the excavation trenches with only two concentrations of three sherds or more (Fig. 3.7). With three exceptions, all came from the parcel 6 and 9 silts overlying the in situ Late Bronze Age levels. The implications of the general pattern and the three exceptions from within the in situ deposits are explored in Chapter 10. One of the metal finds is a La Tène I fibula, again recovered from parcel 6 silts; it was probably deposited on or very near the site to judge from its condition. The other is a Potin class I coin of early first century BC; it was recovered from the fill of a modern water-main trench in Area 27, but probably came from the parcel 6 silts which had been cut through.

Medieval and Post-Medieval finds

Astonishingly few Medieval finds have been recovered during the excavation campaign, and Post-Medieval material is not very frequent (excluding Bovis phase pits and rubble raft). Their occurrence is largely confined to the fills of contemporary river channels encircling the site, parcels 8, 9 and 10, with additions from occasional features, notably the linear ditch bordering the riverside zone of excavation (Table 3.8). Stratigraphically these all tie in well with the overall sequence and some, especially the clay pipes, have provided valuable dating evidence for late alluvial units.

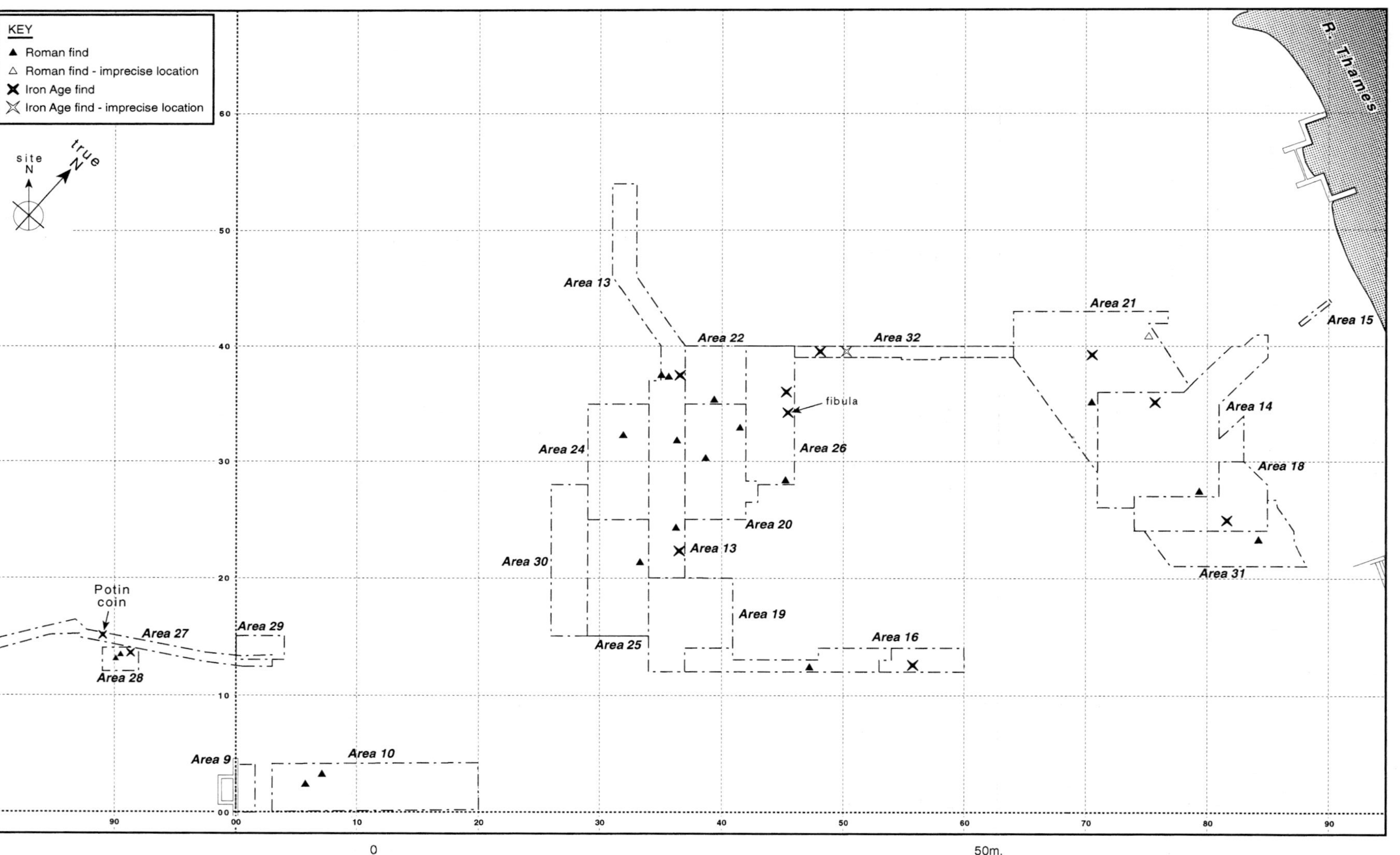

3.7 *Distribution of later Iron Age and Roman finds.*

Table 3.1   *Flint microliths from the Atlantic soil profile and early features (Fig. 3.1); identifications by Roger Jacobi.*

Cat. no.	Context	Grid	Sf no.	Registration no. 1989 10-1	Description
FL30	13.755.2	36/38	226	1001	Microlith, partially made micro-burin: triangle or straight-backed bladelet
FL31	13.774.1	36/31	227	1002	Possible small thick convex-backed microlith
FL32	13.803.1	35/28	202	1003	Slender partially backed point; burnt
FL33	13.846.2	36.30/ 21.26	8	1004	Single-ended 'needle-point'
FL34	19.365.2	35.77/ 19.87	70	1005	Single-ended 'needle-point'
FL35	19.415.2	38/18	68	1006	Scalene triangle, broken
FL36	19.415.2	38/18	69	1007	Scalene triangle, broken
FL37	19.415.2	38/18	109	1008	Scalene triangle
FL38	19.650	37/19	28	1009	Broken microlith
FL39	19.677	36/19	23	1010	Broken microlith, straight-backed bladelet or triangle
FL40	19.677	36/19	24	1011	Single-ended 'needle-point'; impact fracture at tip
FL41	19.677	36/19	25	1012	Scalene triangle, broken
FL42	19.677	36/19	26	1013	Scalene triangle
FL43	20.671	39/34	72	1014	Single-ended 'needle-point'
FL44	20.755	40.03/ 27.15	34	1015	Straight-backed bladelet
FL45	20.767	40.19/ 27.45	77	1016	Straight-backed bladelet
FL46 FL47 FL48 FL49 FL50	20.767	39.76/ 27.28	71 72 73 74 75	1017 1018 1019 1020 1021	Five straight-backed bladelets
FL51	20.832	39.32/ 26.73	90	1022	Straight-backed bladelet
FL52	20.908.2	37/31	313	1023	Single-ended 'needle-point'
FL53	20.908.3	38/32	215	1024	Broken microlith
FL54	20.908.13	37/31	220	1025	Scalene triangle, upper end

FL catalogue numbers continue the sequence from previously published flintwork (Saville 1991; Needham and Spence 1996)

Table 3.2 *Neolithic pottery from parcel 4B (Area 14, riverside zone) and (NP94) from feature exposed in section of modern pit in Area 16; (Fig. 3.3.)*

Cat. no.	Context	Grid	Reg. no. 1989 10-1	Total wt(g)	No. sherds		Comments	
					Body	Feature		
-	14.149	78/36	-	7	1			
-	14.149	80/34	-	8	2			
-	14.151	78/36	-	23	2		One with dense, medium grit	
-	14.151	79/36	-	4	1			
-	14.152	78/36	-	17	1			
-	14.152	79/34	-	22	2			
-	14.152	79/35	-	16	1(+)		?Self-slip	
NP86	14.152	79/36	1026	29	4(+)	2	Small rolled out rim with flattened top	
NP87			1027				Sherd thins, possibly to tapered rim	
NP88	14.152	80/34	1028	3		1	Rolled rim	
-	14.152	80/35	-	1	1			
-	14.153	80/37	-	9	1		Edges only lightly abraded	
-	14.154	79/37	-	4	1		Thin, sandy, lightly gritted	
-	14.154	80/37	-	15	2			
NP89	14.155	-	1029	60		1	Flakey shell-tempered, yet only lightly abraded; burnished surfaces and neat but weak carination adjacent to one break	
NP90			1030			2	Two joining sherds: weakly carinated profile and slightly concave neck; no rim; oblique strokes above carination	
NP91			1031			1	Gently concave neck, thickening towards damaged club rim?	
-	14.163	81/32	-	25	3(+)			
-	14.163	82/32	-	69	16		Some with smoothed or burnished exteriors	
NP92	14.163	82/33	1032	80	3(+)	2	Weak carination and start of neck	Both burnished externally; moderate, fine grit
NP93			1033				Gently concave neck	
NP94	16.877	48/13	1163	2212	46	✓	Restored pot; 30% present	

- NP catalogue numbers continue the sequence from previously published Neolithic pottery (Kinnes 1991; Longworth and Varndell 1996)
- Normal fabric is more or less crumbly with ill-sorted, medium to dense flint

Table 3.3   *Later Neolithic and Early Bronze Age flint arrowheads (Fig. 3.4).*

Cat. no.	Parcel	Context	Grid	Sf no.	Registration no. 1989 10-1	Description
A. *Transverse arrowheads*						
FL55	6D[1]	13.018	34/36	47	1034	trapezoid; near complete
FL56	2B	13.593	35/39[2]	79	1035	long trapezoid; near complete
FL57	2B	22.113	38/39	12	1036	single-barbed; near complete
FL58	2B	24.827	29/30	26	1037	triangular; near complete
FL59	6D/LBA	25.612	29/22	11	1038	triangular; near complete
FL60	LBA/2B	25.635	29/18	39	1039	?asymmetric trapezoid; fragment
B. *Barbed-and-tanged arrowheads*						
FL61	6D	22.001	41/39	8	1040	near complete
FL62	2F	24.836	29/33	86	1041	near complete
FL63	2F/MN	24.849	31/28[3]	303	1042	fragment; battered

[1] Close to interface with in situ LBA deposits
[2] Grid square partly cut by modern pipe trench
[3] Possibly beneath machine compaction zone

Table 3.4   *Beaker pottery (Fig. 3.5); identifications by Ian Kinnes.*

Cat. no.	Parcel	Context	Grid	Reg. no. 1989 10-1	Form/decoration		Fabric
BKP1	2B	13.463	35/36	1043	Base, slightly dished		Buff/grey; sand/grog
BKP2	2B	13.465	34/36	1044	Neck; deep lenticular impressions forming chevron between horizontal grooves and bordered by vertical line (cf. BKP4)		Buff; moderate, fine to medium crushed flint
BKP3	2F 2F 2F 2F 2F 2F 2F 2B/2F 2B/MN[1] 2B	13.540 13.540 13.542 13.542 13.573 13.581 19.187 20.662 20.666 24.826	34/29 35/29 34/24 35/20 35/24 34/26 40/15 38/26 37/28 31/29	1045	Wall Neck Neck Wall Rim Rim Wall Neck Neck Wall	Rim with diffuse internal bevel; beneath rim, three lines of rectangular tooth-comb (lowest slightly blurred), over row of diagonal ovate impressions; neck, pseudo-chevron of paired ovate impressions with rectangular tooth-comb row above, and four below; probable beginning of another chevron row	Dark grey with parts of exterior buff; grog/sparse, fine shell/occasional fine crushed flint (six sherds illustrated)
BKP4	2F	13.573	35/22	1046	Neck; chevron of lenticular impressions (cf. BKP2)		Grey/buff; moderate fine to small crushed flint
BKP5	top MN refuse deposit	13.758.1	36/27	1047	Wall; very thin-walled; two zones of oblique ovate impressions defined by horizontal lines		Buff; sand/light, fine to medium crushed flint
BKP6	MN feature[2]	13.771.1	35/30	1048	Wall; thin; horizontal incised line		Buff; sand/sparse, fine crushed flint (not illustrated)
BKP7	top MN refuse deposit	13.789.1	35/27	1049	Base; rough protruding foot; irregular fingernail rustication; possible vegetal impressions in base		Buff; fine to medium crushed flint/some grog and sand
BKP8	Feature ?tree-root disturbance	13.842.2	36/24	1050	Wall; zone of two rows of rectangular tooth-comb defined by horizontal incised lines, single row above, reserved zone below		Dark grey/ruddy brown; fine to small crushed flint/sand
BKP9	2F	19.187	36/16	1051	Wall; five horizontal rows of rectangular tooth-comb stamp		Light brown/grey; light, fine crushed flint/some grog
BKP10	2F	19.189	34/12	1052	Belly; fine oblique incisions over three horizontal lines, slightly diverging		Dark grey; light, fine to medium crushed flint/sand
BKP11	MN feature[2]	19.192.4	40/18	1053	Wall; horizontal rows of small vertical stabs separated by incised lines		Buff; moderate, fine to medium crushed flint
BKP12	2F/MN	19.265	39/15	1054	Wall, very thin; two rows oblique incisions above two incised horizontal lines		Dark grey; sand/sparse, small flint (not illustrated)
BKP13	MN	19.506.31	36/13	1055	Neck; chevron row of lenticular impressions bordered by horizontal incised lines		Dark grey; sand/ sparse, fine to small crushed flint
BKP14	2F	20.669	41/26	1056	Wall; three rows of weathered comb-stamp above reserved zone (cf. BKP3)		Dark grey; grog/occasional fine crushed flint (not illustrated)
BKP15	2F	20.702	39/25	1057	Rim; tapered to roughly flat top with possible oblique impressions; two rows of deep fingertip rustication on exterior		Grey/buff; light, small to large crushed flint/some grog
BKP16	2B	24.826	31/29	1058	Neck; bone-end impressions, probably in rows		Grey; moderate, fine to large crushed flint/sand
BKP17	2F	24.836	32/29	1059	Wall; four or five ovate tooth-comb rows divided by light horizontal line		Buff/grey/brown; moderate, fine to medium crushed flint

Cat. no.	Parcel	Context	Grid	Reg. no. 1989 10-1	Form/decoration	Fabric
BKP18	2F 2F/MN	24.836 24.855	33/28 32/27	1060	Two wall sherds; incised lattice zone separated by horizontal line from oblique strokes (largely flaked) (cf. BKP19)	Pinky buff; moderate, fine to small crushed flint/sand
BKP19	2F/MN	24.855	32/27	1061	Wall; oblique incisions bordered by horizontal line and reserved zone (cf. BKP18)	Buff/light grey; moderate, fine to medium crushed flint/sand
BKP20	2F/MN	24.855	32/27	1062	Wall; two rows of oblique lenticular impressions	Buff; moderate, fine to small crushed flint

[1] Small patch with effectively no intervening gravel between 2B and Neolithic soil profile
[2] Gravel present in feature fill

Table 3.5 *Early Bronze Age featured pottery (Fig. 3.5).*

Cat. no.	Parcel	Context	Grid	Reg. no. 1989 10-1	Form and decoration	Fabric
EBP1	2F	19.187	36/18	1063	Urn; rim; internal bevel with irregular diagonal twisted cord impressions (cf. EBP2 and EBP6)	Buff/dark grey; grog/sand
EBP2	2F	19.187	37/12	1064	Urn; rim; internal bevel with irregular diagonal impressions; external, two horizontal lines of twisted cord (cf. EBP1 and EBP6)	Buff/dark grey; grog/sand/ occasional crushed flint
EBP3	LBA	20.501	39/32	1065	Base; rough diagonal slashes across angle onto flat base	Grey/light brown; laminar; moderate, fine to medium crushed flint/sand
EBP4	2F/MN	20.748	39/37	1066	Wall; fingertip impressions	Buff; laminar; moderate, fine to large crushed flint
EBP5	2F	22.132	39/35	1067	Neck; two horizontal rows of finger-tip impressions	Light grey/buff; grog/light, fine to medium crushed flint
EBP6	2B	24.827	29/25	1068	Urn; neck; two horizontal and one vertical twisted cord impressions (cf. EBP1and EBP2)	Light orange/dark grey; grog/ sand/occasional crushed flint
EBP7	2B	24.827 24.911	31/33 29/31	1069	Food Vessel; rim; two joining sherds; thick flattened rim with angular cordon beneath external flange	Buff/dark grey; light, fine to medium crushed flint
EBP8	2B	24.830	29/25	1070	Rim; flat top with diagonal twisted cord; exterior, two horizontal lines of twisted cord	Buff/dark grey; grog/sand/ sparse, fine flint
EBP9	2F/MN	24.862	30/29	1071	Wall; chevron in impressed twisted-cord 'maggots'	Dark grey; light, fine to medium crushed flint/sand

Table 3.6 *Later Iron Age and Roman ceramics; identifications by Val Rigby.*

Parcel	Context	Grid	Reg. no. 1989 10-1	Description	Date (centuries)
A. Mid-Late Iron Age					
6D	13.018	-	1072	Probable IA body sherd; sandy ware	-
base LBA	13.032	36/22	1073	LIA foot ring, sandy ware	1st BC
6D	13.417	36/37	1074	Possible MIA; approaching base	?3rd-2nd BC
6D tr	14.004	75/35	1075	Bead-rimmed jar; rim with lid seat	3rd-2nd BC
6D	16.807	56/12	1076	Probable MIA body sherd	?3rd-2nd BC
6D	18.703	80/24	1077	Probable MIA body sherd; sand and vegetal temper	?3rd-2nd BC
6D tr	21.817	70/39	1078	Probable MIA sandy ware; body	?3rd-2nd BC
6D*	26.206	45/36	1079	Probable LIA tazza	?1st BC
6C	28.307	90/13	1080	Storage jar rim	1st BC-1st AD
6D*	32.001	-	1081	Bead-rimmed jar; rim	3rd-2nd BC
top LBA	32.002	48/39	1082	Possible LIA bowl	?1st BC
B. Roman					
6D	10.007	06/04	1083	Parchment ware - Oxford/Verulamium; probable flagon	2nd AD
top LBA	10.010	05/03	1084	Samian - South Gaulish	1st AD
6D	13.011	36/24	1085	Coarse textured grey ware; indeterminate form	2nd-3rd AD
6D	13.012	36/32	1086	Rim of dish; coarse textured grey ware	3rd-4th AD
6D	13.417	36/37	1087	[1]Necked jar rim; might belong to 20.303	2nd AD
6D	13.417	36/37	1088	Storage jar; body sherd; burnt	3rd-4th AD
6D	14.021	79/27	1089	[1]Flagon; same pot as 31.502	2nd AD
6D	16.803	47/12	1090	Grey ware; carinated bowl	Late 1st to mid-2nd AD
6D	20.303	38/30	1091	[1]Cordoned jar shoulder; might belong to 13.417	2nd AD
6D*	20.304	41/33	1092	[1]Cordoned jar neck	Late 1st-2nd AD
9	21.808	-	1093	Likely flue-tile fragment; grooved	1st-4th AD
6D	21.910	70/35	1094	[1]Closed form	Late 1st-early 2nd AD
6D	22.001	39/35	1095	[2]Chalfont ware; closed form	3rd-4th AD
6D	24.701	32/32	1096	Parchment ware; body sherd	-
6D*	25.604	33/21	1097	Probably [2]Chalfont ware; indeterminate form	-
6D	26.202	45/28	1098	Grey ware	2nd-3rd AD
6C	28.307	90/13	1099	Grey ware; base	Late 1st-2nd AD
6C	28.307	90/13	1100	?Roman ceramic fragment; orange	-
6D* tr	31.502	84/23	1101	[1]Flagon; same pot as 14.021	2nd AD

* Low in 6D, thus close to interface with in situ LBA
tr - above 'trough' in riverside zone
[1] 'Staines' ware
[2] For definition of Chalfont see Rigby 1995

Table 3.7 *Iron Age metal finds (Fig. 3.6); identification of coin by John Orna-Ornstein.*

Parcel	Context	Grid	Reg. no. 1989 10-1	Description	Date (centuries)
6D	26.202	45.62/34.08 sf 10	1102	La Tène I fibula, pin missing, but condition fine	5th BC
Modern trench - ?ex-6C	27.-	89.5/15.2 sf 3	1103	Potin class I coin, uninscribed. Obverse - head right, plain border. Reverse - bull butting left, crescent C above, pellet below, exergual line, plain border. Hobbs 1996, 686ff; Van Arsdell 1989,129	early 1st BC

Table 3.8   *Medieval and Post-Medieval finds.*

Parcel	Context	Grid	Reg. no. 1989 10-1	Description	Date (centuries)
A. Medieval pottery and tile					
-	11.-	-	1104	Coarse-ware body sherd, quartz sand and calcareous inclusions, hand-thrown, unglazed; Saxon-Norman	12th AD
8C	11.005	(SE)	1105	Late Medieval floor tile fragment with remnants of plain glaze	13th-15th AD
(?6D)	13.401	-	1106	Pie-crust rim sherd, quartz sand and some flint, hand-thrown; Saxo-Norman (cf. Denham - Farley and Leach 1988)	12th AD
6D	13.417	35/38	1107	Body sherd, quartz sand, frequent mica inclusions up to 3 mm across; perhaps early Saxon (cf. Prospect Park - Williams 1996)	5th-6th AD
B. Medieval and Post-Medieval metalwork					
8C	11.005	-	1108	Contorted copper alloy band, one edge flattened, one thin; likely binding strip; extended length 255 mm, max. width 29 mm	-
9	14.011	79/34	1109	Iron door-furniture, possibly latch; Post-Medieval	-
C. Post-Medieval pottery					
6D	13.017	35/37	1110	Body sherd of white bodied ware with clear glaze	19th AD onwards
PM ditch	18.702.2	86/18	1111	Body sherd of Border ware (Pearce 1990)	17-18th AD
9	21.808	-	1112	Body sherd of Border ware	17-18th AD
PM post hole	31.514	85/23	1113	Collared rim sherd of jar, Post-Medieval red ware with clear glaze inside and out	18th AD
D. Post-Medieval glass					
6D+	13.010	34/28	1114	Tiny body sherd of clear glass	-
6D+	13.010	36/28	1115	Small sherd of green bottle glass; with moulding	18th AD
6D+	13.010	36/29	1116	Flat sherd of clear vessel glass	18th AD
10	13.464	32/51	1117	Complete medicine bottle, bluish	19th AD
PM ditch	14.002.2	-	1118	Sherd of bottle glass	18th AD
PM ditch	18.702.1	-	1119	Two body sherds of green vessel glass	18th AD
PM post hole	18.705	81/27	1120	Two body sherds, brown and clear bottle glass	20th AD
E. Post-Medieval clay pipes					
10	13.420 sf 70	33/47	1121	Bowl and much of thin stem; ribbed	Mid-late 19th AD
10	13.450 sf 72	31/52	1122	Bowl, elongated	Turn 17th/18th AD
10	13.450/464 sf 66	32/50	1123	Bowl; fluting and ?grape-vine	Later 19th AD
10	13.464	32/51	1124	Bowl; fluting and scrollwork	Late 19th AD
10	13.464 sf 89	31/47	1125	Near-complete pipe; leaf 'stitching' along seams; initials 'M' and 'N' on spur	Mid-19th AD
10	13.500	-	1126	Elongated bowl and part of stem	Turn 17th/18th AD
PM ditch	14.002.2	-	1127	Stem fragment	17th-18th AD
9	14.010	77/35	1128	Stem fragment	17th-18th AD
9	14.010	79/33	1129	Stem fragment	17th-18th AD

Parcel	Context	Grid	Reg. no. 1989 10-1	Description	Date (centuries)
9	14.010	79/35	1130	Stem fragment	17th-18th AD
9	14.012	80/33	1131	Stem and foot of bowl; burnt	17th AD
PM ditch	18.702.2	-	1132	Stem fragment	17th-19th AD
PM ditch	18.702.3	-	1133	Virtually complete pipe; fluting and ribbing; initials on spur 'M' and ? 'N'	Mid-19th AD
9	18.710	-	1134	Stem fragment	17th-19th AD
9	31.503.1		1135	Three stem fragments, one with beginning of bowl; likely fairly early form	17th-18th AD
F. Post-Medieval building material					
8B	11.006	(SW)	1136	Two large tile fragments	16th-18th AD
8B	11.006/007	(NW)	1137	Two tile fragments	16th-18th AD
8C	11.005	(Area 11a)	1138	Large tile fragment	16th-18th AD
8C	11.005	(Area 11a)	1139	Large brick fragment	16th-18th AD
6D+	13.010	36/30	1140	Small brick fragment; weathered	16th-18th AD
6D	13.013	35/25	1141	Two tiny brick fragments	16th-18th AD
6D	13.018	35/29	1142	Tiny brick/tile fragment	16th-18th AD
10	13.415	34/43	1143	Small brick fragment; weathered	16th-18th AD
6D	13.417	36/38	1144	Small tile fragment	16th-18th AD
10	13.420	-	1145	Tile fragment	16th-18th AD
10	13.425	-	1146	Small tile fragment	16th-18th AD
10	13.448	-	1147	Brick fragment	16th-18th AD
10	13.448	-	1148	Two pieces tile, one with peg hole	16th-18th AD
10	13.464	-	1149	Small tile fragment	16th-18th AD
10	13.500	-	1150	Large tile fragment	16th-18th AD
PM ditch	14.002.1	-	1151	Brick fragment; very weathered	16th-18th AD
9	14.010	79/33	1152	Small brick/tile fragment	16th-18th AD
9	14.010	79/34	1153	Two small tile fragments	16th -18th AD
9	14.010	79/35	1154	Two small tile fragments	16th -18th AD
9	14.011	79/35	1155	Brick fragment; weathered	16th-18th AD
9	14.011	79/35	1156	Small tile fragment with peg hole	16th-18th AD
9	14.014	80/33	1157	Brick fragment; very weathered	16th-18th AD
9	14.017	80/35	1158	Brick fragment; very weathered	16th-18th AD
9	14.143	82/30	1159	Two tiny brick/tile fragments	16th-18th AD
?9	15.006	-	1160	Large tile fragment	16th-18th AD
PM ditch	18.702.2	-	1161	Brick fragment; weathered	16th-18th AD
6D+*	24.707	33/29	1162	Small brick/tile fragment	16th-18th AD

+ Zone of compression from heavy plant treads - possible intrusions
* Low in 6D, thus close to interface with in situ LBA

Table 3.9   *Charcoal and wood finds from alluvial contexts and selected features (in chronological order); identifications by Rowena Gale.*

Parcel	Season	Context	Grid	Identifications	[14]C dating
1K	89	19.800.1	34/14	1 charc. frag.; *Prunus* sp.[1]	OxA-3580
1E	85	13.549.1	34/44	trunk/branch, *c.* 24 rings; Salicaceae	BM-2550
1E	85	13.549.2	34/44	branch, 8 rings; Salicaceae	
1E	85	13.549.3	34/44	branch, *c.*14 rings; Salicaceae	
1E	85	13.549.4	34/44	incomplete diameter; Salicaceae	
1E	85	13.549.5	34/44	branch, 14 rings; Salicaceae	
1E	85	13.549.6	34/44	branch, *c.* 4 rings; Salicaceae	
1E	85	13.549.7	34/44	branch, *c.* 4 rings; Salicaceae	
1D	89	22.179	37/40	1 charc. frag.; Salicaceae	OxA-3581
1H	89	32.040.1	55/39	branch, 4 cm diam.; *Alnus* sp.	
1H	89	32.040.2	55/39	branch, 6 cm diam.; *Alnus* sp.	BM-2657
1H	89	32.040.3	55/39	3 frags. ?*Alnus* sp.; 1 frag. ?*Alnus/Corylus* sp.; compressed	
LMes/ ENeo post hole and ditch	89	13.003	35/33	several charc. frags, all except one from mature branch or trunk; *Fraxinus* sp.	
	89	20.957.7	37/30	several frags; *Fraxinus* sp.	OxA-3582
2D	84	14.116.1	74/27	water-worn pebble; Salicaceae	
2D	84	14.116.2	76/27	wedge-shaped frag., *c.* 6 rings; Salicaceae	
2E	84	14.120.4	80/27	bark frag.; possibly *Quercus* sp.	
2E	84	14.120.5	80/27	branch, 6 rings (wide); *Alnus* sp.	
2E	84	14.120.6	80/27	burr or knot; probably Salicaceae	
2E	84	14.120.7	80/27	incomplete diameter; Salicaceae	
2E	84	14.121.1-16	80/27	16 pieces; all *Cornus* sp.	
2E	89	31.675	85/24	large piece of trunk; *Alnus* sp.	
2E	89	31.675.2	85/24	5 frags; large branch; *Alnus* sp.; 1 twig; *Cornus* sp. 8 bark frags - unidentified	
2E	89	31.676	85/24	trunk, diameter 22.5 cm, estimated growth 28 years	BM-2662
4B	89	31.665.1	86/25	large branch; *Acer campestre*	
4B	89	31.665.2	86/25	branch; subfamily Pomoideae	
4B	89	31.665.3	86/25	42 frags (≤2 cm diam.); incl. *Cornus*, *Prunus* spp. and Pomoideae	
4B	89	31.665.4	86/25	1 piece; subfamily Pomoideae 8 frags; *Cornus*, *Prunus* spp. and Pomoideae	
4B	89	31.665.5	86/25	104 frags, mostly de-barked (≤2 cm diameter); includes *Acer*, *Cornus*, *Prunus* spp. and Pomoideae	BM-2661
?4B	89	tufa deposit below 31.642	86/24	branch; *Rhamnus Cathartica*	
8C	80	11.016.1	W corner	worked end of pole; possibly Salicaceae (collapsed)	
8C	80	11.016.2	W corner	worked end of pole; possibly Salicaceae (collapsed)	
8C	80	11.016.3	W corner	tangentially cut slice of branch; possibly Salicaceae (collapsed)	
8C	80	11.016.4	W corner	branch with coppice knobbles; unidentified	
8C	80	11.016.8	W corner	branch; *Alnus* sp.; 8 pieces; possibly Salicaceae	
8C	80	11.005	Extension	stake penetrating 11.005; *Quercus* sp.	

Parcel	Season	Context	Grid	Identifications	[14]C dating
?8C	80	11.U/S.5	-	plank with 28 rings oblique to section; *Quercus* sp.	
?8C	80	11.U/S.6	-	segment of post/stake, 13 rings; *Quercus* sp.	BM-2551
?8C	80	11.U/S.7	-	segment of post/stake, 16 rings; *Quercus* sp.	
9	84	14.015.1	78/35	*Larix* sp. or *Picea* sp.	

[1] Identification by Caroline Cartwright

CHAPTER 4

Scientific dating evidence

Radiocarbon sample selection and measurement

JANET AMBERS

Organic objects from the alluvial parcels occupying the Runnymede Bridge site were analysed for radiocarbon at both the British Museum and Oxford Accelerator Laboratories. The samples reported on here come for the most part from deposits other than the two cultural horizons (Middle Neolithic and Bronze Age) and add to those already published (Longley 1980; Needham 1991; Ambers and Leese 1996). They come from a variety of contexts, natural and anthropogenic, which help clarify the chronology of the alluvial sequences. The detailed chronology of the Middle Neolithic and the Late Bronze Age settlement deposits will be covered in other volumes in this series, but two relevant dates are reported here:

1. BM-2773, which helps resolve previous inconsistent dates for the brushwood structure within alluvial parcel 2A, and
2. BM-2659, which is included for the sake of direct comparison with the underlying sample (BM-2661) and hence estimation of the duration of accumulation at this spot.

Results produced at the British Museum were measured using liquid scintillation counting and the methods described in Ambers *et al.* 1989, while smaller samples were analysed by Accelerator Mass Spectrometry at the Oxford Laboratory using the methods described in Hedges *et al.* 1993.

Samples of both wood and bone were analysed at the British Museum. For the wood samples, cellulose was extracted for dating whenever possible. In cases where this proved not to be feasible, because of sample size or nature, a vigorous pretreatment of successive acid and alkali was used. Samples given such treatment are described in Table 4.1 as 'whole wood'. In the case of BM-2435 and BM-2436 separate portions of material from the same context were given whole wood and cellulose treatments. The two results are statistically indistinguishable. Under these circumstances it seems justifiable to combine the figures to give a weighted mean with a lower error term as shown in Table 4.1. Protein was extracted from the single bone sample.

The Oxford samples were all charcoal and were pretreated as outlined in Hedges *et al.* 1993.

Both laboratories maintain continual quality assurance checks (for details see references above; also Ambers and Bowman 1994), with additional participation in international intercomparisons. These checks serve to justify the precision claimed and indicate no laboratory offsets (see, for example, the results of the Third International Intercomparison, Scott *et al.* 1998).

The results are given below in Table 4.1, quoted in the form recommended by Stuiver and Polach 1977, in uncalibrated years BP, corrected for isotopic variation. As far as possible the calibrated age ranges are calculated from the curves of Pearson *et al.* 1986, Pearson and Stuiver 1986 and Stuiver and Pearson 1986, using CALIBM, an in-house British Museum adaptation of revision 2.0 of the University of Washington Quaternary Laboratory Radiocarbon Calibration Program (CALIB, published in Stuiver and Reimer 1987) and the probabilistic method of calibration. For the period not covered by these curves, those of Stuiver and Pearson 1993, and Pearson *et al.* 1993, have been used instead, as calculated by the OxCal program (Bronk Ramsey 1995). The 1986 curves have been favoured because of the doubts about the detail of the 1993 figures voiced at the business meeting of the 15th International Radiocarbon Conference in Glasgow (Bowman 1994). Calibrated date ranges are quoted in the form recommended by Mook 1986, with the end points rounded outwards to five years. An effect of this is to slightly overestimate the date range for the stated confidence level.

The nature of these alluvial deposits made the location and selection of good dating material difficult. All but two of the samples used were of wood or wood charcoal, with the integral problems of age offset and possible curation or reuse frequently discussed (see, for example, Bowman 1990). Most of the samples came from natural accumulations of material within the silt parcels. This has obvious connotations for the interpretation of the results; it is hard to be sure that such material is not fluvially reworked. Efforts were made to lessen these problems; all of the wood samples are from short-lived species, short-life components (branches, twigs, etc.) or from the outer rings of tree trunks. All were selected for apparent 'freshness'; materials used had no abraded edges (other than those caused by beaver gnawing) or other

Table 4.1 *Radiocarbon measurements from alluvial and selected cultural contexts.*

Ref	Date BP	δ¹³C (‰)	Context	Parcel	Material	Possible calibrated age ranges (Cal BC unless stated)	
						68% probability	95% probability
BM-2774	8870±60	-24.9	ERB78 A5 F294 e and f (published context 145). Cluster of branches thought to be naturally accumulated within sediments	1J	Cellulose from *Corylus* branches[1]	8020 to 7900 or 7760 to 7740[4]	8035 to 7855 or 7820 to 7700[4]
BM-2550	7790±80	-27.0	ERB85 13.549 Large section of tree from a small group of wood, none showing signs of working	1E	Cellulose from outer 1.5 cm (after removal of bark) of tree section. Cellulose appearance atypical; sample may include lignin; family Salicaceae[1]	6690 to 6685 or 6655 to 6465[5]	7000 to 6920 or 6890 to 6840 or 6780 to 6420[5]
OxA-3581	8050±100	-27.5	ERB89 22.179 Within layer of marly peat buried deep beneath Neolithic soil	1D	Charcoal; family Salicaceae[1]	7200 to 7180 or 7140 to 7120 or 7050 to 6710[4]	7265 to 6615[4]
OxA-3580	7350±90	-24.6	ERB89 19.800.1 Fill of deep feature underlying silt layer 19.801 and Neolithic soil	1C	Charcoal; *Prunus* sp.[2]	6235 to 6040[5]	6370 to 6000[5]
OxA-3582	5780±85	-24.2	ERB89 20.957.7 Basal fill of linear ditch underlying Middle Neolithic occupation horizon	1C/D	Charcoal; *Fraxinus*[1]	4775 to 4575 or 4555 to 4545[6]	4895 to 4880 or 4845 to 4460[6]
BM-2657	5130±50	-27.8	ERB89 32.040 Two pieces of young roundwood	1H	Whole wood; *Alnus* sp.[1]	4000 to 3935 or 3865 to 3815[6]	4040 to 3785[6]
BM-2773	4760±50	-21.4	ERB78 A4 L3a (published context 120c). Part of brushwood structure with domestic rubbish	2A	Cellulose from *Quercus* sp., branch plus offcuts[1]	3635 to 3505 or 3400 to 3385[6]	3680 to 3375[6]
BM-3039	5100±100	-23.6	ERB84 14.116	2D	Collagen from pig humerus (immature) and cattle 1st phalanx (fused)	4000 to 3780	4250 to 3650
BM-2435	3770±60	-24.8	ERB84 14.121 Bundle of rods and branches, some with beaver gnawing	2E	Whole wood; *Cornus* sp.[1]	2310 to 2130 or 2065 to 2045[7]	2455 to 2425 or 2400 to 2035[7]
BM-2436	3740±80	-23.6	ERB84 14.121 Second sample from same material as used for BM-2435	2E	Cellulose from *Cornus* sp.[1]	2290 to 2035[7]	2455 to 2420 or 2405 to 1950[7]
BM-2435 & 2436	3760±50	–	Weighted mean of BM-2435 and BM-2436	2E		2285 to 2135 or 2060 to 2045[7]	2355 to 2035[7]
BM-2662	3870±50	-24.5	ERB89 31.676 Tree trunk from alluvial silt	2E	Cellulose from *Alnus* sp; around 30 years' growth[1]	2460 to 2310[7]	2480 to 2200[7]
BM-2661	2900±50	-24.9	ERB89 31.665 Dense mat of fresh twiggy material, some bearing beaver gnaw marks	4B*	Cellulose from branches of *Acer, Cornus, Prunus* spp., and sub-family Pomoideae[1]	1205 to 1010[7]	1265 to 935[7]

Ref	Date BP	δ¹³C (‰)	Context	Parcel	Material	Possible calibrated age ranges (Cal BC unless stated)	
						68% probability	95% probability
BM-2659	2720±50	-22.1	ERB89 31.609 Part of stratified LBA occupation deposits vertically above layer yielding BM-2661	LBA	Collagen from cattle metacarpal and humerus[3]	915 to 825[7]	990 to 805[7]
BM-2551	1190±60	-27.4	ERB80 A11 F6 One of two worked stakes from filled channel of relict Southern Loop of Thames		Attempt to extract cellulose failed; whole wood from outer 1.5 cm of heartwood *Quercus*; 13 years growth[1]	720 to 735 or 765 to 900 or 930 to 940 cal AD [8]	685 to 965 cal AD [8]

[1] Identified by Rowena Gale
[2] Identified by Caroline Cartwright
[3] Identified by Dale Serjeantson
[4] Pearson *et al.* 1993
[5] Stuiver and Pearson 1993
[6] Pearson *et al*, 1986
[7] Pearson and Stuiver 1986
[8] Stuiver and Pearson 1986

signs of wear, which would be expected to appear quickly if carried far or exposed for long in water. The beaver-gnawed samples (contexts 14.121 and 31.665) both come from concentrations which are most likely to have accumulated and been sealed shortly after the gnawing event, while the structure in Area 4 (A4L3a) is regarded as in situ.

In addition to the general dating of the sequence given by the measurements, there remain the two specific questions raised above. BM-2773 was dated to help resolve the problem of two apparently divergent Harwell dates for piles from a Neolithic brushwood structure, believed to be of relatively short duration (HAR-6132; 4630±70 BP and HAR-6128; 4920±80 BP; Needham 1991). The calibrated range of BM-2773 at 95% confidence coincides with the overlap in the 95% ranges for the Harwell figures and, together with the archaeological evidence suggesting a short lifespan for the structure, indicates its most likely date.

In the case of the pair of samples BM-2659 and BM-2661, separated by some 0.7 m of channel deposits, Bayesian analysis of the figures using the OxCal program (Bronk Ramsey 1995) indicates a 68% probability of the period of accumulation being 130–330 calendar years, and a 95% probability of it being between 40–420 calendar years.

Archaeomagnetic dating

THE LATE TONY CLARK

The main objective of this work was to ascertain the dates and duration of sedimentary deposition on the site. This was done by measuring the changing secular variation of the geomagnetic field direction as recorded in the sediment columns by the phenomenon of detrital remanence, and matching these to a reference curve mainly derived from lake sediments (Turner and Thompson 1982; Clark 1992).

Magnetic intensity was also measured, with the addition of low frequency magnetic susceptibility on the longer columns. Measurements were also made on a burnt area and two hearths, the latter providing an opportunity to check the calibration curve itself.

In most cases, the sediment columns were collected in 5 cm deep uPVC tubes as a single vertical series of contiguous samples. When possible, the tubes were pushed vertically into 'shelves' cut into the deposit. The samples themselves normally represented a 5 cm depth, but in some columns were slightly shorter because of difficulty of insertion; and in the hardest deposits (mainly dried-out and near the surface) it was necessary to carve the material into pillars which were encapsulated in the tubes with plaster of Paris. The actual sample numbers in relation to depth are shown in Table 4.2. The level of a sample is assumed to be that of its centre, rounded up to whole centimetres. To remove subsequently acquired viscous magnetism, samples were stored for a lengthy period in field-free containers, and also subjected to partial alternating field (AF) demagnetisation if tests on pilot samples showed this to be necessary. These tests also indicate the magnetic stability of the material, which was generally found to be quite good.

Archaeomagnetic measurements of this type are necessarily imperfect because of flow, turbulence and discontinuity effects during deposition, and weathering, bioturbation, slumping and chemical changes afterwards. Rates of sedimentary accretion are likely to be variable and may include static periods or even erosion, effects which are difficult or impossible to assess unless the data are of high quality. The degree of success is therefore variable, especially in the near-surface conditions at Runnymede and most archaeological sites. However, the magnetic susceptibility and intensity records can provide additional markers for the correlation of columns, and checks on

92

continuity and stability of deposition, although their greatest potential value lies in the information they may provide on landscape history: peaks are typically caused by upstream clearance episodes which enhance the magnetic properties of the soil by burning and release it into the river system; or they may be caused by local occupation. Intensity and susceptibility peaks do not necessarily coincide, possibly because intensity may be more dependent upon heavier heating causing a tendency for the development of thermo-remanent magnetisation in the soil. High intensity is also promoted by favourable conditions of deposition leading to a maximum proportion of magnetised particles aligning with the ambient field. In the case of magnetic susceptibility, it may also be possible to ascertain whether high values are inherent in a material or the result of artificial alteration. Research into such possibilities is not pursued here, but the Runnymede material has been saved in case the opportunity arises.

Date spans given for sets of contemporary samples, such as 'column' 7, are, like radiocarbon dates, limits within which the true date could lie anywhere at the confidence level given. The figure alpha-95 (α^{95}) is a measure of the precision of a group of averaged directional measurements: the smaller its value the better. The date spans assigned to columns, however, are best estimates of the actual limiting dates of the usable readings. Error limits are difficult to assign rigorously to these figures because of imperfections in the geomagnetic field direction record contained in the sediment, and uncertainties remaining in the calibration curves themselves. Limits are not better than ±100 years and could be as high as ±150 years at the 68% confidence level. Dates are given in calendar years, based on calibrated radiocarbon dates.

In a number of cases, the estimated date spans do not cover the full length of a column. This is because of a tendency for the uppermost samples to be affected by disturbance and drying out, while the lowest samples have often run into underlying nodular calcareous material which is too coarse, too physically unstable and too weakly magnetic to retain a sensible detrital remanence.

Remanence parameters were measured with a Molspin Archaeomagnetic Spinner Magnetometer; magnetic suscep-tibility with a Bartington MS2 meter with MS2B sensor. All column measurements and preparation information are set out in the summary tables, and the mean figures for contemporary groups in the main text. More detailed information on the archaeomagnetic dating of alluvial sediments, with references, is given in Clark (1992).

Columns through the alluvial sequences

COLUMN 4 (34.0/44.5), ERB85 AREA 13.
Alluvial parcels 1E, 1K
AJC-121, No. of samples: 13

Unlike most of the columns, this was taken as a series of five horizontally paired samples, with a sixth group of three

Table 4.2 *Archaeomagnetic column data.*

m OD[1]	Parcel	No	Declin-ation	Inclin-ation	Inten-sity	χ
Area 13, column 4 (34.0/44.5)						
Treatment: storage in zero field						
13.90	1K	1	[96.85]	[31.02]	52.4	57.5
13.90	1K	2	[89.10]	[61.13]	23.0	
13.70	1K	3	17.53	58.22	68.4	18
13.70	1K	4	16.51	57.01	79.3	
13.50	1K	5	7.07	63.24	44.5	10
13.50	1K	6	12.46	70.30	43.4	
13.30	1E	7	21.28	87.13	21.7	11.5
13.30	1E	8	-29.25	58.56	17.1	
13.10	1E	9	-10.29	60.52	15.7	6
13.10	1E	10	-5.01	65.62	9.3	
12.90	1E	11	-49.2	-37.56	10.3	3
12.90	1E	12	17.56	62.22	8.0	
12.90	1E	13	10.95	63.06	18.7	
Area 14, column 20 (80.5/31.5)						
Treatment: storage in zero field						
13.91	4B	1	39.47	67.08	19.9	6
13.85	4B	2	-30.73	50.66	13.0	4
13.79	4B	3	-36.87	70.11	12.7	4.5
13.73	4B	4	-34.57	61.23	10.8	3
13.67	4B	5	-	43.29	16.8	3.5
13.61	4B	6	-	6.79	2.6	4
13.55	4B/2B	7	-	23.35	3.2	4
13.49	4B/2B	8	-	74.69	7.2	8.5
13.43	2B	9	8.17	74.06	5.0	5
13.37	2B	10	-	9.26	4.2	5
13.31	2B	11	40.47	57.49	9.2	8
13.25	2B	12	13.59	52.68	7.5	7
13.19	2B	13	-19.27	61.20	16.4	7.5
13.13	2B/2E	14	1.88	78.69	11.1	6
13.07	2E	15	9.95	70.17	24.1	7
Area 19, column 31 (34.0/15.0)						
Treatment: 2.0 m T AF						
14.40	1K soil	1	-10.43	66.82	158.2	41
14.35	1K soil	2	5.9	63.19	132.0	32.5
14.30	1K soil	3	25.12	71.43	107.4	28.5
14.25	1K	4	20.16	58.42	107.6	21.5
14.20	1K	5	3.65	75.03	78.3	16.5
14.15	1K	6	12.06	64.26	89.3	17.5
14.10	1K	7	3.60	67.49	112.0	22.5
14.05	1K	8	17.78	68.08	163.3	27

m OD[1]	Parcel	No	Declin-ation	Inclin-ation	Inten-sity	χ
14.00	1K	9	6.06	70.36	148.1	24.5
13.95	1K	10	-6.29	70.24	145.8	29
13.90	1K	11	-0.71	68.74	79.2	19
13.85	1K	12	-2.13	74.97	98.8	16
13.80	1K	13	5.24	68.46	85.9	13
13.75	1K	14	-2.80	70.60	53.2	11
13.70	1K	15	-6.11	65.01	21.9	6.5
13.65	1K	16	-5.74	64.51	9.5	4
13.60	1K/1C	17	6.50	59.90	7.2	4
13.55	1C	18	16.17	56.26	3.1	4
13.50	1C	19	-6.61	53.66	1.2	2
13.45	1C	20	-17.62	68.92	2.4	3
13.40	1C	21	-50.64	61.58	1.8	1

Area 22, column 40 (41.2/40.0)

Treatment: 3.0 m T AF

m OD[1]	Parcel	No	Declin-ation	Inclin-ation	Inten-sity	χ
13.72	1G	1	9.18	66.20	77.7	17.5
13.67	1G	2	31.77	66.11	66.9	14
13.62	1G	3	12.95	66.67	69.7	14
13.57	1G	4	15.62	69.16	93.2	14.5
13.52	1G	5	-2.88	68.07	116.1	15.5
13.47	1G	6	-2.61	67.89	80.8	13.5
13.42	1G	7	-1.23	68.74	118.9	12.5
13.37	1G	8	1.84	69.30	110.8	11
13.32	1G	9	1.83	68.50	117.3	14
13.27	1G	10	-14.76	68.68	124.7	17
13.22	1G	11	-0.48	69.96	78.3	12
13.17	1G	12	-15.22	69.90	43.6	12
13.12	1G	13	37.93	[34.10]	8.6	7
13.07	1G	14	19.25	74.30	8.5	4

Area 32, column 60 (59.3/39.0)

Treatment: 1-16, 29-32, 2.5 m T AF; 17-28, 7.0 m T AF

m OD[1]	Parcel	No	Declin-ation	Inclin-ation	Inten-sity	χ
14.57	6D	1	[-142.35]	84.94	27.9	14.5
14.52	6D	2	[74.25]	78.25	52.9	20
14.47	LBA	3	[-127.89]	61.36	196.8	27
14.42	LBA	4	[-120.23]	[-2.72]	36.8	30
14.37	LBA	5	-10.07	56.12	43.1	18.5
14.32	LBA/?4B	6	[-42.44]	59.49	31.2	13
14.27	4B/2B	7	[-37.55]	72.77	33.3	9.5
14.22	4B/2B	8	29.88	73.71	33.2	10
14.17	4B/2B	9	26.89	84.27	40.3	9
14.12	4B/2B	10	-8.32	73.37	32.1	9
14.07	4B/2B	11	-4.14	68.96	37.8	8.5
14.02	4B/2B	12	-22.11	52.71	24.7	10
13.97	2B	13	[-154.81]	[48.06]	46.1	10
13.92	2B	14	-12.05	73.71	50.3	11.5
13.87	2B	15	2.48	71.56	67.2	14.5
13.82	2B	16	6.54	69.49	69.1	15
13.77	2B	17	20.03	64.04	93.8	15.5
13.72	2B	18	3.92	66.74	83.0	14.5
13.67	2B	19	3.04	69.18	142.0	16.5
13.62	2B	20	-3.91	70.05	96.3	16.5
13.57	2B	21	19.58	73.21	51.8	11
13.52	2B	22	-4.97	69.18	87.3	13
13.47	2B	23	3.33	65.05	60.6	12.5
13.42	2B	24	-14.65	69.06	47.6	11
13.37	2B	25	-4.94	62.08	86.6	14.5
13.32	2B	26	-7.62	69.16	79.8	13
13.27	2B	27	4.31	68.45	44.0	12.5
13.22	2B/2D	28	1.98	67.03	27.6	12
13.17	2D	29	23.35	58.44	5.7	5.5
13.12	2D	30	5.63	65.55	4.48	4
13.07	2D	31	[127.14]	[80.04]	8.1	4.5
13.02	2D	32	-10.42	63.97	10.8	4

Area 28, column 65 (89.8/14.0)

Treatment: storage in zero field

m OD[1]	Parcel	No	Declin-ation	Inclin-ation	Inten-sity	χ
14.65	6C	1	[-49.26]	74.47	24.1	11.5
14.60	6C	2	-7.24	72.84	28.4	10
14.55	6C	3	[89.95]	[-44.00]	238.7	10
14.49	6C	4	-4.64	68.65	24.9	9
14.44	6C	5	[-111.53]	[20.15]	114.6	18.5
14.39	6C	6	22.38	49.08	39.1	9
14.33	6C	7	-9.51	71.37	28.5	8.5
14.28	6C	8	41.94	81.16	21.9	10.5
14.23	6C/7	9	[-54.60]	50.42	57.4	10
14.17	7	10	4.98	73.66	42.0	11.5
14.12	7	11	[2.43]	[-56.00]	106.8	9
14.07	7	12	25.93	59.21	33.7	14
14.01	7	13	28.56	73.01	33.2	9
13.96	7	14	20.11	56.87	278.4	9
13.91	7	15	[-118.82]	62.87	232.8	11.5
13.85	7	16	-5.58	79.16	29.1	12

[1] Level at centre of sample

samples at the base. The vertical interval between the sample groups was 20 cm. The use of more than one sample at each level enabled the consistency of the magnetic behaviour of the material to be assessed, as well as allowing the quality of the readings to be improved by averaging the measurements.

The uppermost pair of measurements were too disparate to be usable, and had to be discarded. Samples 7 and 8 were strongly divergent but gave an apparently sensible mean, and sample 11 was also unusable. The few declination and inclination values obtained (six of each) were broadly comparable with columns 31 and 40, as was the pattern of susceptibility and intensity values. Thus it is tentatively suggested that column 4 belongs to the same period of sedimentation, with a date span similar to that of column 31.

Possible approximate span: 6750–5850 BC

COLUMN 7 (80.0/38.0), ERB85 AREA 14.
Alluvial parcel 9
AJC-122, No. of samples: 3

Taken at a single level (*c.* 13.95 m OD) to test the age of the latest alluvial deposition identified on the site. The mean direction was:

Dec = 23.52°W; Inc = 61.73°; alpha-95 = 10.19° (3 samples)

The measurements were well grouped, but gave a large alpha-95 because of the statistical invalidity of their small number. Assuming some shallowing of the inclination, the distinctive westerly position of the result is in keeping with the following date spans:

68% confidence: AD 1740–1910
95% confidence: post-AD 1700

COLUMN 20 (80.5/31.5), ERB85 AREA 14.
Alluvial parcels 2B, 4B
AJC-123, No. of samples: 15

Too erratic to give a useful date span. There were no distinctive susceptibility peaks, but a step increase in intensity from sample 5 upward. This was followed by more settled directional values, although three of these were much too westerly. These effects may indicate a change from turbulent to a more stable flow which, however, was still strong enough to rotate the settling particles away from the true geomagnetic field direction. The relative proximity of this part of the site to the main river may have been responsible for these effects. Lepidocrocite mottling, signifying post–depositional chemical change, was also severe in this section, but would probably not have dominated these suggested effects.

COLUMN 31 (34.0/15.0), ERB89 AREA 19.
Alluvial parcels 1C, 1K
AJC-126, No. of samples: 22

The magnetic susceptibility and intensity levels in this column closely parallel each other: high at the top, dropping to sample 6, then rising to a peak between 8 and 10 before falling away steadily to very low levels from sample 16 downwards. For dating, the column was only usable down to sample 19. Visual inspection suggested a good fit between the directional, intensity and magnetic susceptibility parameters of this and column 40, and the two are therefore discussed together.

Correlation tests showed that the best overall fit between the two columns was obtained with samples 1–12 of column 40 paired with samples 3–14 of column 31. This gave correlation coefficients of 0.72 for intensity, 0.64 for magnetic susceptibility and 0.83 for declination. Insufficient variation made the inclination values unsuitable for correlation coefficient measurements, but the two curves were very similar, the main change being an overall fall from about 70° to 65° over the overlapping length of the columns.

Magnetic susceptibility and intensity trends were also similar within the individual columns and, as shown by the correlation coefficients, there was a reasonable match between the columns, with fairly continuous maxima between samples 7 and 13 in column 31 and samples 5 and 10 in column 40. Poor results at the lower ends of both columns, necessitating the rejection of readings, were mainly due to very weak magnetisation of the material and its coarse, nodular character.

Probable span (samples 19-1): 6750–5850 BC

COLUMN 40 (41.2/40.0), ERB89 AREA 22.
Alluvial parcel 1G
AJC-127, No. of samples: 14

See column 31 for discussion.

Probable span (samples 12-1): 6500–5950 BC

COLUMN 60 (59.3/39.0), ERB89 AREA 32.
Alluvial parcels 2D, 2B, ?4B, LBA, 6D
AJC-129, No. of samples: 32

The upper six samples contained a broad susceptibility peak reaching a maximum at sample 4, and a narrow intensity peak at sample 3, but gave a poor directional record: effects apparently due to the burnt detritus of occupation. Unsatisfactory results were also obtained from the weakly magnetic and rather loose material from which the lowest four samples were obtained. Coincident susceptibility and intensity peaks between samples 15 and 20, and on a lesser scale at sample 22 and between samples 25 and 26, may represent a quite detailed history of upstream clearance from the end of the sixth to the middle of the fifth millennium BP, unless the enhanced material was derived from the present site. There appears to be a strong westerly deviation of values above sample 15, perhaps caused by flow effects.

Probable span (samples 28-7, parcels 2B and ?4B): 3350–1450 BC

COLUMN 65 (89.8/14.0), ERB89 AREA 28.
Alluvial parcels 7, 6C
AJC-128, No. of samples: 16

A sandy silt near the present surface, its coarseness and a pebble layer suggesting quite rapid deposition. A number of samples throughout the column gave aberrant intensity values, reflecting the presence of flecks of pottery, clay, burnt soil, etc., presumably part of the eroded occupation material observed in the section, and particularly visible just below the main pebble layer at sample 8. The few unaffected readings produced a rather tentative plot which appears to show the distinctive Bronze Age–Iron Age easterly excursion (up to sample 8), followed by the mainly inclination variations of the Roman period accompanied by faster accretion. The high intensity values at samples 11, 14 and 15 suggest on-site activity during the accretion of the lower silt, probably during the Iron Age.

An interpretation that relies less heavily on the few geomagnetic readings that appear usable is that the coarseness of the material, with pebble layers and an unusually uniform magnetic susceptibility sequence, taken with the distribution of anthropogenic material through the column, combine to indicate that the silting was rapid and energetic compared with the other columns. Since the inclusions are of Late Bronze Age date, the inundation and channel aggradation could have occurred at the end of that period or subsequently.

COLUMN 66 (74/37), ERB85 AREA 14.
Alluvial parcel 6D
AJC-124, No. of samples: 8

From river silt immediately overlying latest prehistoric occupation. Eight samples taken in two adjacent columns of four, so that they could be averaged in pairs. Results were too erratic to be interpretable, suggesting much bioturbation and probable churning by river-bank traffic.

Neolithic burnt deposits

AREA 19, ERB88
Neolithic hearths, contexts 19.208 (15 samples) and 19.258
    (16 samples)
AJC-125

These two hearths were stratigraphically contemporary. Magnetic measurements on them have not made any contribution to site dating, but were important in providing data for the refinement of the calibration curves on the basis of related radiocarbon dates (*c.* 3680–3375 BC). Their significance in that study has been summarised in Clark 1992, while the discussion below is more site-specific.

Both hearths were merely small patches of burnt clay thin enough to have broken up with time into a mosaic of separate pieces. These were very convenient for sampling by the disc method (Clark *et al.* 1988) but, although firmly in situ as a whole, many pieces had clearly tilted to varying degrees, reducing the accuracy with which they recorded the geomagnetic field direction. Work at Maiden Castle and elsewhere (Clark 1991; Clark *et al.* 1988) had shown that the fit of results to the calibration curve could be much improved by assuming that flat hearths had originally been level. The direction and degree of maximum tilt were measured in the field and a suitable bedding correction subsequently applied in the laboratory. At Runnymede, this technique had to be used with the added intricacy of separate tilt measurements and corrections on each sample. The equipment used – a Rabone Chesterman adjustable spirit level and a simple compass – could be improved upon.

The result of this approach was that the clustering of the context 19.208 hearth was improved (alpha-95 reduced from 4.87° to 4.01°), while that of context 19.258 deteriorated (alpha-95 increased from 3.52° to 3.81°). It was concluded from these results that much of the irregularity of the second hearth was original, and that the assumption that it had been originally flat was incorrect. A final result was therefore obtained by combining the corrected measurements from 19.208 and the uncorrected ones from 19.258, to give the following mean value:

Dec = 2.93°W; Inc = 67.58°; alpha-95 = 2.59° (31 samples)

This result is discussed in Clark (1992). It is in keeping with a modified zero 5° east of that originally proposed for the lake sediment derived declination curve (Turner and Thompson, 1982), and steeper than the average value of the lake inclination curve over the radiocarbon time-span of the hearths by 10.9°±3.3° at the 95% confidence level. This figure should probably be even larger: it includes an allowance for shallowing of the inclination in the hearth samples by magnetic refraction; in fact the magnetisation of the clay was exceptionally strong and the allowance should probably be larger.

AREA 20 ERB89
Neolithic burnt area, context 20.908.3.
AJC-130, No. of samples: 10

A patch of burnt soil, apparently in situ, partly overlying the fill of a ditch. Stratigraphic correlations suggest it pre-dates the hearths, 19.208 and 19.258, already discussed. The material was soft enough to be collected in tubes in the same manner as the soft sediment columns. It was strongly magnetised, and a test sample produced a sensible result, but most of the samples gave measurements too scattered to be of use, probably because of the wear and tear of subsequent site activity, and possibly some sinkage into the ditch.

Table 4.3   *TL measurements and dose rates of burnt flints; measurements by Nick Debenham.*

Context	Archaeological dose (Gy)	b-value[2] (Gy.µm²)	Alpha count Ms⁻¹ mm⁻²	Potassium oxide[3] K_2O (%)	Dose rates			TL Age (ka BP)
					Alpha[4] (Gy/ka)	Beta[4] (Gy/ka)	External[5] (Gy/ka)	
19.677(a)	4.56 ± 0.29	4.41	0.41 ± 0.05	0.0022	0.246	0.043	0.563	5.36 ± 0.91
19.677(b)	5.03 ± 0.31	3.31	0.67 ± 0.11	0.0024	0.302	0.069	0.563	5.39 ± 1.07
19.677(c)	4.82 ± 0.35	5.42	0.29 ± 0.07	0.0054	0.214	0.033	0.563	5.95 ± 0.98
20.993(a)	4.83 ± 0.18	2.42	0.53 ± 0.08	0.0019	0.175	0.054	0.526	6.40 ± 0.72
20.993(b)	4.15 ± 0.31	2.63	0.33 ± 0.08	0.0020	0.118	0.034	0.526	6.12 ± 0.87
20.993(c)	3.79 ± 0.36	4.19	0.46 ± 0.09	0.0046	0.262	0.049	0.526	4.53 ± 0.75

[1] Crushed and sieved 75-125 µm grain fraction, acetone washed and measured on oiled stainless steel discs
[2] Fine grain measurements (approximately 2-10 µm) on aluminium discs (Bowman and Huntley 1984)
[3] Measured by atomic absorption spectrophotometry
[4] Based on the alpha count-rates and K_2O analyses; uncertainties in the alpha dose rates range from ±19 to ±36%, and in the beta dose rates, from ±12 to ±24%
[5] Using in situ gamma spectrometry; environmental dose-rate uncertainties are approximately 10%

However, three were convincingly grouped, giving the following result:

Dec = 2.76°E; Inc = 59.82°; alpha-95 = 4.50° (3 samples)

Again, alpha-95 is not significant because of the small number of samples. The declination is in keeping with the stratigraphic relationship referred to above, suggesting a date not before about 3950 BC. The inclination is too shallow, probably as a result of compaction of this soft and vulnerable material, together with a rather large magnetic refraction effect due to its strong magnetisation.

Thermoluminescence dating of burnt flint clusters

Among the 'horizon' of cultural material stratified low within the late Atlantic soil profile were two clusters of burnt flint, 19.677 and 20.993. These may have lain in shallow features but, if so, the feature fills were not observable at a higher level and it is more likely that they are effectively surface deposits that migrated downwards along with the other more scattered debris. As excavated this early horizon material was sealed beneath the refuse-rich Neolithic horizon and intervening cleaner soil. This suggested some time interval between the respective occupations but the interval could have been anything between a few decades and two millennia, the latter being the time over which this high bank was a stable surface without significant sediment accretion. The period in question follows the deposition of alluvial parcel 1K, late seventh to early sixth millennia BC, and precedes the Middle Neolithic occupation, mid-fourth millennium BC; it

therefore covers both the Late Mesolithic and Early Neolithic periods. Since few diagnostic finds were recovered from this early horizon, it seemed worthwhile to attempt to date the burnt flint clusters by thermoluminescence (TL) despite the inherent imprecision of the technique.

The work was undertaken by Nick Debenham who conducted the on-site dosimetry and processed flints from each cluster for the measurement. Sheridan Bowman and Mike Cowell advised on the presentation and significance of the data. The results are given in Table 4.3. Since each cluster can be regarded as a single-event deposition, the results may be combined as shown in Table 4.4. These mean results still have large errors but the central dates fall no earlier than the beginning of the range currently accepted for the MN occupation (within the bracket 3680–3375 BC) and the later part of their probability distribution is thus constrained by the stratigraphic evidence. The burnt flint clusters could still date earlier by several centuries, but the TL dating nevertheless places them towards the end of the stratigraphically permissible time-bracket, most likely after 4500 BC.

Context	Dates (BC)	Mean date (BC)
19.677	3370 ± 910	3590 ± 780
	3400 ± 1070	
	3960 ± 980	
20.993	4410 ± 720	3760 ± 810
	4130 ± 870	
	2540 ± 750	

Table 4.4   *Mean TL dates for burnt flint clusters.*

CHAPTER 5

Sediment analysis and soil micromorphology

SUSAN LIMBREY

Studies of soils and sediments exposed in the course of the research excavations are focused on their sedimentary origin, pedological transformation and the contribution of human activity and cultural materials. The materials studied fall into categories which are differentiated by the relative importance of these factors: alluvial deposits, soils and associated cultural layers, 'midden' or 'dark earth' materials, and fills of features.

Field study and sampling

After detailed examination of excavated sections and formal soil description, sampling was carried out for sedimentological and micromorphological study in long columns to cover the full depth of sediments, soils and cultural layers exposed at points where alluvial parcels were fully expressed, and in short columns and single samples (also identified by 'column' numbers) where particular questions were posed concerning origin or nature of individual contexts (see Figs 1.3 and 1.5a for locations). In the columns for micromorphology, Kubiena tins, $10 \times 5 \times 4$ cm, were emplaced with long axis vertical, in slightly overlapping sequences to give complete depth cover. Bag samples for analysis were taken, in parallel columns each normally covering 5 cm depth but adapted to respect visible stratigraphy and horizon boundaries, so that a few samples were deeper or shallower, there were one or two gaps, and in some cases samples overlapped. For single samples, Kubiena tins were placed either vertically or horizontally, as appropriate, and bag samples were taken to cover the context in question. Additional columns were sampled for sediment analysis only.

Field study was done by S. Limbrey and I.D. Máté: columns 15, 21, 22, 62, 63 and 68 were sampled by I.D. Máté; columns 13, 32, 34, 42, 53, 54 and 56 by S. Limbrey, and the remaining columns by site staff.

Analytical procedures

Samples were examined in the laboratory, using low magnification, to amplify field descriptions.

Standard sample preparation and analytical procedures, though without a primary decalcification stage, were used for pH and particle size distribution (Avery and Bascomb 1982), but while sand fractions were determined by sieve, for silt and clay the SediGraph X-ray particle size analyser was used. The silt plus clay was reanalysed after decalcification in acetic acid buffered at pH 5. Organic matter and calcium carbonate content were estimated by weighing after ignition at 450 and 950°C. Analyses were carried out by D. Dungworth and D.A. Smith.

Micromorphology samples were dried through progressive replacement of water by acetone and impregnated with Crystic resin by gradual concentration from dilute solution in acetone, then cured under slow step-wise temperature increase. Sections were cut in the Thin Section Workshop, Department of Agriculture and Environmental Science, University of Newcastle upon Tyne. Micromorphological studies were done by the author.

General points about analytical data

Particle size data are presented in Table 5.1 in terms of sand, silt and clay fractions of the 'fine soil' component (< 2 mm). Except in the clearly identified gravel units, material in grades coarser than 2 mm is almost entirely cultural in origin, even when it occurs in the predominantly sedimentary units. It has not, therefore, been included in tabulated particle size data, but is referred to in the micromorphological descriptions.

Data on pH, calcium carbonate and organic matter are available from archive. The pH values throughout lie in the range 7.8 to 8.5, consistent with buffering by the ubiquitous presence of calcium carbonate. Values for calcium carbonate lie in the range from about 5% in the more strongly developed soils to some 20% in the marl component of the alluvial sediments. The values for organic matter include charred material. Since micromorphology shows that this occurs in variable quantity in almost all the materials studied, the organic matter values do not contribute to assessment of soil characteristics so well as do observations of characteristics of humified residues and fabric colour made in the micromorphological study.

The problems of particle size analysis of calcareous soils and sediments were discussed in Limbrey (1991). In projects undertaken since then, the practice has been adopted of analysing non-decalcified material, this being facilitated by the change in method of analysis to use of the SediGraph, which uses small samples in which a dispersed state can be achieved with the help of ultrasonics and which is stable for the relatively short period required for SediGraph analysis. This obviates some of the difficulties of interpretation, but the multiple origin of calcareous particles still has to be considered, and the avoidance of the decalcification stage in sample pretreatment increases the likelihood of a proportion of the silt fraction being stable aggregates of clay formed by pedogenic processes. Supplementary data was provided by reanalysis of the silt and clay fraction after decalcification. Sand fractions were not decalcified. Sieve analysis of sand fractions and the analysis of silt and clay give subdivisions into coarse, medium and fine fractions for sand and silt. These data are given in Table 5.1. The SediGraph cumulative percentage curves are available from archive. Discussion of the data includes indication of the modal particle size in the sand and silt fractions. In referring to one mode only, it should be noted that this does not imply a unimodal distribution. In all these samples, the greater part of the clay is in the lower end of the size range, which is quantified in total but unsized (the SediGraph analyses were taken down to 0.25 microns). Very fine clay behaves colloidally and in dispersed condition will never sediment out, so the particle size distribution is open-ended and determination of standard moment statistics is not possible. Decalcification in all cases increases the predominance of this very fine clay: in some samples 80–90% is in the less than 0.25 micron grade. To interpret this in sedimentological terms is nonsense: it would imply deposition from still, turbid water almost entirely by infiltration or evaporation. Much more likely is that the clay was either transported as silt-sized aggregates, or flocculated in the calcium–rich aquatic environment during deposition.

The mode in the sand and silt fractions is commonly in the fine sand and coarse, or coarse and medium, silt, but modes in the medium silt alone also occur. Decalcification flattens the mode considerably and often removes it. Apart from cultural layers, and samples within or just above and below the gravel layers, coarse sand content is very low and is usually dominated by mollusc shell fragments and the composite calcite granules produced by Arionid slugs and probably also by earthworms (the identification of earthworm granules, discussed in Limbrey 1992, is still uncertain). Shells and shell fragments, being hollow and thin, may have travelled with finer grade particles, but they may also be autochthonous; where the granules are the only coarse sand, they are certainly so. Clastic particles in the coarse and medium sand include rounded quartz, quartzite, flint, chert, and occasional other rock types. In some samples, tufa fragments are present in the sand grades. The nature of the fine sand and the silt will be discussed under micromorphology.

General points about micromorphology

To avoid repetition, and to explain some of the terms used, the following points should be noted. Descriptions follow the terminology of Bullock *et al.* (1985), but for brevity do not use the complete descriptive scheme. Structure refers to an extension of the macroscopic structure description into the micro range, being the shape, size and arrangement of soil aggregates and voids. Groundmass refers to material within structural units, and is described in terms of colour, relative distribution of coarse and fine material, and patterns of orientation which are apparent under crossed polarisers in birefringence patterns. In those materials which are rich in calcite in the fine silt grade, the birefringence is referred to as crystallic, calcitic. Where there is additive birefringence of clay with its particles aligned, it may give the speckled birefringence of silt-sized aggregates or, if continuous over larger areas, striated birefringence, either reticulate, produced by random alignments, or poro- and grano-striated if the pattern is aligned with the edges of voids or particles. Pedofeatures refers to the redistribution of soil materials to produce infillings, coatings to pores or particles, staining and impregnation, which may be intense enough to form nodules of various form. These redistributed materials may be calcite, or iron and/or manganese oxides, often probably in complexes with organic matter, or clay. Depletion pedofeatures are those produced by losses of these materials. Both iron and manganese impregnations often focus on organic residues. Commonly, the black manganiferous pedofeatures are strongly impregnated aggregate nodules and spreading or elongated aggregates. These have the appearance of micro–organism colonies centred on residues and subsequently mineralised. The iron staining, however, is more diffuse and there is a gradation: staining only, then from weaker to stronger impregnation, forming coatings, hypocoatings and nodules, which are of simpler form than in the case of manganese. More or less pure segregations of iron oxide occur in some samples, as infillings and coatings to voids and also as fragments of such coatings disrupted and embedded in groundmass. The appearance is limpid reddish brown, sometimes finely laminated or fibrous, and birefringence varies, between and within examples, from high (though masked by the mineral's colour) to totally isotropic. The birefringent version is identifiable as goethite. The isotropic form may be referred to as ferrihydrite.

A characteristic of the groundmass which is constantly referred to in these samples needs elaboration, since it has not been discussed in the literature. In materials such as these, in which there is a modal particle size in the fine sand to coarse or medium silt and moderate to high clay content, the related distribution of coarse and fine material is described as porphyric, that is, a fine, continuous groundmass with coarser particles spaced out within it. In these soils and sediments, and in others of similar texture which I have studied (Evans *et al.* 1993) the coarser particles – fine sand and coarse silt – are arranged in strings and clusters forming polygonal reticulate patterns of about 0.5 to 1 mm scale. Depending on the degree to which it is developed, and the amount of the coarser grades present, it varies from open to closed porphyric, that is, the coarser particles are separated or are in contact. The term reticulate porphyric is introduced here. It is suggested that this pattern is a product of faunal activity. The smaller earthworms would avoid these particles, so their casts would contain only the finer soil; alternatively ants, which gather fine soil into aggregates which they carry in their mandibles, would have the same effect. Eventually, the coarser grains are left in clusters and planes (seen as strings in section) with the fine soil between.

Other evidence of faunal activity is more obvious, in the form of mammilated partial infillings to channels; crescentic, partially sorted infillings; and interlaced intercalations of humic silt or calcitic and humic silt (Thompson *et al.* 1990; Limbrey 1992). Since it was obvious in the field that earthworm penetration of the whole suite of soils and sediments has continued to the present, with very conspicuous deep vertical channels, a continuum of features could be interpreted in terms of earthworm activity, from open channels, through partially and completely filled ones with the form of the casting still clear, to coalesced casts and juxtaposed areas of soil of different colour or texture from different layers or horizons. The reticulate porphyric pattern is present in addition to these well-known and more clearly interpreted ones. Where these features dominate the groundmass, the term total faunal structure is used. This is preferable to the total excremental structure of Courty *et al.* (1989), since ants, if they are involved, do not ingest and excrete the soil.

Characteristics of the soils and sediments

In the following sections, the materials studied will be described macroscopically, their particle size characteristics, and their micromorphology, for each of four groups: alluvial parcels, soils and cultural layers, 'dark earth' deposits, and fills of features. Rather than giving full micromorphological descriptions for each sample, which would involve a great deal of repetition, these descriptions will cover whole units, with indications of significant variation within them. Cultural materials are referred to as they occur in each sedimentary unit or soil, but their kind and condition is discussed more fully under 'dark earths'. Transitions between units are accommodated in the discussion of the soils, since it is the alternation of predominance of depositional and pedological processes which characterises the sequence in each column. Thus, the upper part of each alluvial parcel, where characteristics are strongly modified by soil development, are regarded as soil A or B horizons. The slides themselves, and the formal descriptions of them, are available for study on application to the author.

Alluvial deposits

Alluvial parcel 1

This parcel was studied in Area 19, column 63, where sub-parcel 1K was represented in the lower part of the column.

Field description is of strongly mottled clay, light grey, 2.5YR 7/2, with many sharp, fine and very fine yellowish-brown mottles; no structure was developed in section; there were very few medium, fine and very fine macropores, some with decaying roots; no stones, but snail shells were present.

This material became darker upwards, and a medium angular blocky structure developed as it became affected by the soil processes associated with the Neolithic cultural horizon, and some incorporation of stones and archaeological material was apparent.

Analysis of samples 408 and 409 shows the material to be clay-dominated, with silt and fine sand, a slight mode lying in the medium silt. These proportions are little affected by decalcification of silt and clay.

In Area 13, column 15, a component of parcel 1 underlies the pre-Neolithic soil and has many of the same characteristics as the deposit described above, being dominated by comparable gleying phenomena. Field description is of light brownish-grey clay, 2.5Y 6/2, with 30% fine, sharp yellowish-brown ferruginous concretions and 10% black fine, sharp concretions; no stones, no structure, few coarse macropores.

Analytical data for samples 453 and 454 indicate a coarser deposit than in the Area 19 example, with higher silt and sand, the mode in medium silt broadening and skewing into the coarse silt and fine sand. In this case, decalcification of silts and clays yields a 30% relative increase in clay, and shows that most of the medium and coarse silts is calcareous material. The proportions are then close to those in this unit in column 63.

Alluvial parcels 1C and 1K were sampled for sediment analysis but not for micromorphology in column 31, Area 19, in a deep section through the upper levels of 1K represented in column 63 (some 3 m to the north-east), and into marl, 1C.

Data for samples 4 to 25 show strong similarity with 1K in column 63 in the upper few samples, and then greater similarity to the equivalent deposit in column 15 down to sample 18, where a marked increase of fine sand at the expense of clay marks the transition to marl. Decalcification of the silt and clay has variable, but often little, effect on the silt:clay ratios, but does reduce the predominance of coarse silt. These characteristics show that the marl is composed mainly of coarse silt and fine sand-grade calcite grains and aggregates in a matrix of medium and fine silt and clay, which remains more or less constant. Precipitation of calcite, chemically or biogenically, in situ in an environment which was receiving finer alluvial sediment would explain this composition. What we see in alluvial parcel 1K in columns 63 and 15 (and in the upper part of column 31) is the clay-rich sediment after the production of calcite in situ has been somewhat reduced.

Sub-parcel 1H was sampled for sediment analysis but not for micromorphology in column 59, Area 32, which covers a transition from organic mud with sand lenses through clay with sand lenses to clay. The upper part of the column is affected by development of the pre-Neolithic soil and then truncated by the gravel of 2F.

Data for samples 41 to 54, show a loam texture with mostly comparable amounts of silt and clay and a substantial fine sand component. Apart from a markedly sandy incursion and a later coarse silt incursion, there is very little medium and coarse sand, and a mode in the fine sand and coarse silt is skewed towards the sand. Marked fluctuations in sand and clay occur in the lowest levels where the organic mud was encountered. Decalcification of silt and clay strongly increases clay at the expense of coarse and medium silt.

MICROMORPHOLOGY

Microstructure of parcel 1K, in column 63, is angular and sub-angular blocky, with intrapedal vughs, but also with many channels and chambers, in which are loose sub-angular blocky and granular peds. Textural relationships are reticulate porphyric. The small amount of coarse sand, and the medium sand, is almost entirely shell, calcite granules from Arionid slugs or earthworms, and fragments of tufa and calcified plant structures, with only occasional rounded quartz. Fine sand and coarse silt is more quartzitic, but there is also a large amount of calcite, much of it in the form of grey ovoid composite micritic grains in the medium and coarse silt. Groundmass colour is pale yellowish-brown, strongly affected by staining and impregnation by oxides of iron and manganese, which form the dominant pedofeatures. These features range from weakly to strongly impregnated, forming aggregate nodules, irregular patches and strings of coalescing nodules, and hypocoatings to planar voids. Their colour varies from yellowish-brown to black, with black impregnations often separate from yellowish- to dark brown ones. Black impregnations are in places affecting slug granules or shreds of humified organic matter, and the pattern of many of the segregations suggests concentration at loci of organic degradation and the form of colonies of micro-organisms. Iron depletion zones centred on voids complement the staining and impregnation.

In addition to the stainings and impregnations, there are pure ferrihydrite segregations, some of which are in situ as quasi-coatings associated with a depletion zone around a void, but others in the form of fragments which have been disrupted and redistributed.

Calcite pedofeatures are fairly common, in the form of coatings and infillings to intrapedal voids, often in two phases of finer then coarser crystal size. Calcite pedofeatures decrease upwards. Calcite depletion is associated with the iron oxide impregnation, in that the more heavily impregnated areas of fabric, and the ferrihydrite segregations, are weakly calcareous or non-calcareous.

Throughout (to the depth sampled) there are areas of another fabric – coarser, darker, and with fine charred particles and fragments of cultural material. In the lower part, these are small, often crescentic in form, and can be seen to be embedded components of channel fillings in a fully coalesced faunally turbated fabric. With decreasing depth these darker and coarser areas become more clearly faunal channel fills, eventually becoming continuous with the strongly culturally affected soil at the upper boundary of the unit, which will be discussed later.

Micromorphological characteristics of the material in column 15 are again similar to those of the main fabric in column 63, because of the dominance of gley phenomena, but the coarser texture is apparent, and the sand fraction includes an abundance of fragments of cultural material: pot, burnt and unburnt bone, flint, charcoal. It is also rich in humified residues and in fine charred material. Clusters and isolated examples of spores or faunal egg capsules occur. The secondary calcite includes many tufa fragments and crystal aggregates showing elements of structural patterning which may be of semi-terrestrial algal origin or bits of calcified root tissue. In contrast to column 63, there are few calcareous coatings or infillings, and only in the upper part, probably because truncation had affected the deposit to a greater extent in this location.

DISCUSSION

Differences in texture beween the locations sampled reflect variability in depositional regime and a truncation cutting across flat-lying strata. In parcel 1K, any sedimentary structure which may have formed on deposition has been obliterated by faunal turbation. The fabric has been reworked into a total faunal structure. There appear to be several stages of faunal activity, with an earlier one disrupting ferrihydrite coatings and later ones incorporating coarser, darker and more culturally affected soil from above. The development of the strong gley features of depletion and impregnation of iron and manganese oxides post-dates much of the faunal activity affecting the main fabric, which is probably penecontemporaneous with deposition. Penetration by earthworms, however, has proceeded from later surfaces of more stable soil development, right up to the present day.

The origin of secondary calcite in alluvial sediments has been discussed in Limbrey (1992), where deposits having characteristics comparable to those of parcel 1K have been called 'calcite loam'; the multiplicity of organisms which can mediate the crystallisation of calcite is stressed, and reference is made to the abundance in the deposits similar to those being described here of silt-sized ovoid aggregates. They are not rolled chalk particles – no chalk microfossils are present, either within them or free in the sediment. Whatever the organisms involved, the abundance of these pellets and other calcified structures demonstrates clearly the high level of organic activity during the period of accretion of these deposits.

Though the transition from marl to clay was not studied

micromorphologically, it is likely that the very high calcite content of the marl (1C) is a result of chemical and biological precipitation in an entirely aquatic environment, while the clay was subject to seasonally sub-aerial conditions which favoured the more terrestrially based processes.

Alluvial parcel 2

This parcel was studied in Area 13, column 15, in Area 16, column 22 and in Area 19, column 63. Originally there was some uncertainty whether the pre-Late Bronze Age alluvium in column 21, Area 18, was a component of parcel 2 or parcel 4, or included both; it will be dealt with here first for comparison with column 22.

Parcel 2D was not sampled for micromorphological analysis, but was sampled for analysis of particle size distribution, in Area 32, where the lowest five samples of column 60 took in this unit.

Parcel 2F, the gravel, in columns 15 and 63, has a dark greyish-brown, 10YR 4/2, silty clay matrix, with weak sub-angular blocky and moderately developed fine granular structure. In column 22, the gravel has a dark greyish-brown, 10YR 4/2, sandy clay matrix, with patchily developed strong brown, 7.5YR 5/6 to 4/6, iron panning. This matrix was continuous with the underlying soil, the gravel having been incorporated into the soil. A gravel component in the upper two samples of column 59, which is reflected also in the coarse and medium sand figures, accords with stratigraphic correlations indicating that this column too included 2F.

Parcel 2B in column 15 is dark greyish-brown, 10YR 4/2, becoming paler, 5/3 towards the base, silty clay with few to no stones, but snail shells are present; moderately to weakly developed fine granular structure.

While four samples in column 15 (415, 416, 417, 418) are included in 2B, and there is little variation in a loam texture, the lower incorporates part of 2F gravel, and the upper is already strongly affected by cultural material and soil development. Analysis shows that a modal particle size covers the fine sand and the coarse and medium silts. Decalcification of silts and clays has a very marked effect, increasing relative proportion of clay by about 30%, and completely removing the dominance of coarse and medium silt.

In column 22, the greater part of parcel 2 sediments are those of 2B, with gravel and cultural material present at the base of the column. The lowest three samples of column 21 are regarded as 2B (see below). In these columns, the sediments are at a lower altitude than in column 15, and they are strongly gleyed.

In column 22, the deposit was light yellowish-brown, 10YR 6/4, with 20% yellowish-brown, 10YR 5/6, and 10% black very fine, sharp mottles, becoming greyish-brown, 2.5Y 5/2 with 30% black mottling and 10% yellowish-brown, 10YR 5/6 staining in the lower 15 cm. Field texture

was clay, with stones and shell becoming common in the lower 20 cm. Structure was strongly developed prismatic with common fine and very fine macropores and some coarse macropores with black clay linings.

Analysis of samples 444 to 449 shows that, again, a modal particle size lies in fine sand and coarse and medium silt, but this is weakly expressed in the lower three samples, which have more coarse sand and a predominance in the lower silt content of medium silt. Decalcification of silt and clay increases the relative proportion of clay by 15 to 20%, the losses variably distributed in the silt grades.

Field description for column 21 is of a light yellowish-brown, 10YR 6/4, with strong gleying giving variable amounts of fine, sharp, yellowish-brown, 10YR 5/6, and black mottles. In the upper 32 cm (parcel 4B), mottling is 10%, predominantly yellowish-brown, then increasing to 20% yellowish-brown plus about 30% black, and in the lower 15 cm, about 40% mottling, colour intensifying to yellowish-red, 5YR 6/6, with soft nodules and a weakly cemented pan. Field textural assessment is of a clay loam, with a sharp increase in silt in the lower 20 cm and having more shell fragments in the zone of black mottling; there were no stones. No structure was developed in the excavated sections, but there were moderate amounts of fine macropores and a few fine root residues.

The data for samples 425 to 434, again show a modal particle size lying in fine sand and coarse and medium silt, but in the lower three samples, the higher clay content is reflected by low sand and coarse silt, giving a less marked concentration in these grades. The texture change reported above in field assessment appears not to reflect a change in particle size distribution, which is almost the reverse of that implied. Decalcification of silt and clay increases the proportion of clay to a variable extent, by 10 to 20%, mainly at the expense of coarse and medium silt.

The pre-LBA sequence in column 21 was not linked during excavation to the broader alluvial sequence and parcel attribution has to be deduced. Looking at the particle size distribution, this substrate has very consistently high fine sand and coarse and medium silt, until the lower part is reached – samples 432, 433 and 434 – when higher clay content and reduced sand and/or silt makes it comparable to 2B in column 22. It would appear, therefore, that there is a transition, and samples 426 to 431 of column 21 could be attributed to alluvial parcel 4B. This ties in well with the level of the parcel 2B/4B boundary seen in neighbouring sections (Fig. 2.2).

The transition from alluvial parcel 2B to LBA soil in Area 16, column 22, is not strongly marked by change in particle size distribution, but at this level a steady increase in silt and decrease in sand and clay ceases.

In column 15, the LBA has no clear effect on the particle size distribution, and the reworking of LBA deposits and their redeposition over the material in situ again produces a matrix consistent with that below, not changing until a quite dramatic shift from silt to clay in the top level, after the LBA material has died out. (The top three bagged samples were from parcel 6D.)

Alluvial parcel 2B was also sampled for sediment analysis but not for micromorphology in Area 32, column 60. Particle size data indicate a clay with little sand and a substantial silt content, in which there is modal particle size in the coarse and medium silt, extending into fine silt. Decalcification of the silt and clay increases the relative proportion of clay by 15 to 30%, reducing all silt fractions.

MICROMORPHOLOGY

Because of the difficulty of sampling a gravel layer using Kubiena tins, in column 15 only the lower part of 2F was included in the micromorphological study. It is much less strongly gleyed than the parcel 1 deposits below the Neolithic soil, but there is a vertical channel whose loose fill of earthworm casts forms the focus of strong black impregnations. The microstructure is weakly developed sub-angular blocky with many chambers and vughs. Groundmass is calcitic crystallic, reticulate porphyric. Colour is greyish-brown, and there is an abundance of fine charred material as well as much humus staining. Cultural material in small stones and fine sand grades includes charcoal, bone and sherd fragments, and there is burnt and unburnt flint.

Parcel 2B has weakly developed sub-angular blocky stucture and is dominated by channels, chambers and vughs. Channels and chambers have loose granular partial fillings, and crescentic forms of vughs indicate more complete infillings of channels. Groundmass is calcitic crystallic, strongly reticulate porphyric. Colour is light brown with fine, sharp yellowish- and reddish-brown and black composite impregnations, and areas of brown humic staining with black impregnations within them and forming rims to them. Coarse sand and small stones include slug/earthworm granules, shell fragments, burnt and unburnt flint, pot and bone.

In column 21, the deposit has a very uniform microstructure, weakly developed sub-angular blocky with channels and chambers, with granular structure in the fabric partially filling channels, and many vughs. Groundmass colour is light brown, and it is calcitic crystallic. Texture in the lower part is dominated by variably calcitic clay and fine silt, with the coarser components distributed in swirling patterns, giving the appearance of a slurry. In the upper part there is more sand, and the related distribution is reticulate porphyric, which together with crescentic infill to channels and vugh shapes indicate faunal structure which has coalesced. The transition between the two fabrics is by way of vertically orientated zones of coarser and finer texture, with elongated particles of shell and distributions of fine sand and coarse silt vertically aligned. This pattern suggests cracking of the clay-rich deposit and the inwash of the next accretion. After this, the characteristics of faunal structure become dominant. Throughout, the coarse and medium sand is almost all shell fragments, and some complete shells, and slug/earthworm granules, with only very small quantities of rounded quartz and quartzite in the medium and fine sand grades. Fine sand and coarse and medium silt are dominated by grey ovoid micritic calcite aggregates.

The pedofeatures characteristic of gleying have been described above, as they affected sub-parcel 1K. Very similar features are seen in parcel 2 in column 21, the intensity and distribution of impregnations increasing with depth as indicated in the field description. The gley features are partly arranged in vertically orientated zones of about 1 cm width. There is a zone of intense manganiferous impregnation in the middle, and very intense ferruginous impregnation in the lower 20 cm. Segregation of ferrihydrite in voids occurs increasingly with depth.

In column 22, micromorphology is very similar to that in column 21 except that the lower part of the column makes the transition, via a gravel incursion, to the Neolithic soil, and cultural material and darker soil rich in fine charred material occurs in faunal castings within the lower few centimetres of the 2B deposit above the gravel.

DISCUSSION

The very strong similarity of the column 21 material to that of column 22 in macroscopic description and micromorphology, on the one hand, and the distinctive fine sand and coarse and medium silt of the upper part of the putative parcel 2 material as revealed by particle size analysis, on the other, reflect the ways in which soil processes dominate over sedimentary characteristics in both the macro and the micromorphological aspect. The difference in sedimentary regime implied in the attribution of this part of column 21 to parcel 4B, rather than parcel 2B, however, need not be great, since the location is marginal to the channel and the change involved in the transition between the upper fill of the channel and overbank flooding can be minimal.

The various materials of parcel 2 and, in Area 18, parcel 4B, are dominated by the characteristics of 'calcite loam', described above, with a very high representation of probably biogenic calcite, modified by the effects of gleying. The micromorphology suggests that in the deeper examples there is a transition from a wetter environment, in which a slurry structure is evident, to one which is seasonally dry enough for vigorous faunal activity. The vertical zonation of gley features reflects the rooting pattern of plants adapted to a seasonally oxygen-depleted environment.

Alluvial parcel 4B

This unit, as unequivocally identified in the field (see also column 21 above), was sampled in Area 21, column 37, 480 and 481, and column 38, 467 and 468. In both the columns, cultural material from Bronze Age deposits had become incorporated into the 4B deposits within and on the bank of the channel. Analyses show variable texture, close to the lower limit of clay, with high silt and/or sand.

In column 37, micromorphology shows so much cultural material throughout the depth of the sample, that a description of the unit as a sediment must be confined to the observation that the sand component includes quartz and flint sand in coarse and medium sand as well as finer grades, and that there is an abundance of tufa fragments and other calcareous aggregates. Of particular note here are rod forms, up to 2 mm long, composed of calcite grains, graded in size across the rods. Such rods have been observed in most of the other sediments, but less commonly and of smaller size. In column 38, the micromorphological characteristics are very close to those of the Bronze Age soil and the redeposited Bronze Age levels of column 21, and it is therefore covered under that heading.

The probable attribution of the column 21, samples 426 to 431 to parcel 4B is supported by the comparability in particle size between these samples and those of column 37.

Alluvial parcel 6

Parcel 6 was studied in Area 19, column 63, where alluvial material formed a clearly defined layer, and in Areas 13, 16 and 18, columns 15, 21 and 22, where the deposit was strongly affected by the incorporation of cultural material from subjacent layers.

Field description in column 63 is of brown clay, 10YR 5/3, with no mottling, with weakly developed sub-angular blocky structure, 0.5% medium pores and common very fine woody roots. There are abundant shell fragments, and common small (up to 2 cm) angular to sub-rounded flints.

Data for sample numbers 400, 401 and 402 show a modal particle size in the fine sand and coarse and medium silt grades. Decalcification of silt and clay does not have a very marked effect, increasing relative clay content by 4% only in the upper sample, in the second by 15%, at the expense mainly of coarse silt, and in the third only slightly affecting the proportions within the silts.

In column 15, the field description gave silty clay, the same colour but with slight darkening to ped faces, and with granular as well as sub-angular blocky structure, higher porosity and more fibrous roots. Column 15 also showed clay linings and infillings to worm holes.

Again, analysis of samples 410 to 413 shows a modal particle size in the fine sand and coarse and medium silt grades. While analysis puts the texture into the clay, rather than the silty clay, class, the samples straddle the class boundary. Decalcification of silt and clay increases clay by 18%, 24% and 14% in the upper three samples but by only 6% in the lowest. This lowest sample may have included material from the underlying soil and should be regarded as transitional.

In column 21, the material was dark greyish-brown, 10YR 4/2, silty clay with no structure and no stones.

Analysis of samples 422 and 423, puts the texture just into the clay class. A modal particle size lies in the medium silt. Decalcification of silt and clay yields a 24% relative increase in clay, largely at the expense of medium and fine silt.

In column 22, , the material is described as dark greyish-brown to brown, 10YR 4/2 to 5/3, silty clay with weakly developed fine granular structure and common small and few very small sub-angular to rounded flints.

Analysis of samples 435, 436 and 437, shows a modal particle size lying in the medium silt. Decalcification of silt and clay yields 18 to 20% more clay, at the expense of all the silt, but particularly medium and fine grades.

MICROMORPHOLOGY

In column 63, microstructure is angular blocky with porosity of chambers and intrapedal vughs and some fine intrapedal cracks. Fabric is crystallic, calcitic, with reticulate porphyric related distribution very well developed. There are no coatings or depletion features; small black impregnations and nodules are scattered within the fabric. There is a well-distributed scatter of fine black particles and small humified fragments. The only coarse sand is composite calcite granules of slug/earthworm origin, shell fragments and occasional angular flints; medium and fine sand also includes rounded quartz and occasional rounded pot sherd fragments.

In columns 15, 21 and 22, the micromorphological characteristics are considerably modified by the incorporation of archaeological material and the biological processes associated with it. These features are also apparent in the transition in column 63 between the unit described above and the cultural layer below. The characteristics of this transition, and of the whole unit in columns 15, 21 and 22, are so similar that a single description will cover most points.

Microstructure is very porous, being a combination of sub-angular blocky and granular with porosity of vertically orientated channels and large interconnected chambers, with intrapedal vughs and vesicles. Some channels have partial fillings of darker, coarser granular soil, and there are channels completely filled with this soil. Fabric is crystallic, calcitic, with reticulate porphyric-related distribution very well developed, and there are also interlaced humic silty intercalations. There are no coatings or depletion features; black aggregate nodules, strongly impregnated, are common, as are smaller weakly to strongly impregnated rounded brown nodules which appear to be derived particles. Very small black particles are uniformly distributed in the finer material; these are particularly common in column 22. Sand includes common slug/earthworm granules and shell fragments, and occasional rounded sherd and bone fragments. There are occasional small particles of organic material apparently of faecal origin, as discussed below. In column 15, there is a cluster of angular burnt flint fragments. In column 21, which is somewhat lower-lying than the others, a certain amount of gleying is reflected in the development of black aggregate impregnations in a vertical zone which was probably spreading from a channel to which the plane of the section was tangential.

Soils and cultural layers

Neolithic and earlier soil development

The soil and the cultural layers associated with it are described and discussed together. They were studied in columns 15, 22 and 63, where they were part of the columns sampled for extended sequences, and in columns 13, 32, 34, and 42, where they were sampled for study of the soil in relation to the archaeological features of Areas 19 and 24. In those columns sampled only for sediment analysis, the upper three samples of column 31 and the upper two samples of column 59 represent the B horizon of the truncated soil.

In columns 15 and 22, the Neolithic material occurs in heavily gleyed soils at the base of the columns, where field observation suggested that the gravel incursion had truncated the Neolithic surface progressively downslope, so the features of soil development which are evident in these columns are those of gleyed B horizons, with only a bit of the A horizon apparent in column 15. The top of column 59 is comparable to column 22 in this respect. In columns 13, 32, 34 and 63 (all Area 19) the upper horizon of the soil appears to be more complete, albeit modified by human activity, and the transition to the B horizon is seen where samples covered greater depth than the A horizon, in 13, 34 and 63. The top of column 31 overlaps with the bottom of column 63, thus extending the view of the B horizon.

The soil in column 63 is described as: A horizon, 4 cm of very dark grey, 10YR 3/1, clay loam with common clear fine and very fine brown mottles, 10YR 5/3. Structure is moderately developed fine granular. There are common small and very small angular to sub-rounded flints. A clear, wavy boundary makes the transition to a Bg horizon, 13 cm of dark greyish-brown clay, 10YR 4/2, with 20% yellowish-brown and 20% brown mottles, 10YR 5/6 and 5/3. Structure is very weakly developed medium angular blocky. There are few very small angular flints. This horizon gives way to the more strongly gleyed Cg horizon already described under alluvial parcel 1.

Samples 405, 406 and 407 show a fairly high sand content reflecting the cultural material, also present in the stones grade, but apart from this the modal particle size lies in the fine sand and coarse and medium silt grades. Decalcification of silt and clay increases clay content by some 20% in the upper two samples, but only 6% in the lowest, transitional to the underlying material where strong clay dominance leaves little of the calcareous silt.

Column 13 was sampled to expand upon the understanding

of the relationship of earthworm sorting to depositional processes in the formation of a flint line which was expressed to varying degree and complexity in this soil. A TL-dated burnt flint deposit close to column 13 may be part of this horizon. Description of this column is the same as for column 63, with the addition of the flints concentrated approximately 10 cm below the overlying gravel in a deeper, approximately 15 cm, A horizon. Column 32 sampled only the A horizon, and again the description for column 63 applies.

Column 34 was sampled to investigate an area of marked reddening of the soil associated with a Neolithic structure. The soil description is of about 8 cm of a dark brown to dark reddish-brown clay loam, 7.5YR–5YR 3/2, with an admixture of dark reddish-brown and redder peds, 5YR 3/3, merging downwards to dark brown, 10YR 3/3, and then to brown to dark yellowish-brown, 10YR 4/3–4/4. Structure is well-developed prismatic and fine angular blocky.

Column 42, in Area 24, also samples the soil where there was a query as to whether a Neolithic house floor could be identified. The description is of a brown to dark greyish-brown sandy clay loam with fine granular and fine angular blocky structure, with a conspicuous coarse sand component.

Samples 462, 463 and 464, for column 13, 475 for column 32, 476 for column 34, and 477 for column 42, show texture to be clay, the field assessment having suggested a far more sandy class, possibly affected by the presence of charcoal and cultural material. Decalcification of silt and clay strongly reduces the coarse silt fraction, increasing clay. Unfortunately, particle size data is incomplete for column 13, only the lowest horizon, below the flint line, being fully analysed (sample 464). Sand fractions, however, are very similar above and below the flint line, and the total sand, 9% in both, is markedly lower than in the soil in the other columns. Decalcification had very little effect on the one sample fully analysed in this column.

MICROMORPHOLOGY

Basic micromorphological characteristics will be described first in so far as they apply to all samples of the pre-Neolithic soil; the cultural material within the soil will be included. The nature of the soil will then be discussed. Finally, particular queries applying in each column will be addressed.

A horizon: microstructure is very well-developed sub-angular blocky and granular, with many channels and chambers and some intrapedal vughs and vesicles. Some of the channels and chambers are partially and loosely filled with granular peds. Fabric is in two forms: brown, crystallic calcitic and yellowish-brown crystallic, less calcitic, richer in clay, and with speckled birefringence. Each fabric forms separate peds, but there are also peds showing both; commonly the first surrounds the second. Textural relations in both fabrics are reticulate porphyritic, very well developed. Pedofeatures of gleying are common, and increase with depth, with strong impregnations of black and yellowish- to dark brown, brown and black nodules, and black coatings and hypocoatings to pores. There are localised infillings and coatings of ferrihydrite and/or goethite, and disrupted fragments of these coatings embedded in the fabric. There are localised calcite coatings and infillings. Shell fragments and slug/earthworm granules are common. Fine and very fine charred particles are generally distributed through the fabric, especially in the browner, coarser areas, and there are common fragments of humified organic material.

Bg horizon: The yellowish-brown fabric increases in quantity, and becomes less calcareous, and its birefringence becomes reticulate striated. The brown fabric becomes paler, and the texture finer. The features of total faunal structure, while still present, are less dominant.

A distinctive feature of very fine intrapedal porosity in the lower part of this soil, is what appears to be winding burrows, some 15 to 25 microns in width.

Cultural material occurs in all examples of the soil. It includes burnt and unburnt bone and flint, sherd fragments, charcoal fragments and a general distribution of fine charred particles in the silt grade, and small fragments of faecal material. Fragments of daub are recognised by a clay matrix, variably humus-stained and a distinctive content of fine sand rather than the mixed sand and silt clastic material of the soil fabric. Some of the daub is rich in calcite in the form of chalk in sand- or silt-sized particles or as a micritic component of the clay matrix. There are areas of fabric in which concentrations of amorphous humus form irregular patches and often provide a focus for iron oxide concentration. These areas are often lower in both calcite and non-calcitic clastic material than the surrounding soil.

DISCUSSION

The microstructure and the colour are distinctive of this soil profile within the site as a whole, reflecting the fact that it had had a sufficient period for development for there to be a degree of decalcification, clay formation, and the release and oxidation of iron to give the more reddish-brown colour than occurs in all the rest of the soils and deposits. Macro and microstructure are well developed and porosity high. A distinct B horizon is identifiable. The soil is a gleyic brown calcareous alluvial soil (Avery 1980).

The reddened soil of column 34 has all the features of A and B horizons described above plus common peds of dark red soil which do not form a continuous fabric but are interspersed with the basic fabric types. Only heating could produce the reddening, but it is not associated with any more charcoal or fine charred particles than elsewhere in this soil: the fire must have been isolated from the soil in some way, or an upper layer removed from the surface. The reddened

material has subsequently been broken up and incorporated into the soil by faunal activity.

The deeper example of the soil, column 13, with the included flint line and burnt flint deposit, was studied with a view to determining whether the flints had been lowered into the soil by earthworm activity or were buried by soil deposited over the flint-filled feature and surface spread. The whole fabric of the soil is earthworm worked, below, around and above the flints, and there is cultural material throughout including, in the form of very small bone fragments, below the flints. The earthworm hypothesis can therefore be supported; there is no micromorphological evidence to discount the alternative, but it would be necessary to postulate deposition of a layer of soil over the burnt flints, derived from the same soil profile. Since the depth of the soil is variable, and it is partly truncated by the gravel, it cannot be determined whether such deposition had happened or by what means. The question therefore remains open, but from the soil scientist's point of view, giving earthworms the credit is the simpler option.

In column 42, there was no reflection of the putative floor level in either soil structure or fabric. Daub fragments occur, but not in high density, and concentrations of humus are frequent. In column 32, where the soil was refuse-rich and darker than elsewhere, again there are strong humus concentrations and occasional daub fragments. The evidence from micromorphology does not, therefore, discriminate strongly between a house interior and a refuse-rich area outside a house, faunal reworking since occupation having homogenised the A horizon of the soil.

Bronze Age soils

Soils developed on alluvial parcel 2 were studied in the long columns 15, 22 and 63. In columns 15 and 22, the soil is developed on fine alluvium and is not thought to be associated with house floors, and this provides the best opportunity to assess the degree of soil development.

The soil A horizon is dark grey to very dark brown, 10YR 4/1–3/1, clay loam, with no mottles, and with a moderately developed fine granular structure and many fine and very fine macropores. The transition to the subjacent horizon is abrupt, and this horizon has in effect been described under alluvial parcel 2, above; B horizon features are poorly developed and predominantly those of gleying, differentially developed according to absolute level.

Analysis of samples 414, 415, 442 and 443 gives textures close to and on the clay/clay loam/silty clay boundaries. A modal particle size lies in the fine sand and coarse and medium silt. Decalcification of the silt and clay increases the proportion of clay by 26% and 18% in column 15 and by 26% and 16% in column 22. In both cases, decalcification removes part of all the silt grades, but proportionally more in the coarse grade.

MICROMORPHOLOGY

Structure is sub-angular blocky and granular, with channels and chambers, and intrapedal vughs. The structure has the same elements as the Neolithic soil, but on a finer scale and lacking the very large voids. Groundmass is crystallic, calcitic and textural relations are reticulate porphyric. While at a macroscopic scale the colour in thin section is consistently greyish-brown, lacking the higher chroma of the earlier soil, there are throughout two soil fabrics, one which is the same material as the underlying alluvium, and one browner and coarser, the coarser element being richer in fine charred particles. In both, the textural relationships are those of the total faunal structure described earlier, which is extremely well developed in this unit. The combination of the two fabrics can be clearly seen to be entirely the product of infilling of channels by wormcasting, and the transition to the underlying sediments is by way of these same infilled worm burrows.

DISCUSSION

The soil can be classified as a gleyic rendzina-like alluvial soil (Avery 1980). It is apparent that insufficient time elapsed between the abatement of alluvial accretion and the occupation and burial of the surface for earthworms to produce a thoroughly homogenised A horizon, for decalcification to occur, or for iron oxides to contribute substantially to colour. There is no evidence of clay enrichment. There was, however, sufficient time for humus incorporation to gain upon alluvial accretion sufficiently to form a distinct A horizon.

Bronze Age cultural layers

The cultural layers associated with the Bronze Age soil were the distinctive surface in Area 26, column 51, the house interior in Area 30, column 53, the substrate to a possible entrance way in Area 31, column 54, with an adjacent sample for comparison, column 56, and the potentially reworked deposit in Area 21, column 38, 5 m from the in situ material of column 37, which will be described under 'dark earths', below.

In column 51 the soil is described from sample blocks submitted in the absence of a field description, as about 5 cm very dark greyish-brown, 10YR 3/2, with some dark ferruginous mottles, and some stones, merging rapidly to slightly paler, with a yellowish sandy component, and patchy dark staining, with dark reddish-brown mottles; then there is another merging zone of discrete peds of paler silty soil which become the predominant fabric, brown, 10YR 5/4.

Analysis of samples 469 and 470 shows silty clay loam over clay loam, there being a substantial coarse and medium sand component in the lower sample but not the upper and otherwise a very even distribution from fine sand to medium silt, tailing off in fine silt towards a rather low clay by comparison with most of the materials discussed so far.

Decalcification tends to reduce coarse silt preferentially, giving a small increase in clay.

In column 53, the soil is dark greyish-brown, 10YR 3/2, becoming slightly paler, 4/2, towards the gravel at about 15 cm depth, silty clay loam. The structure is well-developed fine angular blocky and it is more or less stone free but there are shell fragments and cultural material.

Analysis of sample 478 puts it on the boundary between clay loam and silty clay loam, and again the silt tails off in the fine grade. Decalcification of silt and clay substantially increases clay at the expense of coarse and medium silt.

In column 38, bench description of sample blocks submitted gives a dark greyish-brown, 10YR 4/2, with a greyer and yellowish coarser component and some white sand, merging in a few centimetres to greyish-brown, 10YR 5/2, with a paler sandy component and fine distinct yellowish-brown mottles, 10YR 6/6 to 5/8. The substrate is the channel fill of alluvial parcel 4B as it rises up to overtop the channel bank, and is therefore in a comparable situation to the putative 4B levels in column 21, but closer to the channel itself.

Analysis of samples 467 and 468 shows clay loam over clay, with very substantial fine sand content in both and high medium silt in the upper sample. Decalcification of silt and clay has little effect in the upper sample, but increases clay by 20% at the expense of medium and fine silt in the lower one.

The samples submitted, columns 54 and 56, were referred to as 'dirty subsoil', and were taken from below the cultural layers in and adjacent to the possible entrance way, with the question as to whether effects of trampling could be distinguished to help with the interpretation. As in the case of column 38, the substrate is a sandy deposit of alluvial parcel 4B.

In column 54, the 'dirty subsoil' is a mixture, on a scale of a few millimetres, of black, greyish and dark greyish-brown, 10YR 6/2, 5/2 and 4/2, and dark brown, 7.5YR 3/2, with light yellowish-brown, 10YR 6/4, and ochreous mottling, 6/6. Texture is variable, with paler sandy material and darker humose clay. Structure was not apparent macroscopically in the sample as examined. There are many fine macropores. In column 56, the mixture of materials is also clear macroscopically, with black, very dark grey, 10YR 3/1, sandy clay and a paler sand component; there are many fine macropores.

Analysis of samples 460 and 461 shows strong similarity to column 38, on either side of the clay/clay-loam boundary, with fine sand and medium and fine silts predominant, but whereas in 460 decalcification has little effect, in 461 it strongly reduces the medium and fine silt, increasing relative proportion of clay by 26%.

MICROMORPHOLOGY

The cultural layers share with the other Bronze Age soils the characteristics described above, but with the addition of both more cultural material and variable amounts of humus staining and concentrations of humified organic residues. In all cases faunal turbation is very strongly expressed: in the reticulate porphyric related textural distributions, in interlaced inter-calations of humic silt, and in the incorporation of casts from subjacent and overlying sedimentary layers into the deposit.

In column 51, the putative floor is possibly represented in the upper part of the sampled layer where soil coloured dark brown by humified material is present, not as continuous areas of groundmass but broken up among fragments of daub and of a mixture of sand and organic material which might also have been used as daub. The Bronze Age daub is more readily recognised as such than the Neolithic version. It occurs in more coherent pieces, and shows a clearer differentiation from the surrounding soil, with a higher sand content and in some pieces clear evidence in the form of particle alignment and lamination of the working of the material.

In column 53, no features distinctive of other than a humic soil rich in cultural material were apparent.

Comparison of columns 54 and 56 did not show discrimination between the entrance way and adjacent area. Both samples were very porous, with very strong evidence of faunal activity. Both were rich in charcoal and contained cultural material, but with more of the former in column 56 and the latter in column 54. Column 56 had a higher concentration of humified residues and associated manganese dioxide concentrations. The possible entrance way showed neither horizontal orientation of stones and other particles nor in situ crushing of charcoal fragments which could be attributed to trampling.

Column 38 has, apart from the textural differences quantified by the particle size analysis, micromorphological characteristics closely comparable to those of the Bronze Age level in column 21, but with a conspicuous content of the 'calcite rod' structures and tufa fragments in the sand grades.

Dark earths

A deep Bronze Age dark earth deposit was sampled in Area 16, column 22. A comparable deposit occurred in Area 21, column 37, where cultural material and a great deal of charcoal were included within the uppermost channel fill of alluvial parcel 4B.

In column 22, field description of the dark earth is black humose clay with no mottles, weakly developed fine granular stucture and few medium macropores containing casts, common rounded to sub-angular flints and chert, and cultural materials.

Analysis of samples 438 to 441 gives a very consistent clay with substantial amounts of sand and silt, very similar to the top of the underlying soil, sample 442. There is a mode in the medium silt grade, but it is broad and skewed to include fine sand and coarse silt, which are also high. Decalcification

of silt and clay reduces all silt grades and increases relative proportion of clay by 20 to 24%.

In column 37, the highest concentration of dark earth and refuse gives a mixed black and dark grey deposit, some 10 cm deep, humose sandy loam with stones. No analyses were carried out at this level in the column.

MICROMORPHOLOGY

In column 22, the microstructure is complex, and varies through the depth of the deposit. In the upper half it is largely weakly developed sub-angular blocky and spongy, with many vughs and a variable amount of chambers and channels, with areas of loose and coalescing granular peds. In the lower half, there is a much more open, better-developed sub-angular blocky stucture, with many chambers and again a granular component. Groundmass colour is greyish-brown, greyness depending on the amount of fine charred particles. These are particularly abundant in the macroscopically black part. They are distributed through the groundmass, in patterns which delineate the wormcast structure evident in the groundmass. Larger identifiable wood charcoal fragments are also common. The groundmass is crystallic calcitic, but less calcitic than the 'calcite loam' of the alluvial sediments. Reticulate porphyric related distribution is very well developed. Flecks and shreds of humified organic material occur, and there is a variable brown humus staining. Pedofeatures include ferruginous nodules, weakly to strongly impregnated. While the fine and medium silt calcite particles are fewer than in calcite loam, calcareous particles and aggregates in many forms and internal organisation range upwards from medium silt through the sand grades. Silt includes numerous phytoliths. Sand includes, apart from the cultural material discussed below, rounded quartz, flint, chert and occasional other rock types, very abundant slug/earthworm granules and rounded tufa fragments.

Cultural materials are very abundant: burnt and unburnt flint, bone and sherds, the fragments varying from silt grade to several millimetres (clearly, also up to the large material recovered in excavation). There are also fragments of what appears to be daub: sand in a matrix of ground-up chalk and organic matter, having a laminated structure. The chalk component is identifiable by its content of microfossils; the organic material is not identifiable, being completely humified.

Faecal material is distributed through the fabric in the form of rounded and irregular pieces, from silt-sized flecks to coarse sand grade. It is identifiable by pale to dark yellow colour, isotropic under crossed polarisers, and fluorescence of variable intensity in ultraviolet light. Within the faecal groundmass are included silt and sand grains, bone, shreds of humified plant material, phytoliths, hairs, spores, pollen grains, possible gut parasite egg cases, and other unidentified structures of plant, animal and micro-organism origin. An attribution to herbivore, carnivore and omnivore faeces is not always clear: some of the distinguishing features referred to by Courty *et al.* (1989) would need to be seen in larger areas and concentrations of the material than occur here, but broad categories are evident. One kind is pale yellow, sometimes with streaky or swirling patterns in the fabric and often having areas of intense reddish-brown staining, sometimes confined to the centre of the piece and having the blobby form of humified micro-organism colonies, or what are clearly brown fungal hyphae; occasionally, bone fragments occur in this kind, and phytoliths are uncommon; this material gives the strongest fluorescence. Another has a stronger ground colour, its fabric has no internal form, and it may contain shreds of humified material, sometimes with detectable plant cell structure, and/or phytoliths, and sometimes with variable amounts of calcite. In rare cases, a few birefringent spherulites (Canti 1997) occur. These two extremes can with little doubt be attributed to carnivores and herbivores respectively, but many bits of probably faecal matter combine these attributes or contain traces of material not identifiable as plant or animal in origin. The omnivores, pigs and people, could be held responsible, and the variety of diet and state of the digestive system would obviously give rise to a wide range of degree of digestion and concentration of residues in the excreta; foxes and badgers are also omnivorous rather than being strict carnivores, and could well forage around a settlement. If the small mammals and birds which surely also contribute their droppings are included, it would be unwise to go further in interpretation.

Evidence for ash, in addition to the abundance of charcoal and burnt cultural material, lies in the presence of calcite masses of varying size and form. Some of these are clearly algal in origin, both tufa fragments similar to those in the alluvial layers and mat-like masses of filaments which could have formed on the midden surface, and much of the calcite is probably derived from the substratum or has the same biogenic origin. There are also, however, structureless masses of fine calcite which are clearly not infill pedofeatures and are probably recrystallised, under the moist conditions of the midden, from original calcite, calcium oxalate and calcium oxide in ash. In some patches, the association with charred material is clear, but commonly there is no clear association between charcoal and calcite. Abundance of phytoliths is probably also a product of ash well distributed in the deposit. While the fine charred particles have strong similarity to the residual carbon commonly seen trapped in fresh ash when it is examined microscopically, they are not necessarily derived only from ash: smoke and soot must contribute masses of this material in and around settlements, and indeed small quantities of these particles are ubiquitous in the alluvial sediments, and higher concentrations occur in all the soils. The particles do not show plant tissue structure even under high magnification, and are probably soot formed from hydrocarbons produced in combustion. Soot may be trapped in the ash, carried away in fine ash particles in smoke, or

deposited on surfaces of all kinds, including people and animals, within buildings.

The structure of the dark earth groundmass shows that it is entirely faunally mixed. The transition to the underlying soil is by way of burrowing and casting, giving a transitional zone in which both soil and midden fabrics are present in infill and cast form, and the process of bioturbation as the midden accumulated must have distributed alluvial material within it and homogenised it. Any original lamination, and concentrations of ash, faeces, or other organic debris, have been lost. The dust, dirt and mud of the settlement will have contributed to the midden, so the high mineral content, and its similarity to that of the surrounding soils, is not surprising.

In column 37, structural characteristics and materials of the deposit are very similar to those in column 22, but the matrix is coarser, and there are more of the pedofeatures of gleying. There is a higher concentration of charcoal, ash, and putative faecal material. Much of the fabric is a mixture on a scale of a few millimetres of ash, with a high content of fine and coarser charred particles, and the faecal material, much of it of the indeterminate, perhaps omnivore, kind. Some of the faecal material contains charred fragments: perhaps people were eating bread with burnt crusts (cf. Lindow Man; Hillman 1986), or charring their roast meat.

Below the level in column 37 at which blackness dies out, there is still a great deal of cultural and organic material, suggesting faeces. The groundmass includes areas of sediment of finer texture, rich in humified organic shreds and charred material and slurry-like open porphyric-related distribution, which appear to fill channels and form laminae within the deposit. Gross mixing processes rather than homogenisation by faunal turbation appear to be at work here.

Fills of features

Two early features were sampled for comparison of their fills with the substrate and information on the way in which they filled up. Columns 35 in Area 20, and 68 in Area 13, sampled the same Late Mesolithic/Early Neolithic ditch, the former at a higher level in its fill than the latter. Column 62 in Area 13 sampled an Early Neolithic pit fill; the sample was taken at a fairly high level in the pit, and the section suggests that soil from the weathering cone would have contributed to its fill.

No analyses were done for column 68; those for column 35 show a particle size distribution comparable to alluvial parcel 1K, especially in column 15, with the consistent low sand content and with the silt/clay proportions central to the range in that deposit. The mode in medium silt is very similar to that in 1K in column 15, and on decalcification of silt and clay there is a strong increase in clay, here at the expense of all the silt grades. Some variation with depth is apparent within the ditch fill, sample 473 being a little coarser, and whereas decalcification leaves less than 10% silt in the other samples, this level has considerably more non-calcite silt. Analysis of the column 62 pit fill again indicates a clay-dominated fill comparable to the alluvial parcel 1K sediments; here decalcification reduces the medium silt more than fine or coarse but leaves much of the silt intact.

MICROMORPHOLOGY

In column 35 microstructure is prismatic and sub-angular blocky, with intrapedal vughs and vesicles and with areas dominated by large chambers and channels. Texture is clay, with much fine sand and coarse silt in the form of dark grey, rounded micritic calcite; coarse and medium sand is entirely slug/earthworm granules and shell fragments or complete shells. No fragments of charcoal, bone, flint or pottery are visible. Textural relationships are open porphyritic but not reticulate, and show flow patterns. Fabric is in two forms: crystallic calcitic, variably speckled birefringence and, in stained areas, less calcitic with speckled and reticulate striated birefringence, with some grano- and poro-striation.

Colour is light brown with 30% iron staining and 10 to 20% black impregnation, the latter especially commonly forming coatings and hypocoatings to intrapedal voids and impregnating calcite granules. Brown humus staining is often the focus for impregnations. Spotty reddish-brown impregnations occur in the stained areas and both colours, but particularly the black, occur in aggregates and lines. Increasingly with depth, there are infill and coatings of calcite, in two grain sizes, commonly micritic followed by sparitic, i.e., fine followed by coarse, but also showing complex gradations and variations in grain size. Some calcite coatings are heavily impregnated with iron and/or manganese oxides. Goethite and ferrihydrite coatings are abundant, some in situ, some fragmented and incorporated. Infilling and coating with dusty clay occurs locally, occasionally subsequent to other pedofeatures.

Column 68 has many of the same characteristics as column 35, with comparable structure and again many large chambers, some having partial infill with very fine angular blocky and granular peds. Textural characteristics and relationships are similar to column 35, but there is a small amount of cultural material in the form of charcoal and small angular flint flakes. Pedofeatures are even more strongly developed than in column 35, and the black and ferruginous impregnations are strongly differentially distributed.

Column 62 has a sub-angular blocky structure with intrapedal vughs and areas of large chambers and channels, but in the lower part of the sample the structure becomes predominantly vughy, the interpedal voids being weakly developed. The textural relationship is open porphyritic, and the coarse and medium sand include angular and rounded quartz, as well as a lot of burnt and unburnt bone

and flint. There is abundant fine, charred material, and some charcoal fragments, and abundant humified flecks and shreds.

The fabric is light yellowish-brown, with areas of iron depletion and staining from yellow to reddish-brown, and areas of humus staining. Impregnations of reddish-brown and black, spotty and as aggregates, are common but do not form coatings and hypocoatings, being mainly in the interiors of peds. There are no calcite pedofeatures. There are, however, abundant segregations of goethite/ferrihydrite, lining or filling pores and as embedded fragments, of fibrous form, sometimes separate from other pedofeatures, but sometimes associated with intense iron staining and impregnation, or humus staining. A bone fragment is partially coated with, and its pores filled with, this material; in places it appears to pseudomorph plant tissue.

DISCUSSION

The relationship of these early feature fills to both the alluvial parcel 1K sediments into which they are cut and the Neolithic soil are illuminated by the micromorphological observations. For the ditch, the texture – with analytical results only for column 35 but micromorphology allowing comparison for column 68 – shows remarkable similarity to the very consistent composition of the substrate, and micromorphology shows that again the coarse and medium sand is entirely of biological origin. The grey micritic calcite granules dominating fine sand and coarse silt grade, which are probably also biogenic, are also common to substrate and feature fills. The textural similarity, therefore, suggests that this feature derived its fill from the immediate surroundings. In the case of the pit, though the texture is again very similar to alluvial parcel 1K, the coarser grades contain not only cultural material but clastic as well as biogenic particles, and there is also much siliceous material in the silt grades. This is a feature also of the Neolithic soil and suggests that both soil and pit fill have input from coarser sedimentary material elsewhere, along with, or as part of, the residues of human occupation.

Both features show the strong gleying affecting the sediment into which the ditch and pit were cut, but more intensely, especially in the case of column 68, from lower in the fill. The intense gleying can be attributed either to greater waterlogging in the feature fills, or to a higher organic content. Texture difference cannot be the cause, but the slurry-like characteristics suggested by the related textural patterns could give an initially less permiable structure. The prismatic and angular blocky structures with the intrapedal porosity characteristic of slaking, though subsequently broken up by chambers giving a much more porous character, would support this. One effect of the intense gleying is the formation of large amounts of the pure iron oxide segregations, and this also suggests a high level of microbial activity, possibly that associated with the roots of plants tolerant of such conditions, but also, in the case of the pit, with organic rubbish. The substrate of these features also has these iron oxide segregations, but in lesser amounts. Comparing the pit fill with other columns in which there is intense gleying associated with refuse deposition – columns 37, 54 and 56 – organic residues form the focus for the gley features at different periods, indicating the dominance of the fill characteristics.

General discussion

In discussion of each of the components of the sediments, the soils which developed on them, and archaeological deposits, common features recur, such as the high level of faunal turbation and the occurrence of gleying processes. The first of these ensures that soil horizon boundaries are clear or gradual (zone of uncertainty of several centimetres) rather than abrupt or sharp (*Soil Survey Field Handbook*, Hodgson 1976), and that even where stratigraphic boundaries appear to be sharp, such as the base or top of gravel layers, soil and sediment materials, together with small particles of cultural material, have been transported across the boundary, both upwards and downwards. Downward transport below the 10–20 cm depth of daily feeding activity by earthworms is via the deeper annual aestivation channels, with some particles being carried down to line aestivation chambers and others falling in as channels collapse and infill. Some of the channels that penetrate the archaeological levels and into underlying alluvial materials originate from later surface horizons, including the recent one. Upward transport occurs when a soil surface begins to be buried and surface-casting earthworms bring soil up and deposit it on the new surface. This continues until there is sufficient depth of overburden and/or insufficient earthworm food in the form of dead roots and unhumified organic residues in the former soil, and earthworms begin to work from a new surface, where plant growth now provides fresh supplies. When feeding, earthworms select against particles larger than about 0.5 mm diameter, whereas for aestivation purposes, they select particles up to about 2 mm diameter. Further, it is larger, rather than smaller, particles which are most likely to detach themselves from walls of old burrows and fall down them. Earthworm sorting can therefore cross boundaries, rather than being confined to the currently active A horizon. An effect that needs to be taken into account is that on phosphate analysis, reported in Chapter 6: enhanced phosphate content may be spread upwards with fine soil transport, and downwards in the form of small fragments of bone or shell, from the level at which it originated. In addition, it should not be forgotten that ant colonies and moles are active vertical transporters of soil materials on floodplains, the more so in periods of reduced flooding but not excluded by seasonal inundation, so earthworm sorting is not acting alone.

The A horizons of the Neolithic and Bronze Age soils are the result of a transition from a regime in which alluvial accretion gains upon incorporation of organic matter, to one in which a stable surface is maintained. Where overbank seasonal flooding occurs, the balance between incorporation of organic matter from each year's plant growth and the amount of new sediment deposited determines the extent to which a distinct A horizon is formed. For the fine-textured alluvial parcels, raw alluvial soils grew upwards, their microstructure showing evidence of the faunal activity which incorporated organic residues, but with insufficient humus accumulation to form an A horizon, until accretion slowed down and eventually flooding stopped. Humus accumulation gained upon accretion, earthworms reworked the same few upper centimetres, and A horizons became differentiated from the underlying raw soil, which then constituted first a C horizon and eventually, in the case of the Neolithic soil, took on B horizon characteristics. In this floodplain situation, the B horizon was seasonally wet enough for gleying to persist, at least for most of the period of soil development, so rather than the deepening of decalcification and clay formation, the soil was limited to a Bg horizon. In the case of the Bronze Age soil, while a distinct A horizon did form, there was not enough time for B horizon differentiation to occur, so this soil remained an A/C profile type.

Faunal activity during and subsequent to occupation has incorporated the residues of occupation into the soil. In the case of the Neolithic soil, residues change the texture and the character of the detrital mineral content. The disintegration of daub, which has a higher and more quartzose sand content than the soil, is an identifiable contributor, together with flint debris, pot sherds and bone. These cultural materials are present as particles which function as components of the fine soil, as well as in the larger fragments recognised archaeologically. The depth of the soil was not increased beyond that normally achieved by an A horizon of a brown earth, however. The organic component of the residues underwent predominantly aerobic humification processes, but high concentration of organic matter resulted in localised anaerobism and segregation of iron and manganese oxides beyond that to be expected in an aerobic soil free of anthropic input. The mottling and segregation produced in the A horizon are distinct from those of gleying due to waterlogging which characterise the B and C horizons of the soil, and which are dependent on absolute height.

It was not possible to establish clearly the relationship of the Mesolithic cultural material to the development of the soil, though the probability that the apparent stratigraphic location was the product of worm sorting is in accord with the evidence of the soil itself that a substantial period of soil development on a stable surface pre-dated the Neolithic occupation.

Development of the Bronze Age soil was of shorter

duration. The A horizon has a colour of lower chroma and yellower hue than the Neolithic soil, showing a lower relative accumulation of iron oxides, and there is no increase in clay relative to the parent material. Combinations of different soil fabrics indicate incomplete homogenisation of the A horizon, with C horizon fabric persisting among more humic soil. No differentiation of a B horizon from the raw alluvial soil condition of the parent material is apparent.

In the areas of dense occupation material, archaeological questions relating to house floors, trampling, etc., could not be answered by consideration of soil properties, since bioturbation had obliterated any evidence of surfaces or compacted levels. The porosity is that of well-developed total faunal structure. As with the Neolithic soil, there was archaeological material in the sand grades as well as the larger fragments, well distributed throughout the A horizon and transported into the transitional zone to the C horizon by earthworm activity. Again, daub provides evidence of structural materials and is one of the ways in which additional mineral material is introduced into the soil.

Where deposition of residues formed 'dark earth', the upgrowth of the deposit and the faunal turbation produced a transition from the normal soil A horizon to what is in effect an anthropogenic soil. In soil classification, where the depth exceeds 40 cm this is an earthy man-made humus soil (Avery 1980). That depth is barely reached in the columns studied here, but the depth of humic soil exceeds the normal depth of daily earthworm feeding activity, and the increasingly dark colour above the initial A horizon reflects both increased organic content and the contribution of charred material. Again, intensification of biological activity where organic input was high has resulted in localised oxygen depletion and intense segregation of iron and manganese oxides.

In those columns in which intepretation of the sediment analysis and micromorphology can be compared directly with the interpretation of biological remains, there is confirmation that the immediate sampling locality is indeed represented by the ecological conditions suggested. For example, Scaife (Chapter 9) suggests that his pollen profiles represent short time-windows. If we look at the particle size analysis for the organic-rich levels sampled for pollen in columns 31 and 59, we see that high sand content, and/or fluctuations in particle size involving pulses of sand, indicate periods of deposition from relatively fast-flowing water over a short time period. Where there is information from molluscs, the interpretations in terms of immediate locality agree very closely with the interpretation of both sediment characteristics and soil development. Starting with the marl, sub-parcel 1C, the molluscan evidence accords with input of fine sediment from gently flowing water, the coarser material being biogenic calcite, which includes shell, produced in situ. As biogenic precipitation decreases, sub-parcel 1K, the molluscs indicate a transition from an entirely aquatic environment to a wet floodplain and then an increasingly well-drained one, culminating with cessation of overbank flooding as the soil develops. Subsequently, in each alluvial parcel, 'site' mollusca indicate very close agreement with the conditions suggested by the nature of the sediments and the development of soil.

The integration of the study of soils and sediments with the stratigraphic analysis made possible by the extensive excavations has meant that a far better understanding of the details of riverine erosion and sedimentation in a small area has been achieved than is possible by the more usual geomorphological studies, which depend on glimpses of chance exposures scattered along a stretch of river bank or by coring. Other archaeologically based investigations of alluvial situations, such as that in the Upper Kennet (Evans *et al.* 1993) have similarly revealed complexity of floodplain structure where small exposures or cores originally suggested uniformity. The situation at Runnymede is particular to a river loop and confluence chosen for occupation presumably because of its morphology, and one cannot necessarily generalise to simpler stretches of the valley.

Table 5.1 *Particle size analysis of soil and sediment samples.*

Sample number	Altitude (m OD)	Coarse sand	Medium sand	Fine sand	**Total sand**	Coarse silt	Medium silt	Fine silt	**Total silt**	**Clay**	Stratigraphic unit
Column 13											
462	14.44-14.50	1.0	3.5	10.0	**14.5**						Neolithic soil
463	14.34-14.40	1.0	2.0	7.0	**10.0**						Neolithic soil
464	14.24-14.30	tr	1.5	8.0	**9.5**	8.0	18.0	11.0	**37.0**	**53.5**	Neolithic soil/1K
Column 15											
410	14.74-14.79	2.0	3.5	11.5	**17.0**	11.0	15.5	10.0	**36.5**	**46.5**	6D
411	14.69-14.74	1.0	3.5	11.5	**16.0**	18.5	17.0	8.5	**44.0**	**40.0**	6D
412	14.64-14.69	1.0	3.0	13.0	**17.0**	18.0	15.0	8.5	**41.5**	**41.5**	6D
413	14.58-14.63	2.0	4.0	12.0	**18.0**	14.0	17.0	8.5	**39.5**	**42.5**	6D/LBA dark earth
414	14.55-14.60	3.0	5.0	13.0	**21.0**	19.0	15.0	8.0	**42.0**	**37.0**	LBA dark earth
415	14.51-14.56	2.5	7.5	15.5	**25.5**	15.5	15.5	8.5	**39.5**	**35.0**	LBA/soil
416	14.46-14.51	2.0	7.0	14.0	**23.0**	13.0	16.0	8.0	**37.0**	**40.0**	2B
417	14.43-14.48	2.0	7.5	14.5	**24.0**	13.5	16.5	8.5	**38.5**	**37.5**	2B
418	14.38-14.43	4.0	7.5	12.5	**24.0**	14.0	15.0	7.5	**36.5**	**39.5**	2B/F
419	14.34-14.38	7.5	8.0	12.5	**28.0**	12.5	12.5	7.0	**32.0**	**40.0**	2F
420	14.28-14.34	8.5	9.5	10.0	**28.0**	14.0	13.5	6.5	**34.0**	**38.0**	2F/Neolithic soil
421	14.22-14.27	3.5	7.5	10.5	**21.5**	9.5	12.5	7.0	**29.0**	**49.5**	Neolithic soil
453	14.19-14.24	1.5	4.5	11.0	**17.0**	11.5	15.0	8.5	**35.0**	**48.0**	1K
454	14.12-14.17	1.0	3.5	11.0	**15.5**	11.5	15.5	9.0	**36.0**	**48.5**	1K
Column 21											
422	14.39-14.44	2.5	5.5	12.5	**20.5**	15.5	17.5	9.0	**42.0**	**37.5**	6D
423	14.34-14.39	2.0	7.0	13.0	**22.0**	14.5	16.5	9.0	**40.0**	**38.0**	LBA soil
424	14.29-14.34	1.0	5.0	15.0	**21.0**	17.5	17.5	8.5	**43.5**	**35.5**	LBA soil
425	14.13-14.18	tr	3.5	14.5	**18.0**	19.5	19.0	9.0	**47.5**	**34.5**	4B
426	14.09-14.14	0.5	4.0	14.0	**18.5**	20.5	21.0	6.5	**48.0**	**33.5**	4B
427	14.04-14.09	0.5	2.0	12.5	**15.0**	22.5	21.0	10.0	**53.5**	**31.5**	4B
428	13.99-14.04	0.5	2.0	11.0	**13.5**	18.5	21.0	10.5	**50.0**	**36.5**	4B
429	13.94-13.99	1.0	2.5	14.0	**17.5**	18.0	18.5	9.0	**45.5**	**37.0**	4B
430	13.89-13.94	1.0	4.0	14.5	**19.5**	16.5	16.0	9.5	**42.0**	**38.5**	4B
431	13.86-13.91	tr	4.0	17.0	**21.0**	18.0	16.5	8.5	**43.0**	**36.0**	4B
432	13.81-13.86	0.5	7.5	11.0	**19.0**	13.0	14.0	4.0	**31.0**	**50.0**	2B
433	13.76-13.81	tr	3.0	8.5	**11.5**	11.5	19.5	10.0	**41.0**	**47.5**	2B
434	13.71-13.76	0.5	3.0	8.5	**12.0**	12.5	15.5	9.0	**37.0**	**51.0**	2B
Column 22											
435	14.50-14.56	1.5	6.5	10.0	**18.0**	13.0	18.5	10.5	**42.0**	**40.0**	6D
436	14.48-14.53	1.5	6.5	13.0	**21.0**	12.0	17.5	8.5	**38.0**	**41.0**	6D
437	14.44-14.49	1.5	7.0	12.5	**21.0**	12.5	17.0	8.5	**38.0**	**41.0**	6D
438	14.39-14.44	1.5	8.5	13.5	**23.5**	14.0	15.5	8.0	**37.5**	**39.0**	LBA dark earth
439	14.34-14.39	1.5	10.0	14.5	**26.0**	11.0	15.0	8.0	**34.0**	**40.0**	LBA dark earth
440	14.29-14.34	2.0	10.0	15.5	**27.5**	11.5	15.5	8.0	**35.0**	**37.5**	LBA dark earth
441	14.24-14.29	2.5	11.0	15.0	**28.5**	13.0	13.5	8.5	**35.0**	**36.5**	LBA dark earth
442	14.18-14.23	2.0	9.5	13.5	**25.0**	13.5	15.5	8.5	**37.5**	**37.5**	LBA soil
443	14.13-14.18	2.0	6.5	10.0	**18.5**	13.5	17.5	10.0	**41.0**	**40.5**	LBA soil

Sample number	Altitude (m OD)	Coarse sand	Medium sand	Fine sand	Total sand	Coarse silt	Medium silt	Fine silt	Total silt	Clay	Stratigraphic unit
444	14.06-14.11	0.5	4.0	6.0	**10.5**	14.0	20.5	11.5	**46.0**	**43.5**	2B
445	14.00-14.05	0.5	3.0	6.5	**10.0**	11.5	19.5	11.0	**42.0**	**48.0**	2B
446	13.95-14.00	2.0	4.0	9.0	**15.0**	11.0	19.5	10.5	**41.0**	**44.0**	2B
447	13.90-13.95	1.0	3.5	10.0	**14.5**	11.5	18.5	10.0	**40.0**	**45.5**	2B
448	13.82-13.87	2.0	5.5	7.5	**15.0**	7.0	12.5	8.0	**27.5**	**57.5**	2B
449	13.77-13.82	3.5	8.5	7.5	**19.5**	4.5	11.0	7.0	**22.5**	**58.0**	2B
450	13.71-13.76	3.5	9.0	8.5	**21.0**	6.5	13.5	8.0	**28.0**	**51.0**	2F
451	13.65-13.70				Not analysed						Neolithic soil
452	13.60-13.65	4.0	7.0	9.0	**20.0**	9.5	17.0	10.5	**37.0**	**43.0**	Neolithic soil
Column 31											
1	14.39-14.44	1.5	5.0	9.5	**16.0**	17.5	9.0	5.5	**32.0**	**52.0**	Neolithic soil
2	14.34-14.39	0.5	2.0	6.5	**9.0**	9.0	16.5	6.5	**32.0**	**59.0**	Neolithic soil
3	14.29-14.34	0.5	1.5	3.5	**5.5**	15.0	13.0	9.5	**37.5**	**57.0**	Neolithic soil
4	14.24-14.29	0.5	1.5	6.0	**8.0**	10.5	8.0	6.0	**23.5**	**67.5**	Neolithic soil
5	14.19-14.24	8.0	3.5	7.0	**18.5**	6.5	6.0	3.5	**16.0**	**65.5**	1K
6	14.14-14.19	1.5	1.0	10.5	**13.0**	11.5	11.5	9.5	**32.5**	**54.5**	1K
7	14.09-14.14	0.5	1.0	10.0	**11.5**	8.0	11.5	10.0	**29.5**	**59.0**	1K
8	14.04-14.09	3.0	2.0	7.0	**12.0**	5.5	9.0	8.0	**22.5**	**65.5**	1K
9	13.99-14.04	0.5	2.5	14.5	**17.5**	13.5	16.0	12.5	**42.0**	**40.5**	1K
10	13.94-13.99	0.5	2.0	13.5	**16.0**	18.0	11.5	8.5	**38.0**	**48.0**	1K
11	13.89-13.94	0.5	3.0	10.0	**13.5**	19.0	18.5	13.0	**50.5**	**36.0**	1K
12	13.84-13.89				Not analysed						1K
13	13.79-13.84	tr	2.0	11.0	**13.0**	11.5	15.0	11.0	**37.5**	**49.5**	1K
14	13.74-13.79	2.0	2.0	13.0	**17.0**	11.0	11.0	8.5	**30.5**	**52.5**	1K
15	13.69-13.74	tr	0.5	15.5	**16.0**	13.0	13.5	9.0	**35.5**	**48.5**	1K
16	13.64-13.69	tr	0.5	20.0	**20.5**	14.5	11.5	8.0	**34.0**	**45.5**	1K
17	13.59-13.64	tr	2.0	25.0	**27.0**	15.0	10.5	7.5	**33.0**	**40.0**	1K/1C
18	13.54-13.59	0.5	2.0	31.5	**34.0**	15.0	10.0	7.5	**32.5**	**33.5**	1C
19	13.49-13.54	tr	2.0	32.0	**34.0**	16.5	11.0	7.5	**35.0**	**31.0**	1C
20	13.44-13.49	1.5	2.5	32.5	**36.5**	28.5	13.0	8.5	**50.0**	**13.5**	1C
21	13.39-13.44	tr	5.0	37.5	**42.5**	12.0	8.5	5.0	**25.5**	**32.0**	1C
22	13.34-13.39	tr	4.0	33.0	**37.0**	18.0	11.0	7.5	**36.5**	**26.5**	1C
23	13.29-13.34	0.5	6.5	36.5	**43.5**	22.0	11.5	7.5	**41.0**	**15.5**	1C
24	13.24-13.29	2.0	8.5	39.0	**49.5**	12.0	7.5	5.0	**24.5**	**26.0**	1C
25	13.19-13.24	tr	8.0	34.5	**42.5**	15.0	9.5	6.5	**31.0**	**26.5**	1C
26	13.14-13.19				Not analysed						1C
Column 32											
475	14.43-14.53	3.0	5.5	9.5	**18.0**	10.0	12.5	7.5	**30.0**	**52.0**	Neolithic soil
Column 34											
476	14.25-14.35	1.0	3.5	8.5	**13.0**	13.5	18.0	14.5	**46.0**	**41.0**	Neolithic soil
Column 35											
471	14.06-14.11	1.5	2.0	4.5	**8.0**	8.5	14.0	8.0	**30.5**	**61.5**	Neolithic fill
472	14.01-14.06	1.0	3.0	6.0	**10.0**	9.5	14.0	8.5	**32.0**	**58.0**	Neolithic fill

Sample number	Altitude (m OD)	Coarse sand	Medium sand	Fine sand	**Total sand**	Coarse silt	Medium silt	Fine silt	**Total silt**	Clay	Stratigraphic unit
473	13.96-14.01	1.0	3.0	11.0	**15.0**	10.5	14.0	9.0	**33.5**	51.5	Neolithic fill
474	13.91-13.96	2.0	3.5	8.0	**13.5**	8.0	15.0	8.0	**31.0**	55.5	Neolithic fill
Column 37											
480	13.94-14.00	0.5	1.5	9.5	**11.5**	16.5	21.5	14.5	**52.5**	36.0	LBA/4B
481	13.84-13.90	0.5	2.0	13.0	**15.5**	9.5	15.5	12.5	**37.5**	47.0	LBA/4B
Column 38											
467	14.28-14.37	1.5	8.5	21.0	**31.0**	9.5	24.0	12.0	**45.5**	23.5	LBA/6
468	14.20-14.28	0.5	7.5	26.0	**34.0**	3.5	8.0	17.0	**28.5**	37.5	LBA/4B
Column 42											
477	14.43-14.48	6.0	10.5	9.5	**26.0**	15.0	9.0	5.5	**29.5**	44.5	Neolithic soil
Column 51											
469	14.34-14.44	0.5	2.5	16.0	**19.0**	13.0	19.5	13.0	**45.5**	35.5	LBA dark earth
470	14.27-14.34	6.0	8.0	14.0	**28.0**	16.0	14.5	12.0	**42.5**	29.5	LBA soil
Column 53											
478	14.56-14.66	1.5	6.0	13.5	**21.0**	16.5	22.5	10.0	**49.0**	30.0	LBA soil
Column 54											
461	14.09-14.14	2.0	9.0	18.5	**29.5**	4.5	16.0	11.0	**31.5**	39.0	LBA soil
Column 56											
460	13.91-13.96	2.0	13.0	17.0	**32.0**	9.5	16.0	10.0	**35.5**	32.5	LBA soil
Column 59											
39	13.67-13.72	7.5	6.5	8.5	**22.5**	14.5	11.0	8.0	**33.5**	44.0	2F
40	13.62-13.67	3.0	7.0	18.0	**28.0**	11.5	8.5	8.0	**28.0**	44.0	2F/Neolithic soil
41	13.57-13.62	1.0	3.5	18.0	**22.5**	8.5	18.5	11.0	**38.0**	39.5	1H
42	13.52-13.57	tr	2.5	20.5	**23.0**	17.5	9.5	6.0	**33.0**	44.0	1H
43	13.47-13.52	tr	1.0	18.5	**19.5**	32.0	14.5	10.5	**57.0**	23.5	1H
44	13.42-13.47	tr	1.0	20.5	**21.5**	19.0	15.5	8.0	**42.5**	36.0	1H
45	13.37-13.42	tr	2.5	29.0	**31.5**	12.5	11.5	7.0	**31.0**	37.5	1H
46	13.32-13.37	tr	1.0	17.0	**18.0**	15.5	12.5	7.5	**35.5**	46.5	1H
47	13.27-13.32	tr	1.0	12.5	**13.5**	19.0	12.0	9.5	**40.5**	46.0	1H
48	13.22-13.27	1.0	12.5	21.0	**34.5**	10.5	10.0	8.0	**28.5**	37.0	1H
49	13.17-13.22	0.5	11.0	22.0	**33.5**	7.5	8.0	4.5	**20.0**	46.5	1H
50	13.12-13.17	tr	6.0	15.5	**21.5**	9.5	11.0	9.5	**30.0**	48.5	1H
51	13.07-13.12	tr	2.0	21.0	**23.0**	11.0	11.0	7.5	**29.5**	47.5	1H
52	13.02-13.07	tr	0.5	8.5	**9.0**	9.0	10.0	7.5	**26.5**	64.5	1H
53	12.97-13.02	1.0	2.5	14.0	**17.5**	13.0	11.5	10.0	**34.5**	48.0	1H
54	12.92-12.97	0.5	5.0	15.0	**20.5**	3.0	11.0	10.5	**24.5**	55.0	1H
Column 60											
27	13.52-13.57	0.5	1.0	4.0	**5.5**	8.5	6.5	5.0	**20.0**	74.5	2B
28	13.47-13.52	0.5	1.5	5.5	**7.5**	14.0	21.0	11.0	**46.0**	46.5	2B
29	13.42-13.47	tr	1.0	5.5	**6.5**	13.0	14.0	9.5	**36.5**	57.0	2B
30	13.37-13.42	tr	1.0	5.0	**6.0**	15.0	14.0	9.5	**38.5**	55.5	2B
31	13.32-13.37	tr	1.0	3.5	**4.5**	7.5	11.5	8.5	**27.5**	68.0	2B
32	13.27-13.32	0.5	2.0	6.0	**8.5**	11.0	12.0	8.0	**31.0**	60.5	2B

Sample number	Altitude (m OD)	Coarse sand	Medium sand	Fine sand	**Total sand**	Coarse silt	Medium silt	Fine silt	**Total silt**	Clay	Stratigraphic unit
33	13.22-13.27	tr	0.5	4.0	**4.5**	13.5	13.5	9.5	**36.5**	59.0	2B
34	13.17-13.22	tr	1.0	5.5	**6.5**	11.0	9.5	7.5	**28.0**	65.5	2D
35	13.12-13.17	0.0	tr	4.0	**4.0**	1.0	3.0	5.0	**9.0**	87.0	2D
36	13.07-13.12	0.0	tr	10.0	**10.0**	2.5	10.0	10.0	**22.5**	67.5	2D
37	13.02-13.07	tr	1.5	24.5	**26.0**	15.5	11.0	6.5	**33.0**	41.0	2D
38	12.97-13.02	tr	1.0	15.0	**16.0**	21.0	17.0	9.0	**47.0**	37.0	2D
Column 62											
455	13.78-13.88	1.5	3.0	12.0	**16.5**	8.5	16.0	8.5	**33.0**	50.5	Neolithic fill
Column 63											
400	14.67-14.72	2.0	5.0	10.5	**17.5**	11.0	14.5	8.5	**34.0**	48.5	6D
401	14.62-14.67	1.5	4.5	11.5	**17.5**	12.0	17.0	7.5	**36.5**	46.0	6D
402	14.57-14.62	1.5	5.0	14.0	**20.5**	11.0	15.5	9.0	**35.5**	44.0	6D
403	14.52-14.57	2.0	6.0	13.5	**21.5**	14.5	15.5	8.5	**38.5**	40.0	LBA soil
404	14.47-14.52	2.0	6.0	14.0	**22.0**	15.0	17.5	7.5	**40.0**	38.0	2F/Neolithic soil
405	14.42-14.47	5.0	7.5	11.0	**23.5**	13.0	13.5	7.5	**34.0**	42.5	Neolithic soil
406	14.36-14.41	5.5	7.0	9.0	**21.5**	11.0	13.0	7.5	**31.5**	47.0	Neolithic soil
407	14.32-14.36	1.5	4.5	8.5	**14.5**	5.5	7.0	4.0	**16.5**	69.0	Neolithic soil
408	14.26-14.31	1.0	2.5	6.0	**9.5**	7.5	12.0	7.0	**26.5**	64.0	1K
409	14.20-14.25	0.5	2.5	7.0	**10.0**	2.5	10.5	3.0	**16.0**	74.0	1K
Column 64											
1	14.01-14.21	0.5	1.0	8.0	**9.5**	7.0	10.5	7.5	**25.0**	65.5	1K
2	13.83-14.01	0.5	2.0	11.0	**13.5**	10.5	10.5	7.0	**28.0**	58.5	1K
3	13.71-13.81	tr	3.0	33.0	**36.0**	11.5	10.5	7.5	**29.5**	34.5	1C
4	13.51-13.61	tr	1.5	20.5	**22.0**	4.0	16.5	14.0	**34.5**	43.5	1C
5	13.31-13.41	0.0	0.5	18.0	**18.5**	18.0	18.0	10.5	**46.5**	35.0	1C
6	13.16-13.21	tr	1.5	26.5	**28.0**	8.0	13.5	11.5	**33.0**	39.0	1C

CHAPTER 6

Soil phosphate analyses of column samples

MIKE COWELL

Introduction

An extensive soil phosphate survey was planned as part of the investigation of the alluvial and occupational history of the Runnymede Bridge site. Occupation-related phosphate may enter the soil in a variety of forms and by a variety of pathways. These include: burials; human excreta; food with vegetable matter, animal bones and related debris either as a scatter or in discrete middens; animal dung perhaps concentrated by corralling (Clark 1977) or used in construction; and concentrations of vegetable material by human or natural processes. Phosphate in solution, derived from a variety of animal or vegetable sources, may also be concentrated by secondary processes, for example by run-off into ditches (Craddock *et al.* 1985). Once in the soil, the predominantly organic phosphate is converted to inorganic forms and largely fixed as iron or calcium compounds depending on the soil type and pH. In this form there seems to be little opportunity for solution movement although other forms of disturbance, such as the activities of earthworms, may cause displacement vertically (and, less significantly, laterally) (Craddock *et al.* 1985). Unfortunately, although a simple quantitative measurement of phosphate will identify zones of phosphate concentration or enhancement, it is not possible to determine the origin of that phosphate from such a simple measurement; this may only be inferred from the context.

The preliminary stages of the soil phosphate survey at Runnymede Bridge involved the collection and analysis of a series of column samples with generally uniform vertical spacing. In total, twenty-two sets of column phosphate samples were taken over the period 1984–9, all from excavated sections at the site. Two of these sets (archive only) formed part of a pilot scheme to provide a general evaluation of the correlation of soil phosphate with occupation horizon and to determine the typical concentration of phosphate at the site; they were not individually related to the site datum and are not reported on here. The remaining columns were taken from levelled sections across the site, usually over the full range of the exposed section but this varied from section to section depending on the stratigraphy, with accurate sampling at intervals of 2 or 3 cm. Such comprehensive column sampling for phosphate is unusual; most site surveys have relied solely on gridded or linear sampling of the presumed occupation horizon. Extensive column sampling was carried out at the flint mine complex at Grimes Graves, Norfolk (Sieveking *et al.* 1973) but the vertical sample intervals were much coarser (at 20 cm) than for Runnymede Bridge.

The main objectives of the column sampling were to clarify the correlation of the concentration of phosphate with the stratigraphy at different parts of the site and to determine the precision with which subsequent grid sampling should be carried out. The latter were to be taken from occupation horizons with the intention of investigating the density changes across the site and answering particular questions about the functions of certain areas and features. Evidence from the columns should show the extent to which phosphate has moved vertically through the columns. This will assist in the evaluation of the significance of variations across a particular gridded surface.

Most of the columns were sampled throughout a profile including upper silts, containing Late Bronze Age material, which seem to have been subjected to seasonal and repeated flooding. This may have caused sorting and redeposition with probable loss of some occupation debris and vertical settling of the remainder. This so-called 'flood-reworked' or redistributed horizon (Needham 1992) overlays a variable thickness of in situ LBA material, apparently undisturbed by this process. The column sampling was also intended to investigate the detailed distribution of phosphate through these deposits and into the underlying soil or sediments. Where the Neolithic occupation horizon was present beneath, and was excavated, columns were generally extended right down to the underlying soil.

Analytical procedure

The analytical procedure employed followed that described by Craddock *et al.* (1985) for on-site phosphate surveys involving acid extraction and colorimetric measurement of phosphate using the method devised by Murphy and Riley (1962). However, since the analyses in this instance were conducted off-site in a laboratory environment, the opportunity was taken to improve the reliability of the procedure and measurements without greatly adding to the processing time.

The bulk soil samples (about 50–100 g) were first air dried, broken up by hand, then sieved (2 mm mesh) to remove larger stones and other debris. Sub-samples of 1 g were taken from these and transferred to 25 ml beakers. Then 10 ml of 2M hydrochloric acid were added in stages to each beaker and these were heated on a hotplate to about 80°C for about 1 hour. Distilled water was then added to make up to 25 ml and, after cooling, an aliquot of 100 μl from each was transferred to 20 ml flasks. These were then filled with diluted molybdenum-blue reagent (Craddock *et al.* 1985), left for 30 minutes and the absorbance of the blue phospho–molybdenum complex measured at 882 nm using a UV/Visible colorimeter. Standard phosphate solutions were used to construct calibration curves and determine the concentration of phosphorus in the original soil sample. Replicate measurements indicated that the precision of the analytical technique is about ±10% but the overall reproducibility of the procedure, including the sub-sampling of the bulk soil, is probably in the range of 10 to 20%.

This analytical procedure extracts and measures a major, but not necessarily consistent, proportion of the inorganic phosphate content of the soils. The soils at Runnymede are calcareous, with a pH of about 8, and thus the dominant inorganic phosphate component would be expected to be calcium phosphate (Bethell and Máté 1989).

Results

The individual phosphate concentration data for the soil samples are not listed here (although they are available from the author on request); instead, the results for the twenty sets of column samples are plotted in Figs 6.1–6.4 with phosphorus concentrations expressed in units of mg/100 g of soil (dried weight) and the sampling level according to the site datum (m OD). Common scales have been used for all plots for ease of comparison and some relevant contextual information has been included on the figures to show the relationship between phosphate and occupation level. In general, plots show the results for only one column. However, Figs 6.3 and 6.4 both include a combined plot for closely adjacent columns; in one case this is effectively a continuation of the same column with minimal overlap (24/25) but in the other most of the levels are duplicated for comparison (57/58).

COLUMN 1A (AREA 13, GRID 34.0/30.0, FIG. 6.1)

This column includes post LBA deposits and in situ LBA. The isolated high value at the base of the sequence is supported by similar levels for some continuation samples not reported here.

COLUMN 9 (AREA 16, GRID 49.5/14.0, FIG. 6.1)

This column runs from post-LBA through in situ LBA and underlying alluvium to a possible stripped Neolithic surface.

It was sampled from the cleaned wall of a modern pit or soakaway and hence it is possible that there has been some more recent phosphate contamination or movement. As in other columns, there is a gradual increase in phosphate concentration through the reworked deposits peaking at in situ deposits at about 14.5–14.4 m OD. There are then decreased concentrations but occasional higher values above background.

COLUMNS 13A AND 13B (AREAS 13 AND 19, GRIDS 37.0/20.0 AND 37.2/20.0, FIG. 6.1)

These are two closely adjacent columns which may be considered together. Both commence with the post-LBA layers. However, whereas 13a terminates just below the in situ LBA, 13b continues through and below the Neolithic levels. There is good correlation between the two columns with similar trends and concentrations.

COLUMN 16A/B (AREA 13, GRID 37.0/35.6, FIG. 6.2)

This is a combined sequence with samples taken in two successive seasons. Sub-column 16a runs from the post-LBA levels to just below the LBA in situ deposits, then 16b continues below this to the Neolithic occupation. There is reasonably good correlation of phosphate concentration at the junction and the sequence as a whole differentiates the occupation horizons.

COLUMN 17 (AREA 13, GRID 34.3/44.0, FIG. 6.2)

This column is close to the edge of the channel which crosses the north-west end of Area 13. It runs from post-LBA (perhaps even Post-Medieval) through possible in situ LBA and down to Neolithic layers.

COLUMN 18 (AREA 14, GRID 75.9/36.0, FIG. 6.2)

This column runs through the post-LBA to in situ LBA and shows similar trends to other columns for this sequence.

COLUMN 23 (AREA 16, GRID 55.0/12.0, FIG. 6.2)

This column also runs through reworked and in situ LBA layers and shows similar trends to column 9 for this sequence.

COLUMNS 24 AND 25 (AREA 16, GRIDS 60.0/12.0 AND 60.0/12.3, FIG. 6.3)

These columns are closely adjacent (0.3 m) and were intended to serve as a single column. Column 24 runs down into the in situ LBA; column 25 picks up at a similar level and then continues through intervening silts to Neolithic layers. There is good correlation at the junction of the two columns. In common with 16a/b they show clear demarcation of the occupation levels.

COLUMN 36 (AREA 21, GRID 69.9/43.0, FIG. 6.3)

This column runs through reworked and in situ LBA into sediments beneath. There is a gradual increase through the reworked and in situ deposits, and high levels apparently below the in situ LBA zone. This might imply some vertical movement of phosphate which is discussed below.

COLUMNS 44 AND 45 (AREA 25, GRIDS 29.0/19.5 AND 29.0/15.5, FIG. 6.3)

These columns were taken within a round house and run through LBA-occupation soil. Column 44 is positioned just inside the wall line of the building, whereas column 45 is within the interior. Both columns show a single peak in phosphate concentration which coincide at 14.63 m OD. The peak value is highest in column 44, although there is no significant difference between the mean concentrations of the columns as a whole.

COLUMNS 49 AND 51 (AREA 26, GRIDS 45.7/40.0 AND 46.0/32.6, FIG. 6.4)

Both columns run from post-LBA through in situ LBA to the underlying soil. Column 51 was not accurately levelled but, assuming that the LBA occupation layer is at a comparable position to that in column 49 (they are about 7 m apart), the two columns may be interrelated. Both columns show a single peak in phosphate concentration corresponding with the LBA layers, but it is a higher concentration for column 51 and presumably reflects different occupation intensity.

COLUMNS 57 AND 58 (AREA 31, GRIDS 85.2/21.0 AND 86.2/21.0, FIG. 6.4)

The columns are adjacent samplings covering post-LBA and in situ LBA dark earth deposits near a suspected entrance way. The occupation deposits here show a lateral colour change with column 57 sampling browner deposits and column 58 blacker deposits. Both columns show peaks in phosphate concentration but this is much broader for column 57 than column 58. In addition, column 58 (sampling the blacker deposits) has very much higher concentrations at the peak with 345 mg/100 g P as opposed to only 135 mg/100 g. This would indicate that the blacker deposits represent larger amounts of occupation debris.

COLUMN 61 (AREA 32, GRID 59.9/40.0, FIG. 6.4)

This column mainly spans a thick deposit with reworked LBA material over a possible remnant of in situ LBA. There is a very gradual increase in phosphate down the column, a single high value which may correspond to some in situ LBA, followed by a return to low concentrations.

Discussion

It can be seen (Figs 6.1–6.4) that the expected correlation between phosphate and occupation horizon is clearly confirmed for most of the columns across the site, although the detail of this for all the individual sections will not be elaborated here. The phosphate peak recorded for the in situ LBA horizon in the columns (see Table 6.1) is not, however, uniform across the site; presumably this is a reflection of the type and intensity of occupation activity and will be explored in more detail by the gridded survey. The peak is broad (or variable) and occupies several samples in some columns (e.g., as in columns 9 and 36; Figs 6.1 and 6.3), but for others the peak is virtually confined to a single sample with sharp falls either side (e.g., as in columns 23 and 51; Figs 6.2 and 6.4) . Clearly, sampling at the vertical precision adopted for the columns was also necessary for the gridded survey to monitor significant three-dimensional changes.

One of the clearest correlations between context and phosphate is that shown for columns 24/25 (Fig. 6.3) which, although separated horizontally by 30 cm, can be considered as one continuous column. Particularly distinct peaks in phosphate correspond to the in situ LBA and Middle Neolithic horizons. Apparently less well correlated are columns 9 (Fig. 6.1), 17 (Fig. 6.2) and 61 (Fig. 6.4).

In column 17 a high phosphate concentration was obtained for the part of the profile at the top of the gravel bed, immediately below a diffused LBA zone. Since the high phosphate trend is not confined to a single sample it is presumed to be representative of that part of the section. This particular column is adjacent to the western loop of the

Table 6.1 *Phosphate peak concentrations at the LBA in situ occupation horizon (in grid order from site West to site East).*

Column	Area	Grid	P (mg/100 g)
45	A25	29.0/15.5	150
44	A25	29.0/19.5	195
13a	A13	37.0/20.0	380
1	A13	34.0/30.0	380
16	A13	37.0/35.6	260
51	A26	46.0/32.6	314
49	A26	45.7/40.0	215
9	A16	49.5/14.0	380
23	A16	55.0/12.0	630
24/25	A16	60.0/12.0	705
36	A21	69.8/43.0	475
18	A14	75.9/36.0	320
57	A31	85.2/21.0	135
58	A31	86.2/21.0	345

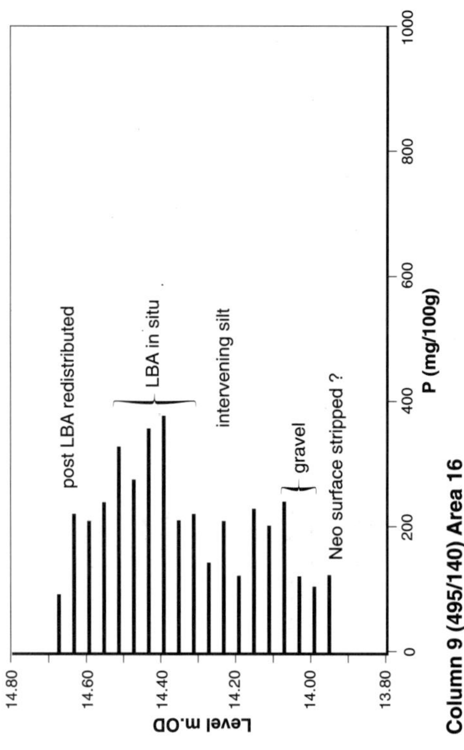

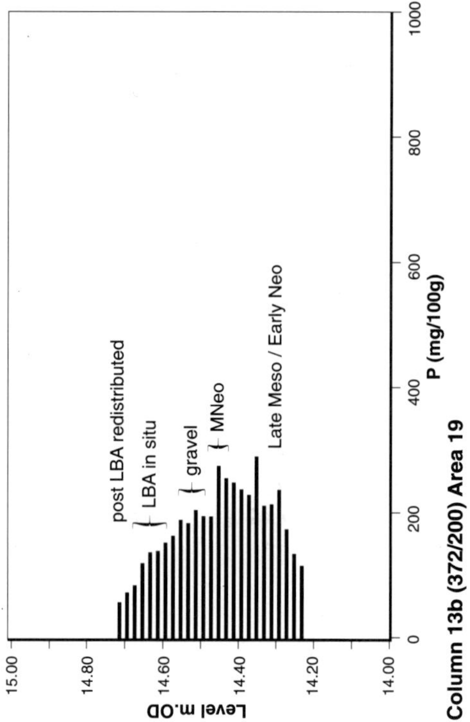

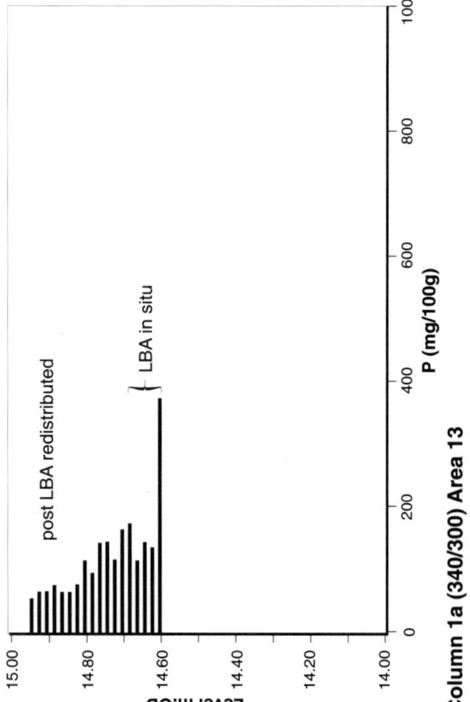

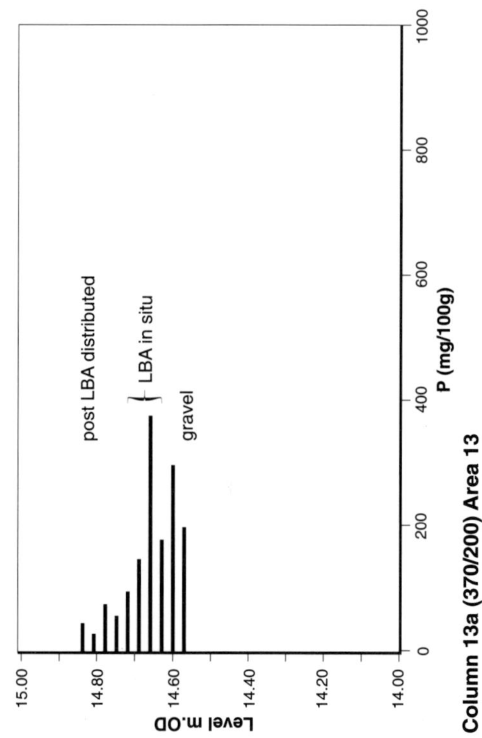

6.1   *Phosphate measurements: columns 1a, 9, 13a, 13b.*

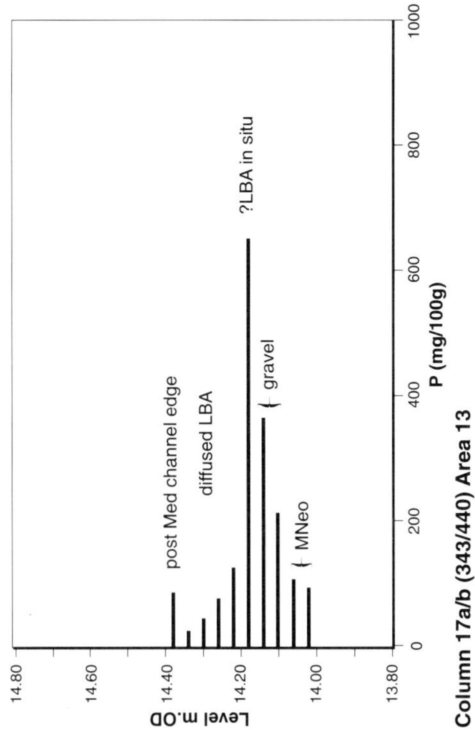

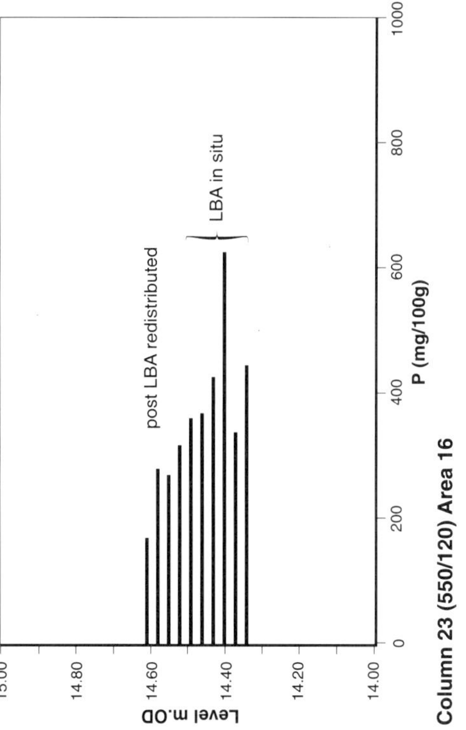

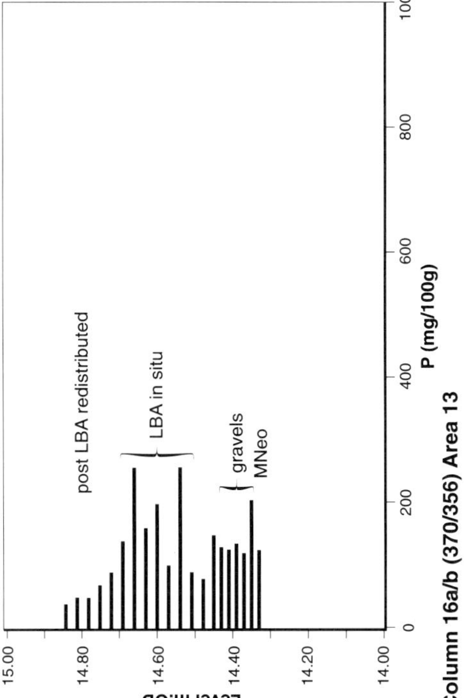

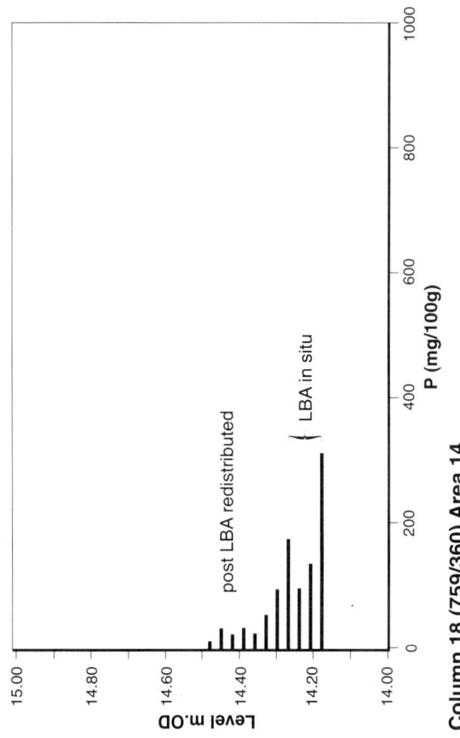

6.2 *Phosphate measurements: columns 16a/b, 17a/b, 18, 23.*

121

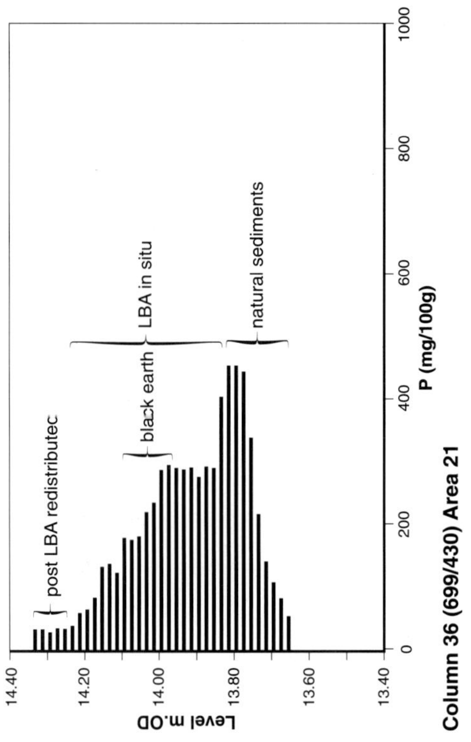

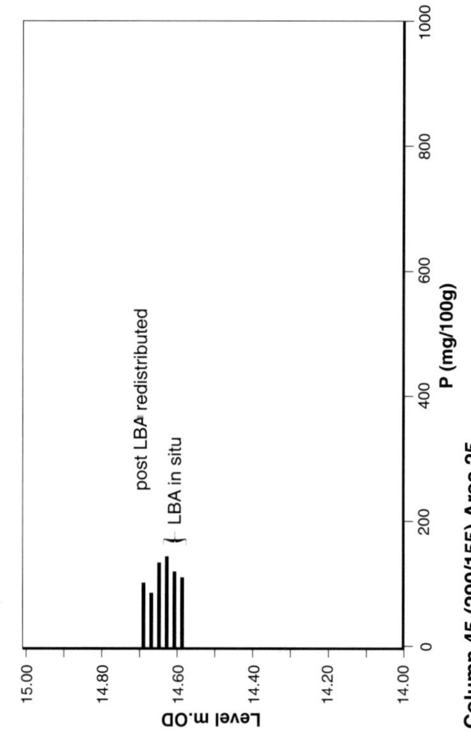

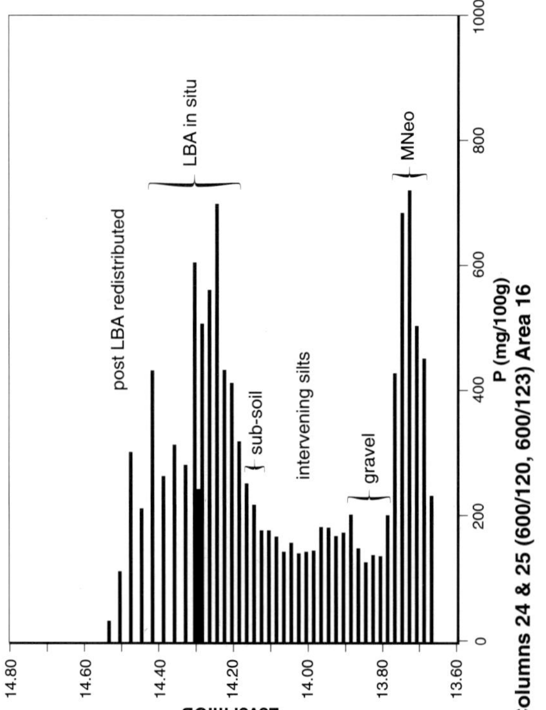

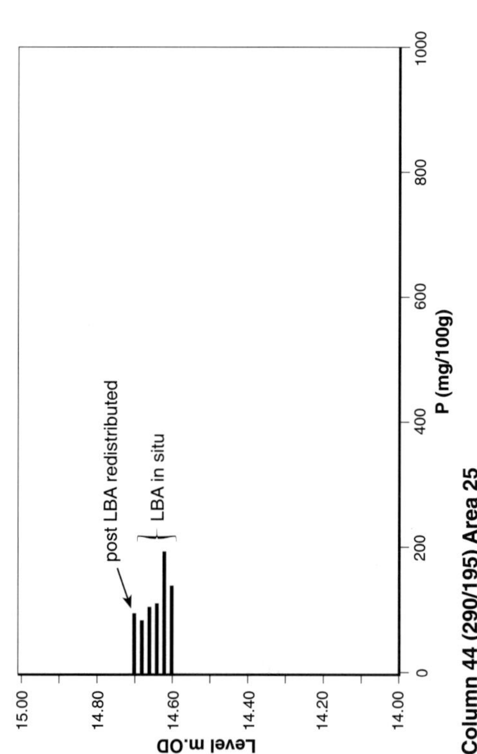

6.3   *Phosphate measurements: columns 24/25, 36, 44, 45.*

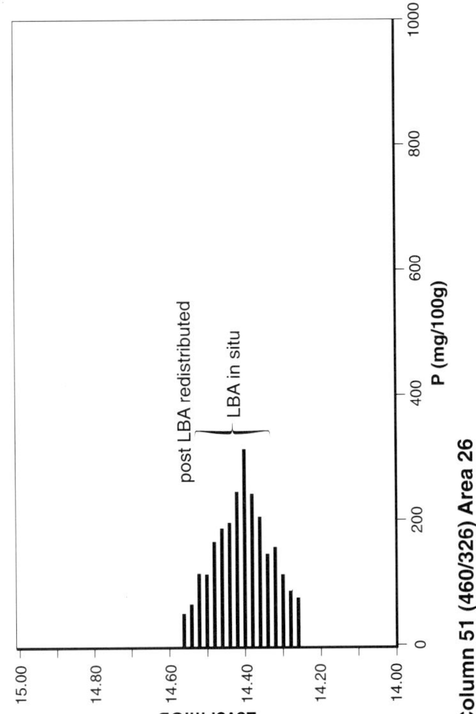

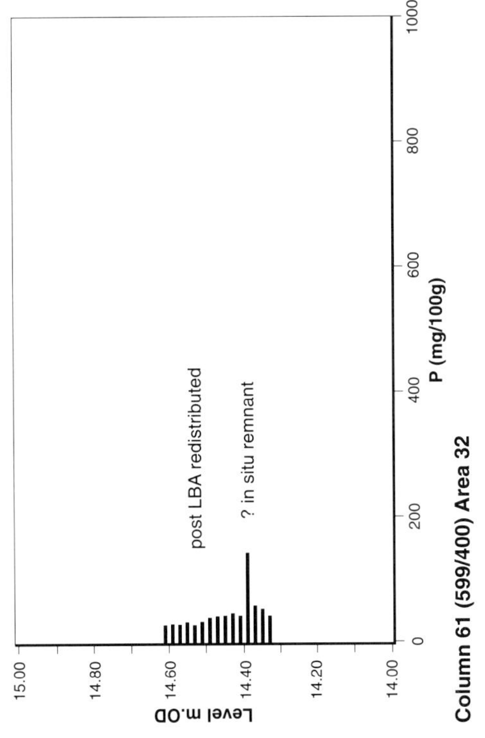

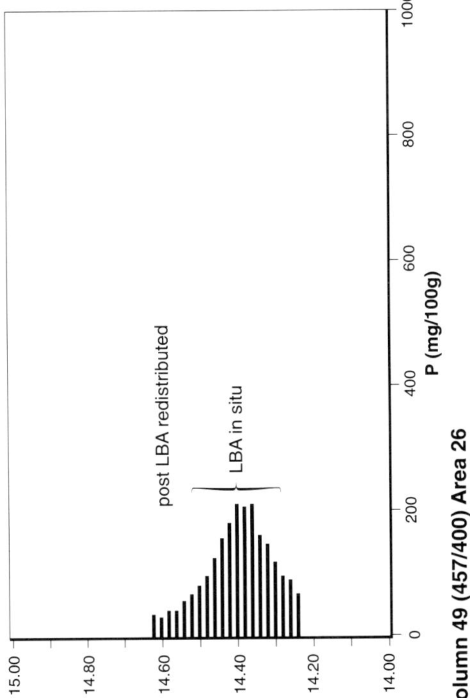

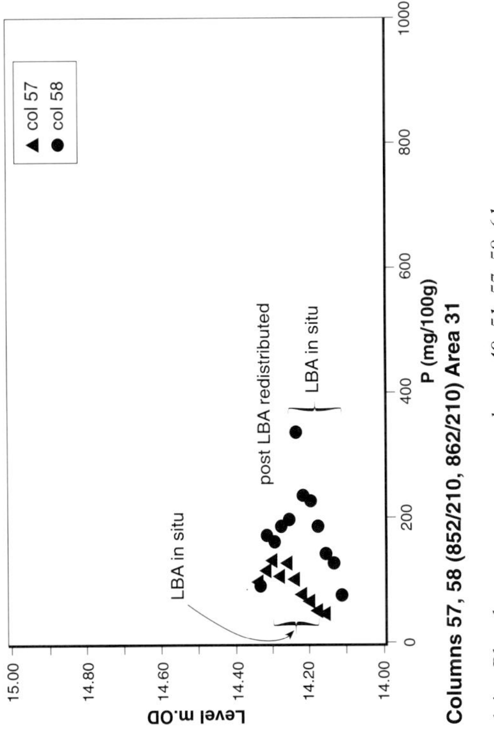

6.4   *Phosphate measurements: columns 49, 51, 57, 58, 61.*

Post-Medieval river channel where the LBA evidence is somewhat diffused. It is possible that the high phosphate reading relates to a localised small pocket of occupation or other debris on the gravel.

In column 61 (Fig. 6.4), an isolated high value was found at 14.39 m OD within a sequence of reworked LBA deposits which otherwise shows a gradually increasing phosphate down the section. It is possible that this represents a remnant of the original LBA soil profile or is the result of a random inclusion of a piece of high phosphate debris, a consequence of taking relatively small samples from inhomogeneous material.

As noted above, column 9 (Fig. 6.1) was sampled from the wall of a modern pit or soakaway and the possibility of recent phosphate contamination cannot be ruled out. The gravel and underlying silts had low phosphate levels suggesting that any Neolithic surface deposits had been stripped by fluvial action, but there are higher values in the intervening silts below the LBA deposits. This enhanced phosphate may be due to some vertical movement from the overlying LBA horizon. There is evidence from other columns (for example, 24/25, 36, 49 and 51) to support this as a possibility. This is unlikely to be solution transportation and is more probably due to disturbance by faunal activity, for example, by earthworms. Such was the explanation given for high phosphate levels detectable in the surface soil at Cat's Water, Fengate, where occupation horizons had been covered by later clay (flood alluvium) deposits (Craddock 1984).

In column 36 (Fig. 6.3) the lowest five samples lie within the undisturbed natural sequence, yet the phosphate concentration is already rising towards the base of the LBA deposits (from about 50 to 220 mg/100 g). It is also noteworthy that the peak in phosphate values, between 13.76 and 13.86 m OD, corresponds with a dirty sand virtually barren of cultural debris, rather than with the overlying dark earths rich in varied debris which might have been expected to have yielded the highest phosphate values. In column 24/25 (Fig. 6.3) the intervening (sterile) silts between the LBA and the Middle Neolithic horizons remain high in phosphate, about 150 mg/100 g, as was the case in column 9. It is possible that some disturbance of the upper part of the Neolithic occupation layer in column 24/25 (which is not represented in column 36), together with downward movement of phosphate from the LBA horizon, has produced a level higher than the intervening 'background'.

In the absence of appropriate control samples for the sterile levels in this part of the sequence the background can only be assumed to be comparable with the lowest phosphate concentrations encountered in the columns analysed, i.e., 50–100 mg/100 g. Columns 49 and 51, as with column 36, show fairly high, although decreasing, phosphate levels in the natural sequence below the LBA layer. This movement of phosphate, which is not unexpected, is small but will need to be taken into account when interpreting the results of the gridded samples.

A related feature of the column data is the apparently 'flood-reworked' or redistribution horizon overlying the in situ LBA. Those columns which include a good profile of reworked layers (1, 9, 13b, 16, 18, 24/25, 61) exhibit a consistent pattern of gradually increased phosphate as the in situ occupation horizon is approached. It is noticeable that the slope of this increase is generally less steep than the distribution created by the downward movement below the occupation horizon. The trend in phosphate concentration of the reworked layers is consistent with a mechanism involving flooding, sorting and redeposition, with consequent loss of some occupation debris and vertical grading of the remainder. However, subsequent earthworm activity must presumably have modified the structure somewhat. Yet, clearly, the levels in the reworked horizon remain above 'background' (about 50 mg/100 g at the surface) and could be recognised as being derived from an occupation layer by their phosphate concentration.

Conclusions

As expected, the phosphate concentrations in the columns are generally well correlated with the occupation contexts. The peak levels are well defined and indicate that grid sampling must be precise to monitor significant changes over a large area. There is nevertheless clear evidence of some movement of phosphate vertically within the section, perhaps by the action of earthworms rather than solution transfer. This does not detract from the general correlation of phosphate with the occupation horizons, but may give rise to slightly elevated levels in known sterile levels. The phosphate distribution within the suspected flood-reworked levels is consistent for all the columns analysed. The trends are compatible with the suggested mode of formation and the detailed data available here may help to explain similar features on other sites.

CHAPTER 7

Molluscan evidence

MARY EVANS AND JOHN EVANS

Work on molluscs from river valley deposits of Pleistocene and Holocene age has shown that, although the assemblages are often mixtures of species from a wide range of ecologies, it is possible to reconstruct the past environments of both the land and aquatic habitats (Sparks and West 1970; Kerney 1971; Alexandrowicz *et al.* 1984). The results are usually of site or local significance, especially where the assemblages are autochthonous, and of narrower relevance than insect or pollen data; valuable information about lateral spatial variation can thus be obtained. From allochthonous deposits, especially those of river channels, results are of wider spatial significance although lower in resolution. The molluscs from the earlier rescue excavations at Runnymede (Evans 1991a) confirmed their usefulness at this locality specifically, so it was sensible to do a more detailed study of the deposits of the later research excavations.

Methodology

All the analyses, identifications, the primary histograms and report are the work of Mary Evans, with help from John Evans for *Pisidium* identification and the report.

Sampling, analysis and identification

Soil and sediment samples were provided. They were taken as columns, each sample usually 5 cm thick (see Figs 1.3 and 1.5 for locations).

Sampling strategy aimed to obtain as long and complete a sequence as possible and also to examine lateral variation between different altitudes and topographical and depositional contexts, for example in relation to bank, levee and trough, and overbank floodplain and channel deposits and soils. Sampling in columns provides a vertical control on location even though it may lead to some parts of the sequence being missed, e.g., in lenses and wedges of sediment which are not on the column line. The sample thickness of 5 cm reflects experience of this sort of deposit, the fact that there is mixing, and that finer intervals are unlikely to reveal more detail; the aim in any case was to obtain a general picture of trends, not micro-sequences within the soils and sediments. Finer sampling was applied to the soil profile in column 13 and, in retrospect, would have been useful at the

top of other soil profiles, especially in column 10, parcel 2B. Furthermore, sub-sampling took place in that not all samples were analysed, but usually only alternate ones, since vertical changes, especially in the channel deposits, were often minimal.

None of the samples had been pretreated except column 55 where they had been sieved for insects. Air-dried weights were usually 0.5–1.0 kg; although these weights yielded large numbers of shells, they were often necessary to get a representative number of land snails which were often in low abundance, especially in the channel deposits.

For comments on identification problems, see Evans (1991a). *Vallonia pulchella* and *V. excentrica* broken shells and juveniles are impossible to separate and are recorded as *Vallonia* spp. (not *costata*); most are probably *V. pulchella*. *Euconulus fulvus* is the aggregate species, since both *E. fulvus* and *E. alderi* are present but difficult to separate.

Presentation of the results

All the results are archived as tables of shell numbers, with *Pisidium* numbers being for the maximum of left or right valves as relevant. Most sequences (except columns 12 and 40, which were very short) are presented as histograms of shell numbers, again with the maximum of left or right valves of *Pisidium*. In some of the histograms where not all samples were analysed (columns 6, 27–29, 52 and 60), the gaps between samples are reduced.

Graphical presentation implies some interpretation. Thus, since the aim of the work is to identify past environments and since interpretation is mainly through analogy with present-day species ecology, the order of species in the histograms is an ecological one. However, patterns in the data were also considered at this stage.

First, there is a basic division into aquatic and land species.

Aquatic species are those demanding submergence under water for their everyday living, even though some can survive in mud during periods of dormancy (probably *Bathyomphalus contortus*, *Anisus leucostoma*, *Lymnaea truncatula* and *L. stagnalis*), or even out of water in conditions of high humidity – especially *Lymnaea* spp., *Pisidium personatum* and *Planorbis planorbis*.

Land species include a distinct wetland group which,

although essentially terrestrial in that they generally live out of water, demand very high atmospheric humidity and are generally absent from drier land habitats, especially in conditions of high pH. *Lymnaea truncatula* can live under water, but is generally found 'out of water, on the mud at the edge of rivers, streams and ditches' (Boycott 1936, 143). Some taxa, notably *Zonitoides nitidus* and the Succineidae (Bishop 1981 for the latter), can live under water for several days at a time, but generally do not breed so.

True land species are those which live in drier land habitats, although many are not excluded from wetlands.

Following the procedure of Kerney (1971), the aquatic and land (the latter including wetland) species are presented as separate histograms (or as a single diagram where numbers of one group are very low).

The ratio of aquatic to land species is shown in the land or total diagram (Figs 7.1–7.24) and in the summary diagram (Fig. 7.25) expressed as percentages of total shells.

Lesser groups of species have been established on the basis of present-day ecology, and their behaviour (presence–absence and relative abundance) in the assemblages themselves. The order of the species in the histograms from left to right is equivalent to the order of the groups from 1–11 respectively. The numbers attached to these groups refer specifically to the assemblages from Runnymede in this study and are not equivalent to Sparks's (1961), Evans's (1991a) or any other numbered groups.

AQUATIC GROUPS (1–5)

1. *Ancylus fluviatilis* and *Theodoxus fluviatilis*. On stones in fast-flowing rivers (or in lake littorals where there is wave action).

The presence of *Ancylus* from the start of the sequence (column 31, Sn 39) and the absence of *Theodoxus* until the Atlantic (column 40), could be chronological rather than ecological, reflecting different rates of biogeographical spread. An incomplete survey of the literature shows the following: in earlier work at Runnymede, *Ancylus* predominates over *Theodoxus* in the Neolithic (SS1) sequence (Evans 1991a); at Staines, *Ancylus* is abundant from 8360 ± 100 BP but *Theodoxus* is absent until after 6655 ± 55 BP (Preece and Robinson 1982); at Anslows Cottages, in the lower Kennet valley, *Ancylus* predominates in the pre-Bronze Age levels, while *Theodoxus* does so in the late prehistoric to Saxon ones (Evans 1992); at Apethorpe, Northants, *Ancylus* is present by pollen zone V (Boreal), but *Theodoxus* is absent even into zone VIII (Sub-Atlantic) (Sparks and Lambert 1961); in Berkshire, *Ancylus* is recorded from Lateglacial pollen zone III, but *Theodoxus* is absent until zone VIIb (Holyoak 1983).

2. Other distinctive river species, *Pisidium amnicum*, *P. henslowanum*, *P. supinum* and *P. moitessierianum*. These (except *P. supinum* which he does not classify) are in Sparks's (1961) group 4 – moving-water species – and like *Ancylus* and *Theodoxus* they are characteristic of rivers, although not

excluded from lakes. At Runnymede they are notable for their absence from the earliest, eighth millennium BC, deposits (column 31, Sn 31–39) while being present subsequently.

In the case of *P. amnicum* and *P. henslowanum* this early absence is unlikely to be biogeographical (or regionally chronological) since both were present from an early stage in the Weichsel and Weichsel Lateglacial in eastern and southern England, including the Thames valley (Allison *et al.* 1952; Gibbard *et al.* 1981; Holyoak 1982; Kerney *et al.* 1982; Gilbertson 1984). On the other hand, *P. moitessierianum* is not known from before pollen zone V in Berkshire (Holyoak 1983), although a more detailed survey of the literature is needed before this can be confirmed.

3. All other river species, being *Valvata piscinalis*, *Bithynia* spp., *Physa fontinalis*, *Lymnaea auricularia*, *Sphaerium corneum*, *Hippeutis complanatus*, *Pisidium subtruncatum*, *P. milium*, *Planorbis carinatus*, *Armiger crista*, *Gyraulus* spp., *Lymnaea peregra*, *Pisidium nitidum*, *P. hibernicum*, *Acroloxus lacustris* and *Unio* spp. These behave similarly, although some – *V. piscinalis*, *Bithynia* spp., *Armiger crista*, *Gyraulus albus*, *Lymnaea peregra* and *Pisidium nitidum* – are more abundant than the rest. All occur in a wide range of habitats, and embrace Sparks's groups 2, 3 and 4.

4. A possible swamp and/or floodplain pools group – *Planorbis planorbis*, *Lymnaea palustris*, *Valvata cristata*, *Bathyomphalus contortus*, *Pisidium pulchellum*, *Sphaerium lacustre*, *L. stagnalis*, *Pisidium casertanum*, *P. obtusale*, *P. personatum*, *Anisus leucostoma* and *L. truncatula*. These species can occur in other habitats, but when some or all go together in antipathetic distribution to other aquatics, as in columns 4 and 31, then a different kind of environment from river channel is indicated. Of course, each species may be responding positively to different conditions through a sequence, as in column 4 with *V. cristata* which may be a floodplain pool or swamp species in the lower part (Sn 41–44) and a river-channel species in the upper (Sn 36–40). Equally, the shells of one species from a single assemblage may conflate two populations, one from river and one from floodplain-pool/swamp habitats.

5. *Lymnaea truncatula*. The association of *L. truncatula* with *Perforatella rubiginosa* in late second millennium BC and later deposits probably signifies mudflat, since this is the main habitat of this two-species association today (I. Killeen, pers. comm.); it may also occur in floodplain pools and swamps (as in group 4), and as a marsh/damp grassland species (group 6).

WETLAND GROUPS (6–8)

6. *Lymnaea truncatula*. In addition to being an aquatic species, *L. truncatula* occurs in marsh and damp grassland; such a habitat may, where overgrazed, have a lot of bare soil and be flooded in winter (Robinson 1988). It may seem odd having two separate groups for one and the same species but this is

the way we should be thinking in interpreting the molluscan data: as with *Valvata cristata*, discussed above under group 4, one species may occur in two or more distinctive habitats throughout a sequence, and both separately or at the same time.

7. *Zonitoides nitidus* and *Carychium minimum*. These two species behave similarly, although *Z. nitidus* probably indicates more shaded or structured conditions than *C. minimum* (Evans 1991a, figs 119 and 122). They are often abundant, especially where the humidity is constantly high, living under stones and fallen branches and among grass and sedge tussocks; highly structured habitats (i.e., those in which there are numerous lateral and vertical variations of plant layering and architecture) and shade are particularly suitable for *Z. nitidus* ('carr' in Bishop 1981), and the species can also live under water. *C. minimum* is more wide-ranging, additionally favouring less structured habitats ('herbaceous fen' in Bishop 1981). As wetland species, they fall into the same category as *Vallonia pulchella* and the Succineidae, but are generally found in more shaded (and perhaps damper) places.

8. Succineidae. Species in this family probably like light (Bishop 1981), living in places with high humidity, often on monocotyledon, e.g., flag iris, leaves or under stones and at the base of vegetation. *Vallonia pulchella* (group 11) could well be placed here.

LAND GROUPS (9–11)

9. Structure-loving species, *Aegopinella nitidula* to *Vitrea* in the diagrams. Although none of the species individually is confined to shaded habitats (and we have found every one except *Cochlodina laminata* in modern grassland, although *Balea* admittedly on the Isle of Harris in the Outer Hebrides), when several occur together, and especially where Clausiliidae are present, a more diverse habitat than grassland is indicated. In a strictly dry land context, this would indicate some trees and bushes, the Clausiliidae (*Balea*, *Clausilia*, *Cochlodina*) especially liking trees. In a wetland context, though, as at Runnymede, this could also mean large sedge tussocks (up to 1 m high) since, as discussed by Bishop (1981), such a habitat can support dry land species in the tussocks and aquatics on the ambient fen surface. However, none of Bishop's faunas were as rich in this group as those at Runnymede, and woody vegetation is generally implied here.

The absence of *Pomatias elegans*, *Aegopinella pura* and *Acanthinula aculeata*, which are usually present and often common in Holocene archaeological deposits from southern England, may signify the lack of a particular requirement, perhaps a combination of dryness and high lime content. Even without knowing what this requirement is, one can use the absence of these species to suggest a local origin for the Runnymede Bridge assemblages. The absence of *Ashfordia granulata* is also noteworthy since it is often common in damp woodland and sedge habitats today.

10. *Vitrina pellucida* to *Vallonia costata* in the diagrams.

These species occur in a range of habitats and are less fussy about woodland than are the previous group. Some, like *Nesovitrea hammonis*, are often common in wetlands. *Vallonia costata* comes here, not necessarily with the other *Vallonia* as an index of openness (although it could do); it is another species, like *Valvata cristata*, which would be worth plotting among two or more ecological categories in histograms. *Cochlicopa*, *Trichia* and the Limacidae (and maybe others) probably occur as damp grassland species, especially in conjunction with the next group (although in impoverished grassland on their own), and this is a distinctive association (Evans 1991b).

11. Open land, generally treeless. *Vallonia excentrica* to *Helicella itala* in the diagrams. *H. itala* probably equals dry calcareous grassland, and the shells may come from some distance (although note the comments about certain absences from group 9 in this respect). The other species are not necessarily confined to dry habitats, and *V. pulchella* probably signifies quite damp grassland, especially when associated with the wetland taxa, Succineidae and *Carychium minimum*.

The relative proportions of the groups are only a guide to the general range of environments. Many species have a range of habitats, and species could be placed in two or more groups. On the other hand, in one locality the choice may be narrower, with some species living in specific habitats. To reconstruct an ancient habitat in detail from its molluscan fauna can only be done by analogy with modern faunas from clearly circumscribed habitats and where alternative habitats for such faunas can be shown to be unknown. Usually this is impossible because of changes from one locality to another in modern faunas for the same set of physical conditions (Bishop 1981).

In historical work there are likely to have been changes in the preferred habitat of species through time, especially with those which can occupy a wide range of habitats. This difficulty has been indicated on several occasions in the previous section concerning the delimitation of the groups. Thomas (1985) also refers to it in his discussion of 'intermediate' land species, suggesting that they be assigned to distinct ecological groups only after these have been defined from a consideration of the more stenotopic species of an assemblage. This problem arises especially with the species of group 4, the floodplain-pool and swamp species, which occur in habitats ranging from river, through swamp, floodplain pools (permanent or seasonal), to almost terrestrial. *Valvata cristata* is particularly flexible in this respect, and its relevance at different stages in the sequence is often open to several interpretations.

Taphonomy

Many of the assemblages from Runnymede are made up of species from widely different habitats, and most include land

and aquatic species. In effect, any assemblage having two or more species reflects different habitats simply by virtue of the ecological principal of competitive exclusion, but wetland and river environments display this palimpsest more than most. Taphonomic processes and problems, scales and procedural stages of interpretation relevant to the environmental significance of molluscs from river channels, floodplains and wetlands generally are discussed in Bishop (1981), Robinson (1988), Evans (1991a, 1991b, 1992, 1995) and Evans *et al.* (1992). The only other work specifically on the taphonomy of river molluscs (Briggs *et al.* 1985), is not really relevant to Runnymede since it deals with cold-climate braided environments and simpler faunas.

In assemblages from moving-water conditions, there is always a mixture of species from different habitats because of water transportation. This is especially the case with river-channel deposits, even during the life of the animals, but it also obtains in assemblages from overbank alluvium. In the Runnymede assemblages, there are practically no instances where fully aquatic shells are absent, although they are as low as 1% of the total fauna in some cases (Fig. 7.25).

Redeposition of old shells is an extension of this problem, and it is likely, particularly in the land deposits, that a small aquatic component and a small wetland component are always of this category.

Where flooding and alluviation are slight or absent, another taphonomic problem arises. This is post-depositional faunal mixing (biopedoturbation) and it occurs especially in the upper parts of standstill phases and soils where, because of the gentle or complete absence of sedimentation, the deposits are often aerobic and biologically active. Molluscan sequences in such contexts of, say, open land giving way to woodland (as in column 10, Sn 9–14), may accordingly be blurred and not as clear-cut as they were in life (see Chapter 5 for discussion of this).

Mixing of species from widely different habitats can also be related to the life processes rather than the transportation of shells. For example, where different habitats occur on the valley bottom at different times of the year, especially dry land in summer and flooding in winter, two distinct communities of species – a dry land one of, e.g., *Vallonia*, *Trichia* and *Cochlicopa*, and a wetland one of, e.g., *Lymnaea truncatula* and *Anisus leucostoma* respectively – may be able to complete their life-cycles in the same place (Robinson 1988). Mixing of the shells in the subfossil assemblages in this situation is then a function not of transportation but of seasonal behavioural overlap. Another source of mixing relating to the life environment is in highly structured environments of, e.g., large sedge tussocks or trees and bushes in permanently aquatic conditions which, as Bishop (1981) has shown, can produce communities of true aquatics on the swamp or fen surface and true land species on the sedge tussocks and woody vegetation (as discussed above, group 9). At death and in subfossil assemblages, these two communities are mixed. So a mixed assemblage of aquatics

and land species is not necessarily an index of water transportation, although it is suspected in the case of most of the Runnymede assemblages – in view of the high species diversity of both the land and aquatic assemblages – that lateral transportation and mixing of shells from true land and aquatic habitats has indeed taken place.

For the land species, and at the extreme autochthonous end of the taphonomic spectrum, because the animals are slow moving and not widely dispersed, information is relevant to the very local or sampling spot of the assemblages. It is not directly comparable to the insect data which is of wider significance, and even less so to the pollen. This needs to be understood when comparing the information of these other groups with that of the molluscs.

Presentation procedure

In the following descriptions/discussions, the sequences of assemblages have been placed into chronological order as indicated in Chapter 2, the ordering and naming being based on 'alluvial parcels', 'soils' and 'occupation layers'. These are chronological units and are used here in the same way as stage names, not solely as litho-/pedostratigraphical units. Pollen zone names, specifically Boreal, Atlantic and subdivisions of these, and archaeological periods are used to qualify these. Finer subdivisions are based on context, specifically marl, channel fill, overbank alluvium, soil or occupation, and on topographical location – altitude and land configuration (see Chapters 2 and 10 for details). Sample groups/columns within each unit are sometimes more or less contemporary, varying in their horizontal configurations rather than vertically, so that differences in the fauna – e.g., in parcel 2B between column 60, Sn 40, and column 10, Sn 14 – are of lateral spatial rather than chronological significance.

The next section is essentially a brief commentary on the molluscan data, indicating the main environmental deductions in conjunction with other biological, soil/sediment and topographical data. 'Site' refers to the environment of the sampling site, 'aquatic' refers to the environment of the river channel and a few other aquatic habitats including reed swamp and mudflat, and 'land' refers to the land environment, including that of floodplain and wetland habitats.

Environmental comments are supported by the relevant molluscan data in brackets. Where there is no information for 'aquatic' or 'land' or where the site is clearly river channel or land, no entries are made.

Assessment of the 'site' environment in terms of aquatic or land is based on the proportions of aquatic to land shells in the assemblage, and although no precise numerical criterion is used, on the whole, where there is over 25% (and often up to 90%) aquatics, an aquatic site environment of specifically river channel is considered likely. However, account must be taken of the comments made above about

1 *Area 19, section RH5 (1989) from NE. The marl (parcel 1C) is seen along the bottom of the profile, overlain by parcel 1K silts and the mid–Atlantic soil profile. The gravel of parcel 2F is clearly seen high up and overlain by thin deposits of 2B and the Late Bronze Age. The early cut feature is visible towards the left side. (cf. Fig. 2.6).*

2 *Early cut feature in Area 19, section RH5 stratified within parcel 1K silts and partly excavated (1989; 19.800).*

3 *Areas 20, 22 and (left edge) part of 13 after full excavation of the Neolithic soil profile. The early ditch runs away from bottom left, terminating just across the baulk in Area 13 (1989; viewed from SE). The prominent oval pit crossing it is later. In the background section RH14 may be seen (Area 22).*

4   *Area 32, section RH17 (1989) after sample column 59 had been taken (cf. Fig. 2.13). The shelf to the right of the column is the Neolithic land surface, which was overlain by gravel 2F (visible in foot of section). The palaeobank and overlying spill of gravel dip away to the left of the column. The concrete block at the top is near the base of a modern pit.*

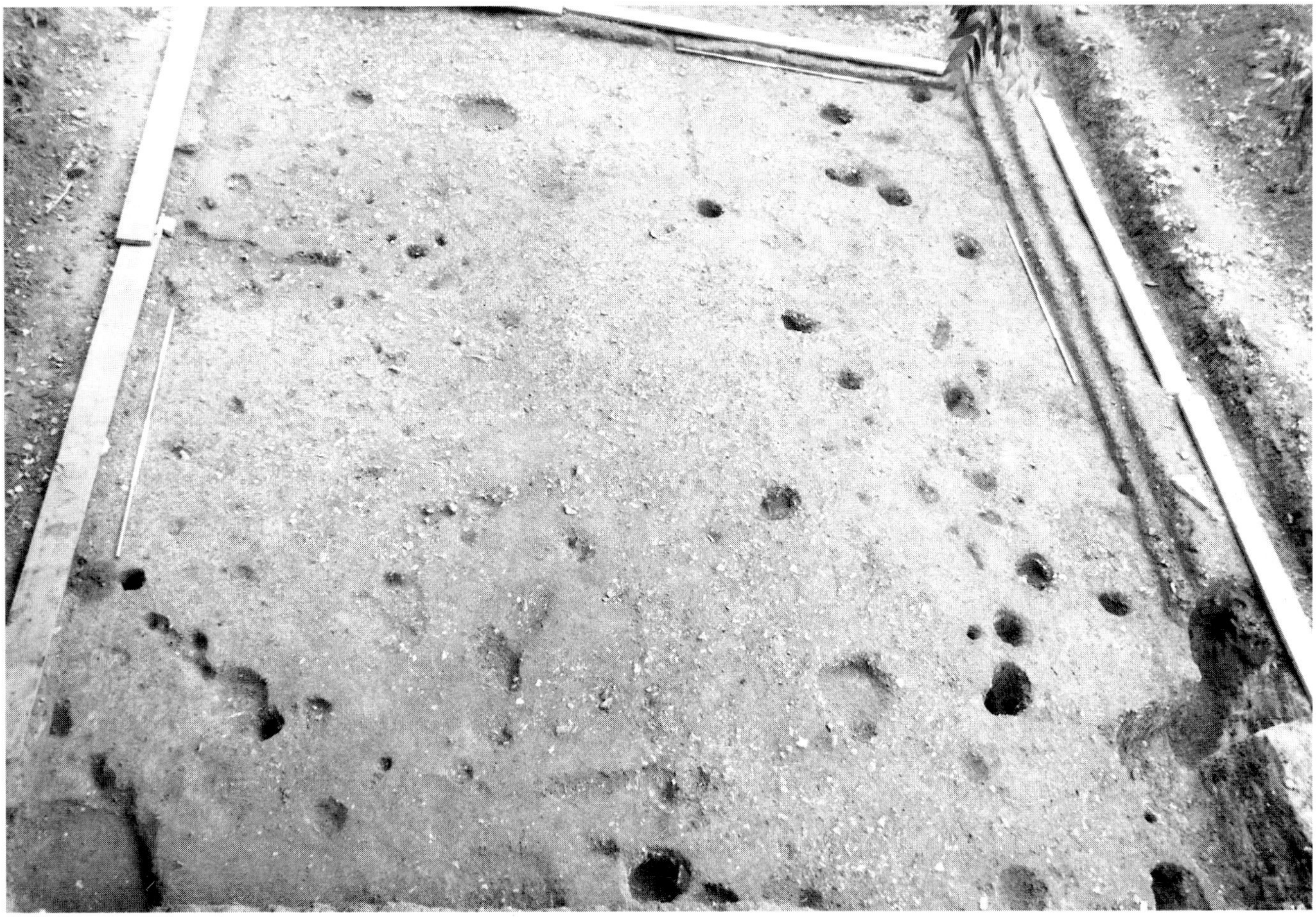

5   *Surface of the gravel bed (parcel 2F) exposed over most of Area 19 on the high spine (1986; from N). Excavated features are Late Bronze Age (most) and modern.*

6 *Area 14, section RH30 (1984). The parcel 2D/2E erosion line is centre bottom, amid dark silts. Paler overlying silts belong to parcels 2B and 4B, with grey Late Bronze Age soil and overlying 6D silt at the top (cf. Fig. 2.24).*

7   *Area 14, cluster of beaver-gnawed branches stratified in parcel 2E (1984; Context 14.121; from SW).*

8   *Area 14, sections RH22–24 (1985; from E). The foreground shows an excavated part of the palaeobank at the parcel 2B/4B interface, also visible in the section just behind (RH24). To the right of the ranging pole is the Post-Medieval cut bank, abutted by parcel 9 silts. Further back the trench shows the surface into which Late Bronze Age features penetrated, having been covered by dark earth deposits visible in the section behind (RH23). These are overlain by pale silts (parcel 6D) and more recent rubble (above walkway; RH22) (cf. Figs 2.17–2.19).*

9 *Area 23, section RH43 (1987, from NE). Gravel bar and overlying sands of parcel 5, cut by erosion line (to left) and overlain by parcel 6A and 6B silts (cf. Fig. 2.28). The sloping cut at the right-hand side is the wall of the current M25 floodway.*

10 *Post-Medieval cut bank running obliquely across north corner of Area 14 (1984, from E; prior to extension trench). Late Bronze Age dark earth and features can be seen in the cut face and are yet to be excavated across the trench. Excavated features are Industrial Age. The Post-Medieval cut revealed the top of the underlying gravel bar of parcel 4B (cf. Fig. 2.21).*

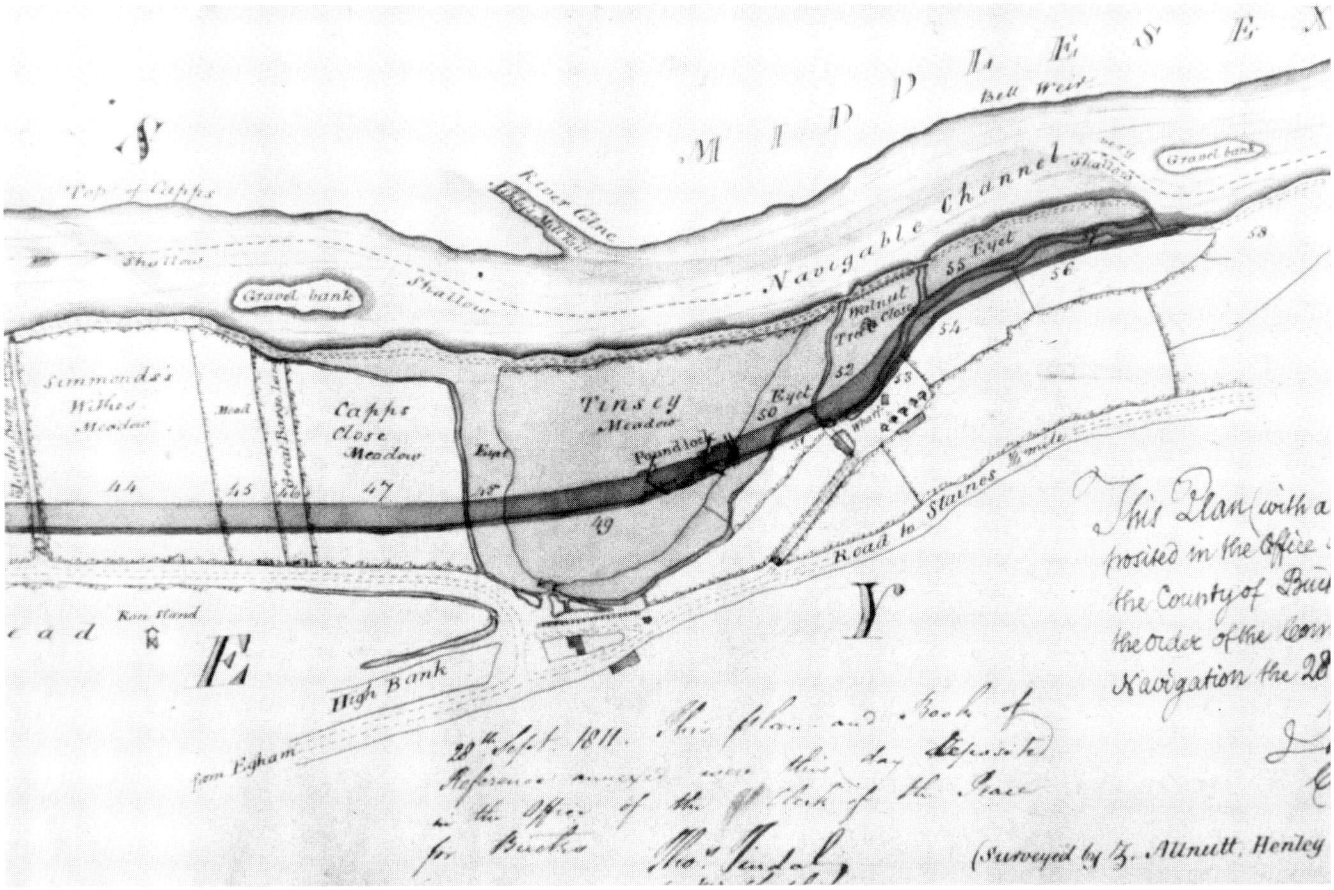

11 *Z. Allnutt's 1811 plan of the proposed new cut at Runnymede. The cut would have bisected the Medieval/Post-Medieval 'island' of* Tyngeyt *(Tinsey meadow) and thus the Late Bronze Age site. In the event the plan was rejected in favour of improving the extant river channel. (Courtesy of Buckinghamshire Record Office, Aylesbury).*

12 *A pre-1903 postcard showing Bell Weir lock and moored craft, from the foreshore alongside the site (close to Area 15). The lock is that rebuilt in 1867.*

13   *The Angler's Hotel and the western flank of the site, photographed during a flood in 1915. The damaged fence line follows the western arm of the by now defunct Southern Loop, which then swings to front right.*

14   *View from SW across Tinsey Mead in more clement conditions sometime around 1915. The boathouse stood close to Area 21, while the site of the building on the far right lies under the A30/M25 embankment.*

highly structured environments of sedge tussocks in permanent standing water and of floodplains where fully aquatic and fully land faunas may alternate within a short time. Additionally, aquatic shells are usually washed onto the floodplain during flooding, although it may be possible to identify such assemblages partly by the low abundance of aquatics and partly by the paucity of shells which do not float, notably *Pisidium* valves (e.g., contrast the channel with the overbank assemblages in columns 5 and 36). Often in assessing the site environment, the nature of the sediments and soils, and the topographical and archaeological contexts, rather than the molluscan assemblages themselves, are used.

The 'aquatic' environment refers to the river channel, unless otherwise stated, even if the site environment is overbank alluvium. Assessment of the aquatic environment from overbank alluvial assemblages derived from the river channel is difficult because they are often sparse and thus lacking in low-abundance species. They also often lack *Pisidium* valves because they are not buoyant enough to be transported far from the channel.

Assessment of the 'land' environment from channel deposits is equally difficult because it is never certain how many different habitats, and of what scales, are represented. It can thus only be done in general terms.

In order to define the land and wetland habitat changes more clearly, a diagram has been constructed (Fig. 7.25) in which the wetland and land groups are plotted in four groups – wetland (groups 6–8), structure-loving species (9), catholic species (10) and open-country species (11) – each as a percentage of the total wetland and land species (groups 6–11).

Only those assemblages which had sufficient numbers of shells for percentage calculations were used (usually more than 150 and only three below 50). Only a selection of the post-LBA assemblages was used. In the interpretation of the land environment on the right side of the diagram, the first column is for vegetation, the second for soil moisture, and the third for flooding.

In the vegetation column, the ascription of an assemblage to woodland or open land is based on an assessment of the relative abundance of group 9 to group 11 species respectively. Although no specific percentage values have been used, the situation is usually fairly clear-cut as to which of the two groups predominates. Note should be made of the likelihood of faunal mixing in soils and slowly accumulating sediments, since assemblages in these contexts may originally have been better defined in terms of the relative proportions of different ecological groups than they are now.

'Open' refers to land in which non-woody plants predominate – i.e., vegetation which shows a considerable reduction in height during winter. 'Woodland' implies more or less closed woody vegetation. 'Open woodland' implies woody vegetation with gaps for herbs.

In the soil moisture column, 'wetland' is usually indicated where groups 6–8 reach over 15%, although in comparison with dry chalkland conditions, practically all the situations here are wetland. 'Wetland' refers to land where there is virtually 100% relative humidity at the land surface and among the non-woody vegetation for most of the time of the life-cycle of the relevant mollusc species. In some cases this may just be for a part of the year. The species of groups 6–8 are a clear indicator of such conditions.

Within the major blocks of dry land and wetland there are some minor trends which are not indicated, e.g., drying in column 4, Sn 41–42, followed by wetting in column 31, Sn 29. In some cases these may encompass temporally overlapping sediments.

'Flooding' is indicated in clear overbank alluvial contexts (as shown by topographical and sedimentary properties) by the presence of fully aquatic shells. Up to 3% is usually considered as no flooding or very little flooding.

Results

Parcel 1C: early Boreal, marl
Column 31, Sn 31–39 (Figs 7.1 and 7.2)

Site: intimate association of aquatic and land environments (fine sediment and its sheet-like, not dipping or lensed, disposition are not indicative of a river channel; absence of *Theodoxus* and river species of group 2; high numbers of aquatic and land shells, but very low diversity of latter).

The aquatic part of the environment was probably permanent (high numbers of aquatic shells), although drying cannot be ruled out; it was flowing (high numbers of aquatic and land shells, *Ancylus*; allochthonous alluvial material), but with gentle sedimentation (absence of *Theodoxus* – but see above for possible chronological reason for this – and absence of group 2 river species; in situ precipitation of calcite); mudflats/bare ground (*L. truncatula*) and pools/swamp (group 4, especially *V. cristata*, *A. leucostoma* and *P. personatum*). A lake edge with some wave action is a possibility, but lake marl assemblages usually lack land species totally.

The land part was wetland (very high numbers of wetland groups 6–8, very low numbers of all other land groups), treeless, with grass/sedge tussocks (low diversity land fauna compared to all later ones, lacking group 9 species and *Vitrina*, *Nesovitrea* and *Trichia* in group 10).

All components of the fauna – aquatic and land groups – and the sediment are distinctive by comparison with all later ones. Gently flowing water, with emergent vegetation, e.g., reeds ('reed swamp') and sedge tussocks ('sedge fen'), and some mudflat are the most likely overall conditions which would satisfy all the data, becoming more swampy with less open water towards the top (increase in group 4). The pollen from the base of the marl (Sn 39 and below) indicates a valley-bottom environment of freshwater and reed swamp (Chapter 9). The sediments are in agreement with this complex origin.

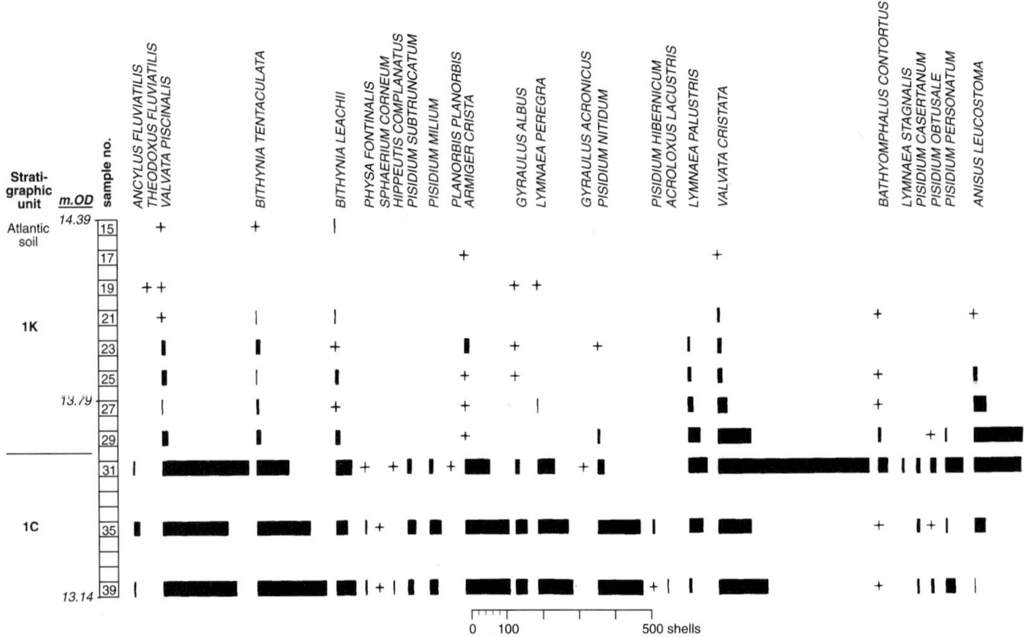

7.1 *Column 31, Area 19, aquatic molluscs.*

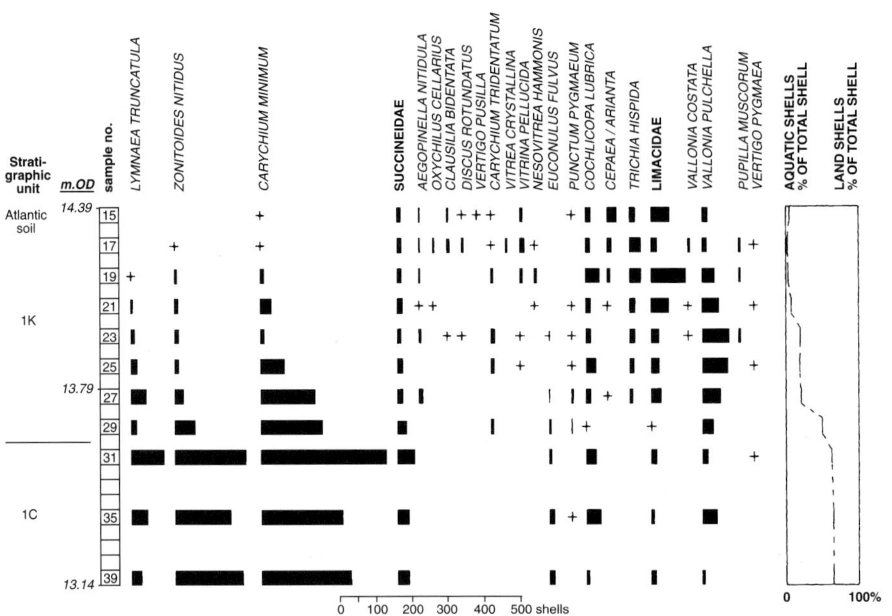

7.2 *Column 31, Area 19, land molluscs.*

Aquatic and land: as for site. (Under 'site' to emphasise close spatial association.)

Noteworthy records: the records for the two *Bithynia* spp. are useful, both being firmly established by the very start of the sequence, at almost 8000 BC (9000 BP). *B. tentaculata* is recorded sporadically from the Devensian Lateglacial – Bingley Bog (Keen *et al.* 1988), Skipsea Withow (Gilbertson 1984) and Nazeing (Allison *et al.* 1952) (although some of these records are probably dubious; see also Holyoak 1982) –

but is not firmly established until the early Holocene – pollen zone IV (pre-Boreal) at Bingley Bog, pollen zone IV/V at Hawes Water (Sparks 1962) and 9600 BP at Llangorse (Walker *et al.* 1993), in the last, as at Runnymede, prior to the *Corylus* rise. For *B. leachii* there appear to be no records prior to that of 8360 ± 100 BP at Staines (Preece and Robinson 1982), pollen zone VIb/c in Berkshire (Holyoak 1983), and zone VI at Apethorpe (Sparks and Lambert 1961). As already discussed, the absence of *Theodoxus* at this early stage may be of regional significance, although the

130

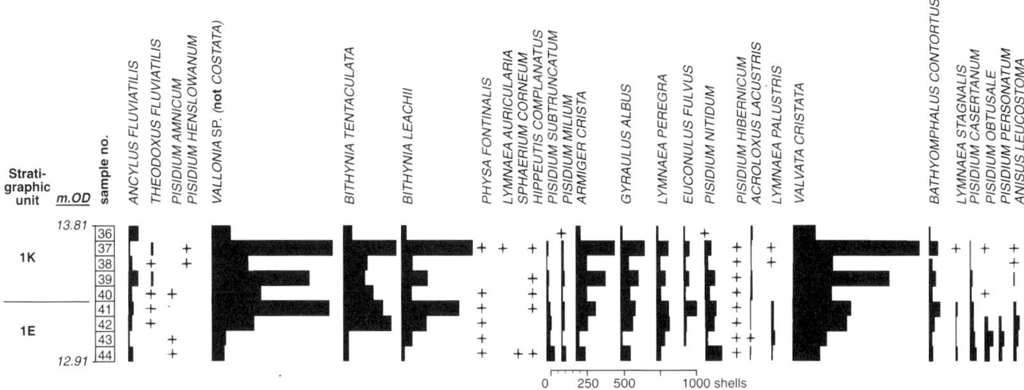

7.3   *Column 4, Area 13, aquatic molluscs.*

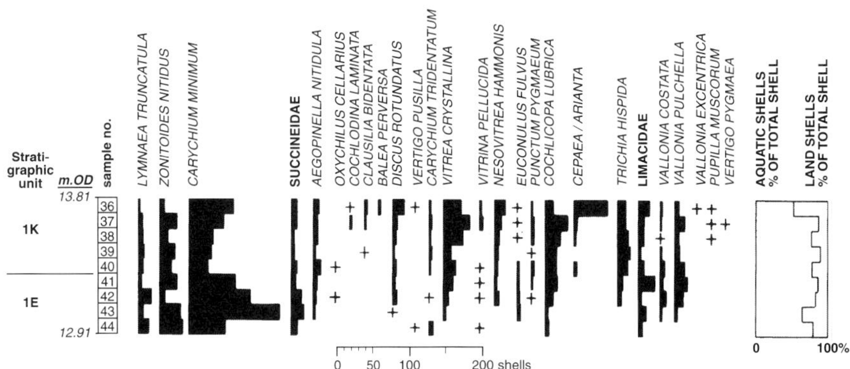

7.4   *Column 4, Area 13, land molluscs.*

species tends to favour more stony, uneven substrata than presented by the marl. *Planorbis planorbis* is another species of interest whose absence until the very top of parcel 1C may be of regional significance, with records from only Nazeing for the Devensian Lateglacial zone III (Allison *et al.* 1952), and then the early Boreal at Nazeing, zone VI at Apethorpe (Sparks 1961), the *Corylus* rise at Llangorse (Walker *et al.* 1993) and zone VIIa in Berkshire (Holyoak 1983).

For the land fauna, although sparse, there is nothing that is out of place with an early Boreal date, and indeed, all the species were present in southern Britain by the end of the Devensian Lateglacial period.

Parcel 1E: late Boreal–early Atlantic, channel fill

Column 4, Sn 43–44 (late Boreal); column 4, Sn 41–42 (early Atlantic) (Figs 7.3 and 7.4)

Site: river channel, permanently aquatic (relatively coarse sediments, presumed to be occupying a cut through the parcel 1C deposits; aquatic shells predominate); fast flowing, well aerated, richly vegetated and with a diversity of habitats; swamp/pools (group 4) as in the upper part of parcel 1C (Fig. 7.1). There was an increase in the influence

of the river and a decrease in swamp/pools across the Boreal/Atlantic boundary.

Aquatic: as site.

Land: wetland, open, but more diverse than before; becoming drier across the late Boreal/early Atlantic divide (decrease in groups 6–8, increase in dry land species (groups 9–11).

Noteworthy records: *Theodoxus* appears for the first time in the early Atlantic deposits. In the land fauna, relevant age indicators are *Discus rotundatus* in the late Boreal and *Oxychilus cellarius* in the early Atlantic (Kerney *et al.* 1980).

Parcel 1K: early Atlantic, fine alluvium

Column 31, Sn 21–29 (Figs 7.1 and 7.2)

Site: floodplain (very fine sediments – much finer than the marl of parcel 1C; sheet-like disposition; seasonally subaerial – soil evidence, Chapter 5; predominance of land over aquatic shells); more or less treeless; initially (Sn 27 and 29) with swamp and/or pools (group 4 species, especially *L. palustris*, *Anisus* and *V. cristata*) and wetland, possibly 'sedge fen' (groups 6–8), becoming drier (decrease in groups

131

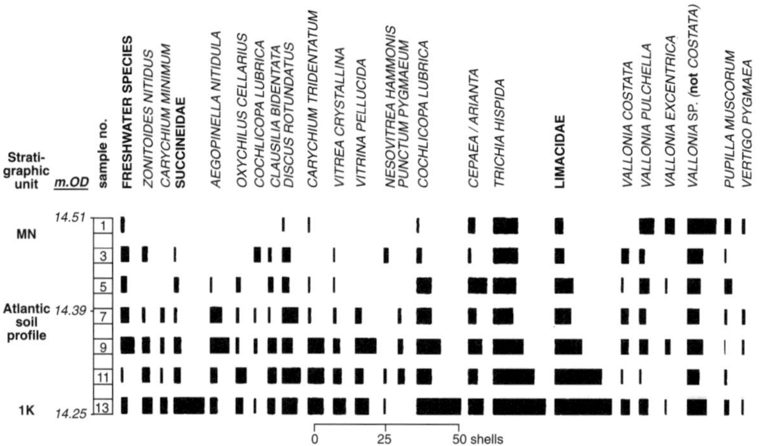

7.5 *Column 13b, Area 19, total molluscs.*

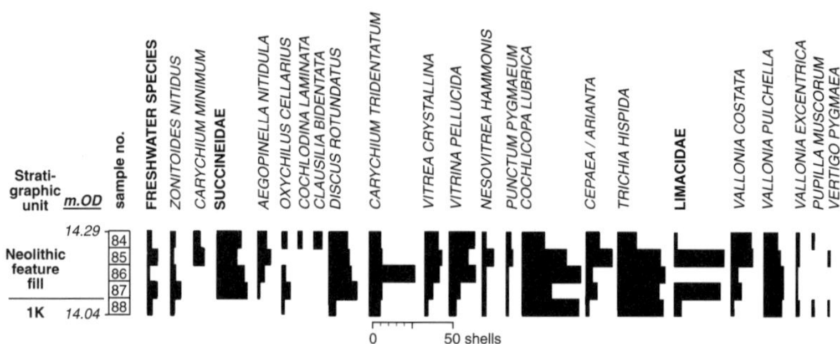

7.6 *Column 41, Area 24, total molluscs.*

4–8) but still flooded, and ultimately grassland (*Vallonia pulchella*). The association of groups 6–8 with *Cochlicopa*, *Trichia*, Limacidae and *V. pulchella* is typical of wet grasslands on floodplains (Evans 1991b).

Aquatic: river, as in parcel 1E, but with an overall lower habitat diversity (no group 1 and 2 species, and fewer group 3, although this could be a reflection of the small sample) and less standing water (declining swamp/pools, group 4, element).

Column 4, Sn 36–40 (Figs 7.3 and 7.4)

Site: gradual transition from channel to overbank flooding (*Pisidium* valves abundant, except in Sn 36). Floodplain (sediment type and disposition; the shells alone would indicate a channel aquatic environment since there is a much higher aquatic component than in any other overbank alluvium); much more frequently flooded than column 31, Sn 21–29 (higher proportion of aquatics), probably because of the lower altitude (by 0.3 m) and different location (slope rather than levee); wet (groups 6–8) woodland (group 9, including three species of tree snail), but some open ground.

Aquatic: as in parcel 1E, and substantially more diverse than in parcel 1K, column 31 (perhaps also related to the different locations); little or no standing water (low swamp/pools, group 4, element, except *V. cristata*).

Parcel 1G: Atlantic, channel fill
Column 40, 0–5, 10–15 cm (Table 7.1)

Site: river channel (sediment type and disposition; aquatic shells, 94–95%); environment as before.

Land: wet woodland, some open ground.

Late Mesolithic–Early Neolithic, soil and features on high bank (mid- to late Atlantic)
Column 13b, Sn 5–13 (Fig. 7.5); column 31, Sn 15–19 (Figs 7.1 and 7.2); column 41, Sn 84–87 (Fig. 7.6); (column 41, Sn 88 is not considered because its context is unclear)

Site: floodplain, although flooding slight to absent; damp open woodland. The autochthonous nature of these deposits (very low aquatics; soil features) makes this phase of open woodland convincing; open land species at 7 to 16% do not show a downward fall-off in abundance, so they are unlikely to be derived from higher deposits by faunal mixing.

Parcel 1H: Early Neolithic, channel fill to overbank alluvium
Column 59, Sn 62–66 (channel fill), Sn 54–60 (overbank alluvium) (Figs 7.7 and 7.8)

Table 7.1 *Numbers of snails from two samples in column 40, Area 22.*

Species	depth: (m OD)	
	12.97-13.02	12.87-12.92
Theodoxus fluviatilis	22	28
Valvata cristata	1215	1004
Valvata piscinalis	2511	1664
Bithynia tentaculata	1425	976
Bithynia leachii	813	1008
Carychium minimum	152	74
Carychium tridentatum	0	8
Lymnaea truncatula	9	4
Lymnaea palustris	3	4
Lymnaea stagnalis	5	3
Lymnaea peregra	307	142
Anisus leucostoma	2	14
Bathyomphalus contortus	279	162
Gyraulus acronicus	285	252
Gryaulus albus	183	144
Armiger crista	260	178
Hippeutis complanatus	4	6
Ancylus fluviatilis	147	66
Acroloxus lacustris	47	12
Succineidae	26	32
Cochlicopa lubrica	95	32
Vertigo pygmaea	0	1
Vallonia costata	14	6
Vallonia pulchella	36	20
Punctum pygmaeum	2	0
Discus rotundatus	20	14
Vitrina pellucida	10	0
Vitrea cystallina	31	14
Nesovitrea hammonis	12	6
Aegopinella nitidula	18	2
Zonitoides nitidus	69	74
Euconulus fulvus	0	2
Clausilia bidentata	3	2
Trichia hispida	42	16
Cepaea/Arianta	7	8
Sphaerium corneum	4	3
Pisidium casertanum	72	94
Pisidium milium	63	31
Pisidium subtruncatum	135	67
Pisidium hibernicum	20	9
Pisidium nitidum	212	173
Total land shell	546	312

Site: sequence from river channel (Sn 62–66), through damp, open floodplain (Sn 58–60), frequently flooded, to damp grassland (Sn 54–56); occasionally flooded. There is no swamp/pool component, so the transition from channel to floodplain was sharp (and by implication rapid).

Aquatic: fast-flowing, well-oxygenated river; environment as before. No significant difference, apart from overall abundance, between the river environment as represented in the channel deposits and that in the later floodplain deposits.

Land: open throughout (group 9 species sparse), and probably grassland (*V. pulchella* abundant, especially in Sn 54–60).

Middle Neolithic, occupation soil
Column 13b, Sn 1–3 (high bank) (Fig. 7.5)

Site: floodplain, hardly flooded, and more or less dry; some woody vegetation in Sn 3; Sn 1, grassland.

Column 10, Sn 19 (damp flanks; note that this sample was partly from deposits below the MN occupation) (Fig. 7.13)

Site: river channel (mainly aquatics).

Column 12, Sn 20–21 (damp flanks)

Site: Sn 21, too few shells for interpretation; Sn 20, grassland (but note that this sample is partly from the overlying alluvium, context 16.932).

Parcels 2D and 2E: Late Neolithic, channel fills
Column 5, Sn 71–76 (2E), 77–85 (2D) (Figs 7.9 and 7.10); column 6, Sn 89–97 (2E) (Fig. 7.11); column 60, Sn 48–51 (2D) (Fig. 7.12)

Site: fast-flowing river, as before; little or no swamp. No significant difference between parcels 2D and 2E. Mud flats in 2E.

Land: wetland, open, with sparse woody vegetation; perhaps less open (fewer group 11 species) than the Early Neolithic floodplain of parcel 1H (column 59, Sn 54–60). 2E shows more land influences than 2D.

Parcel 2B: Early–Middle Bronze Age, overbank alluvium
Low-lying area
Column 5, Sn 56–69 (Figs 7.9 and 7.10)

Site: floodplain, with flooding frequent in the lowest part (Sn 69), slight to none in the upper part (Sn 59–67); very damp woodland with a little open land (Succineidae and *Vallonia* throughout); possible pasture (Sn 65–69) (high *L. truncatula*) on the Robinson (1988) model, but more likely to be mudflat in view of the generally shaded vegetation.

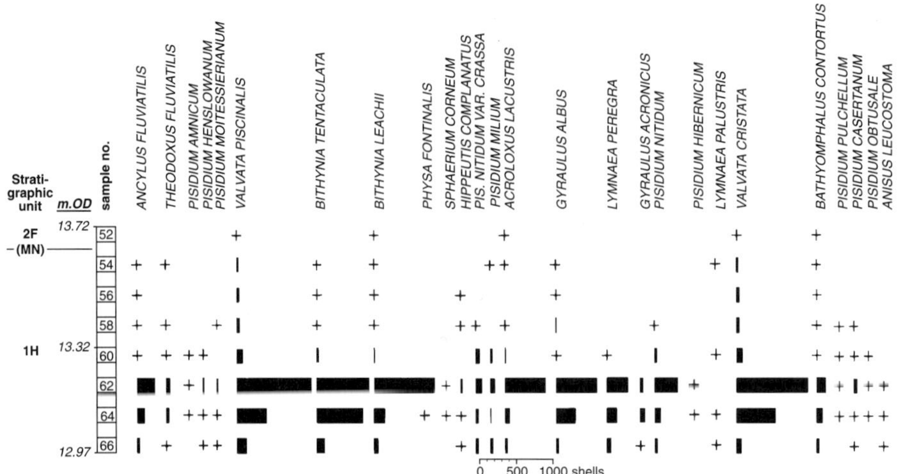

7.7   *Column 59, Area 32, aquatic molluscs.*

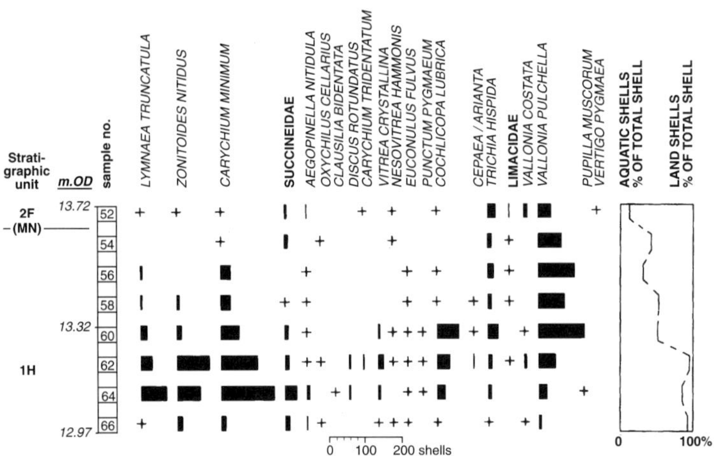

7.8   *Column 59, Area 32, land molluscs.*

Aquatic: river environment possibly impoverished by comparison with that of the Late Neolithic (parcel 2D and 2E), but this could be due to the small sample size of the floodplain assemblage.

Column 60, Sn 40–44 (Fig. 7.12)

Site: floodplain, with flooding decreasing to insignificance at the top; very damp, open grassland (high *Vallonia pulchella*), possibly pasture (high *L. truncatula*) on the Robinson (1988) model.

Aquatic: possible mudflats, Sn 44 (*L. truncatula*), as an alternative to the pasture interpretation.

HIGHER GROUND
Column 10, Sn 9–14 (Fig. 7.13)

Site: floodplain, slight flooding in the lower part to none in the upper; vegetational sequence from open land (Sn 14–12), through a clear grassland phase (Sn 11), to woodland

(Sn 9–10). This is a very distinctive sequence, in the faunal changes (drop in *V. pulchella* and Succineidae, rise in the structure-loving species of group 9), in the autochthonous nature of the material, and in the location at the top of a layer just prior to a sharp change to occupation deposits.

Parcel 4B: Middle–Late Bronze Age, channel fill

Gravel bar, column 8, Sn 63–65 (Figs 7.14 and 7.15); lensed channel silts, inlet, and beaver evidence, column 55, Sn 1–8 (Figs 7.16 and 7.17); higher location than previous two sequences, coarse sand lenses, column 8, Sn 53–61 (Figs 7.14 and 7.15); column 36, Sn 77–83 (Figs 7.18 and 7.19)

Site: fast-flowing river (rich fauna of aquatics; low numbers in column 55, Sn 4 due to shingle lens).

Aquatic: as site; some mudflat in column 8, Sn 63–65, column 55, Sn 1–8, and column 36, Sn 77–83 (*L. truncatula* and

134

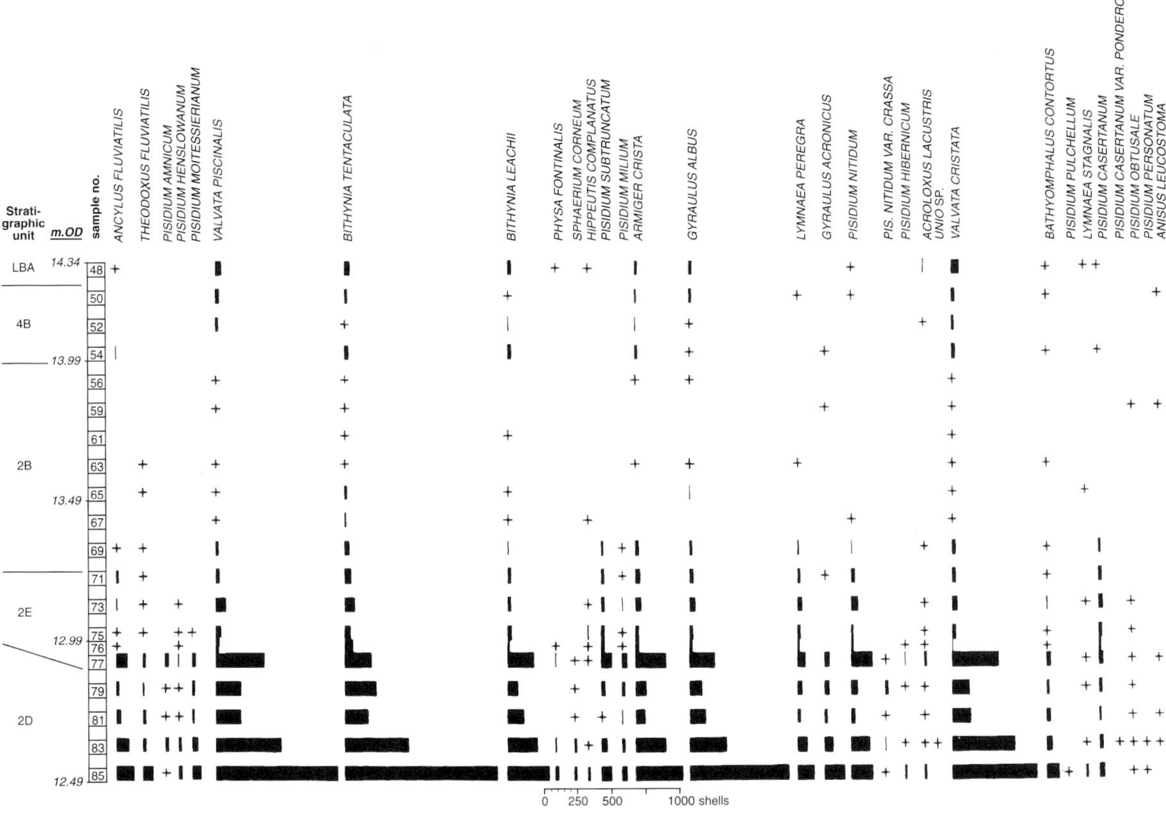

7.9 *Column 5, Area 14, aquatic molluscs.*

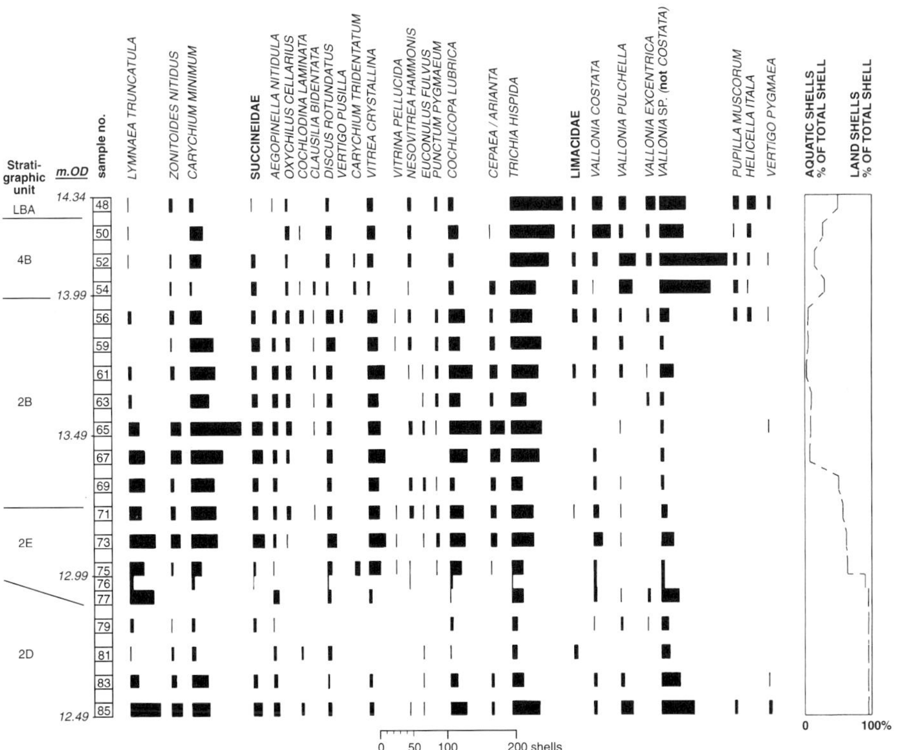

7.10 *Column 5, Area 14, land molluscs.*

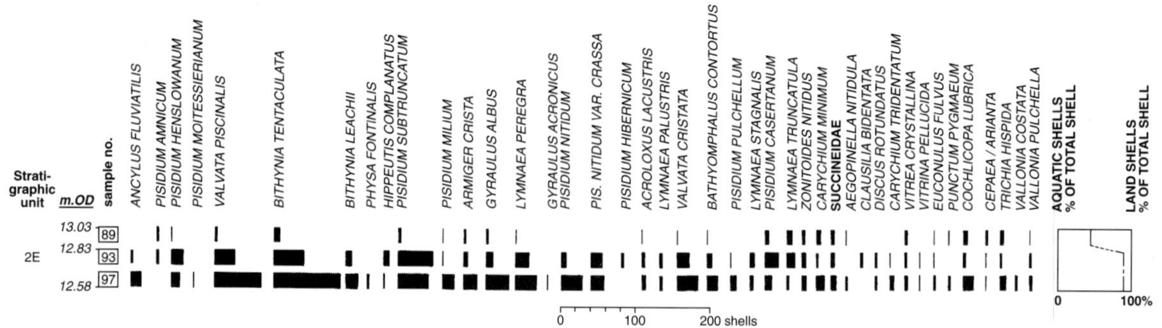

7.11 *Column 6, Area 14, total molluscs.*

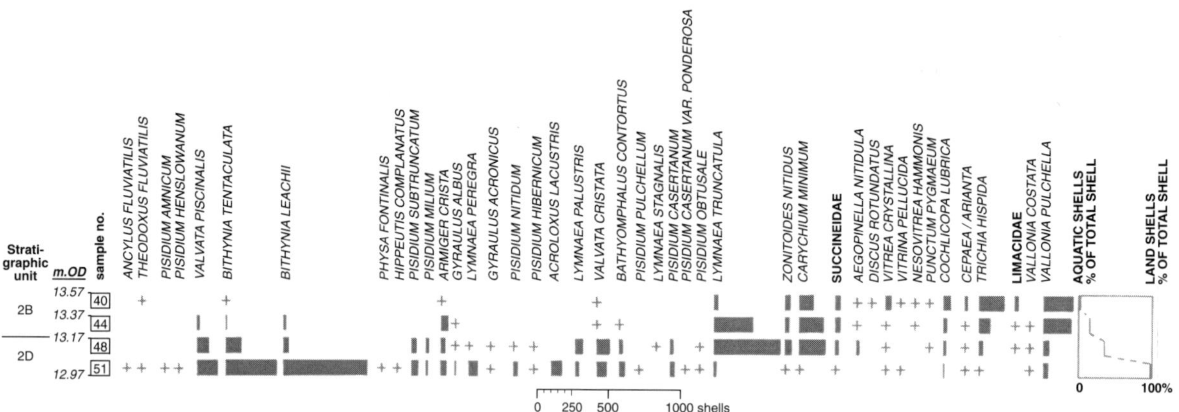

7.12 *Column 60, Area 32, total molluscs.*

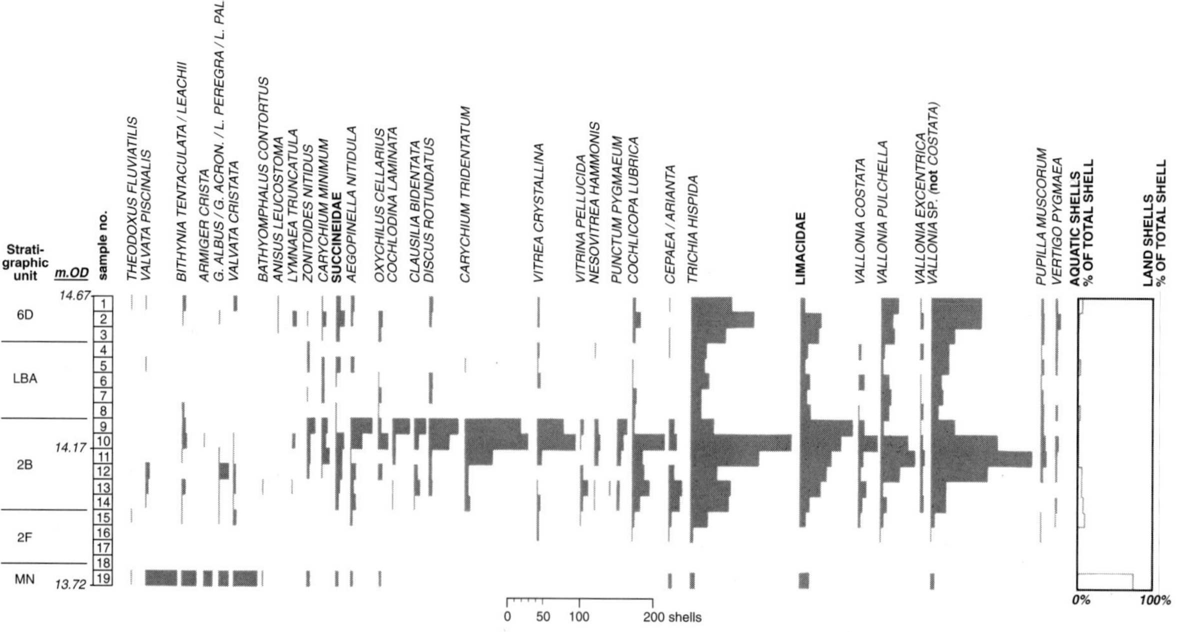

7.13 *Column 10, Area 16, total molluscs.*

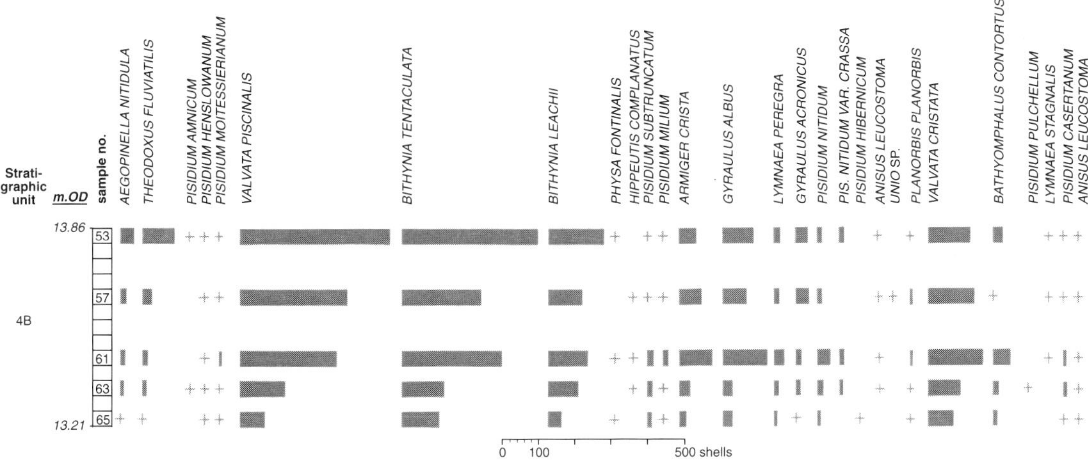

7.14   *Column 8, Area 14, aquatic molluscs.*

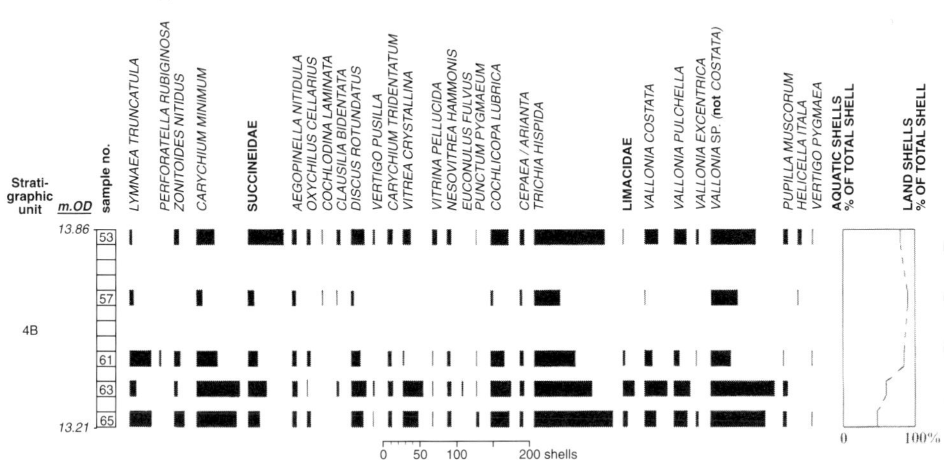

7.15   *Column 8, Area 14, land molluscs.*

P. rubiginosa); swamp in column 55, Sn 1–8, and column 36, Sn 77–83.

Land: wetland; open land with some woody vegetation.

Parcel 4B: Middle–Late Bronze Age, overbank alluvium
Column 5, Sn 50–54 (Figs 7.9 and 7.10)

Site: floodplain with frequent flooding (relatively high numbers of aquatics); open land, marsh and grassland, relatively dry (*H. itala*, *P. muscorum*, *V. pygmaea*).

Aquatic: less diversity by comparison with earlier river environments, with a loss of plant-rich habitats, and more slowly flowing water (reduction in molluscan diversity, especially some phytophilic and fast-flowing-water species although, as in previous cases, this may be an artefact of the low numbers of shells).

Late Bronze Age, occupation deposits and soil

FLANKS/LEVEE
Column 10, Sn 3–8 (Fig. 7.13); column 5, Sn 48 (Figs 7.9 and 7.10)

Site: floodplain with practically no flooding in column 10, but flooding in column 5; relatively dry, open land, probably grassland. Low numbers of land shells in column 10 may be due to in situ LBA occupation.

Aquatic: same comments apply here as for MBA overbank alluvium, 4B, with low molluscan diversity being a function of reduced river diversity and/or low numbers of shells.

TROUGH
Column 36, Sn 71–75 (Figs 7.18 and 7.19); column 11, Sn 24–28 (Fig. 7.20)

137

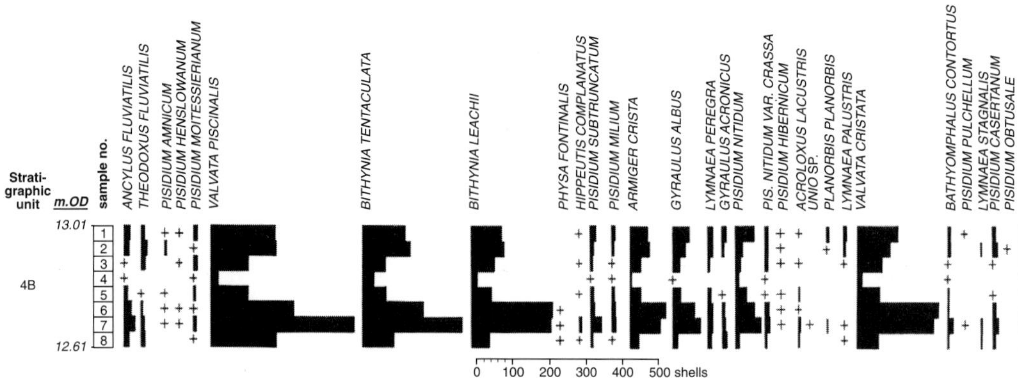

7.16 *Column 55, Area 31, aquatic molluscs.*

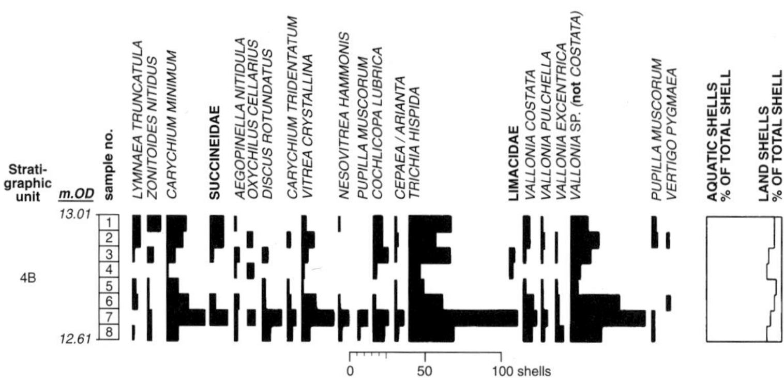

7.17 *Column 55, Area 31, land molluscs.*

Site: floodplain with overbank flooding, reduced in column 36, Sn 71; relatively dry open land, probably grassland. Low numbers of land shells in column 36 may be due to in situ Bronze Age occupation; relatively high numbers of land shells in column 11 suggests weaker influence of LBA occupation in this area.

Aquatic: as for flanks/levee.

Parcels 5, 6 and 7: post–Late Bronze Age, channel fills to overbank alluvium

CHANNEL FILLS
Column 52, Sn 10–18 (parcel 7) (Figs 7.21 and 7.22); column 27, Sn 5–7 (parcel 5); column 28, Sn 9–11 (parcels 5–6A); column 29, Sn 17–19 (parcel 6B) (Figs 7.23 and 7.24)

Site: fast-flowing, well-oxygenated river.

Land: wetland, probably open grassland, marsh; some bare ground and/or mudflats (*L. truncatula* and *P. rubiginosa*).

OVERBANK ALLUVIUM
Column 10, Sn 1–2 (parcel 6D) (Fig. 7.13); column 27, Sn 1–3 (parcel 6A) (Figs 7.23 and 7.24); column 29, Sn 13–15 (parcel 6B) (Figs 7.23 and 7.24); column 52, Sn 2–6 (parcel 6C) (Figs 7.21 and 7.22)

Site: floodplain with overbank flooding; bare ground and/or mudflats; open land, relatively dry, probably marsh.

Aquatic: as previous comments for overbank alluvium contexts, e.g., Late Bronze Age occupation deposits and soil.

Discussion

JOHN EVANS

The past environmental significance of the molluscan data cannot be considered fully out of the context of the other data. In the preceding descriptions, the data of localised contexts such as topography, sediment and soil disposition and properties, and other biological groups have been used. However, in addition, the spatio-temporal dynamics of the river channels as well as wider aspects of the valley bottom environment – such as sea-level change and its effect on ponding, flooding and alluviation, climate and atmospheric properties and their effect on hydrology and the precipitation of calcium carbonate – need to be taken into account. Major questions such as 'Why is there a major change from calcium carbonate precipitation to clastic sedimentation in the eighth millennium BC?' and 'Why did the channel shift 55 m ENE during five millennia?' need to be addressed.

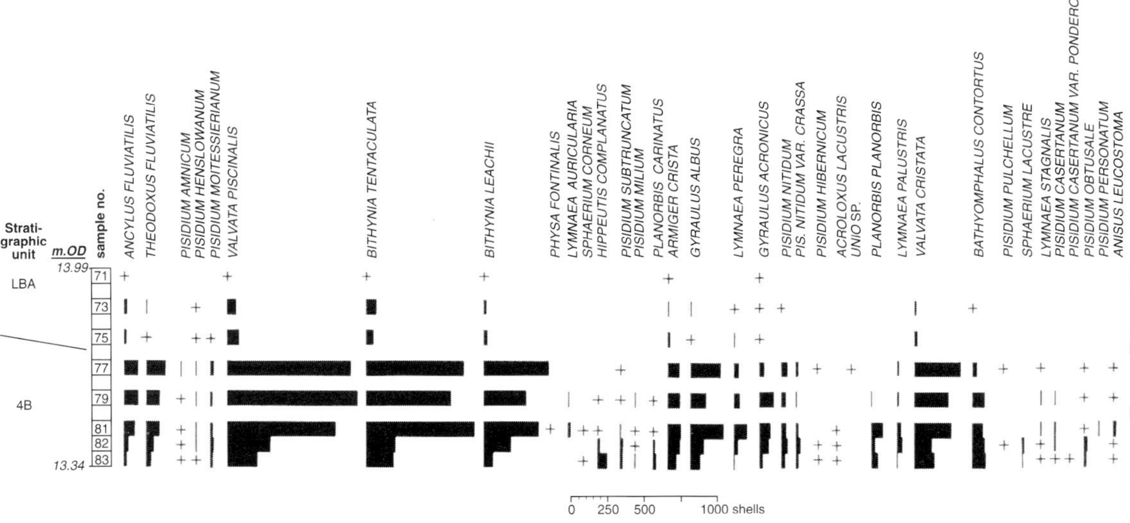

7.18 *Column 36, Area 21, aquatic molluscs.*

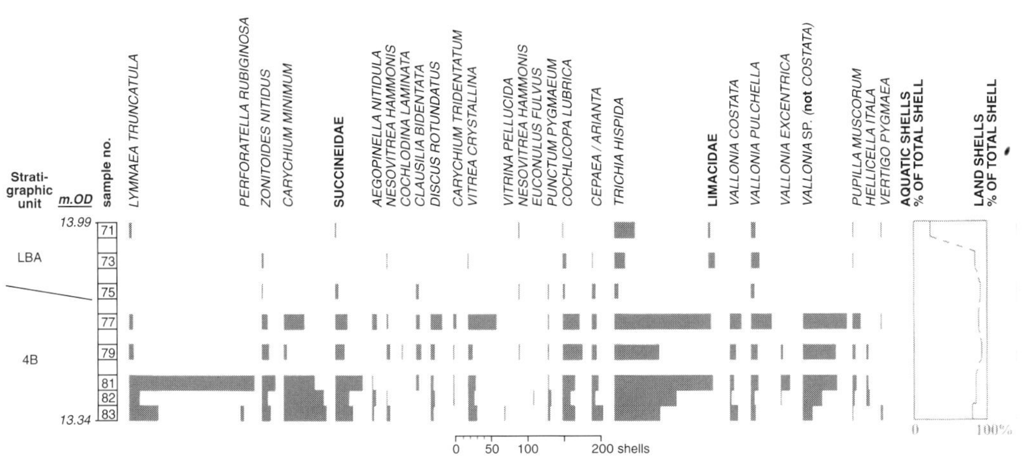

7.19 *Column 36, Area 21, land molluscs.*

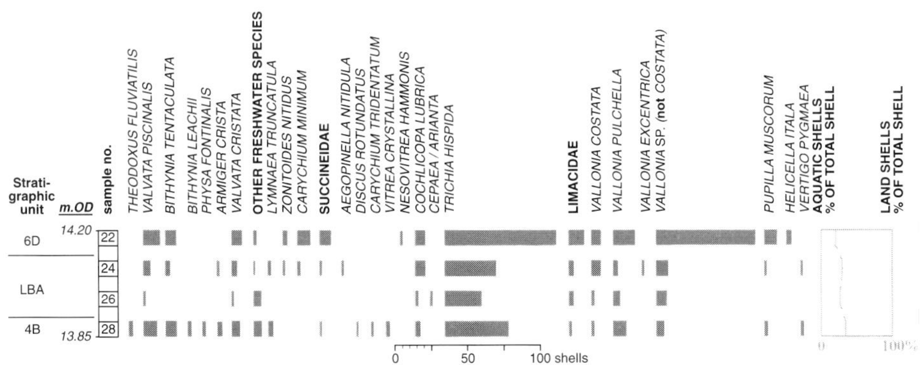

7.20 *Column 11, Area 18, total molluscs.*

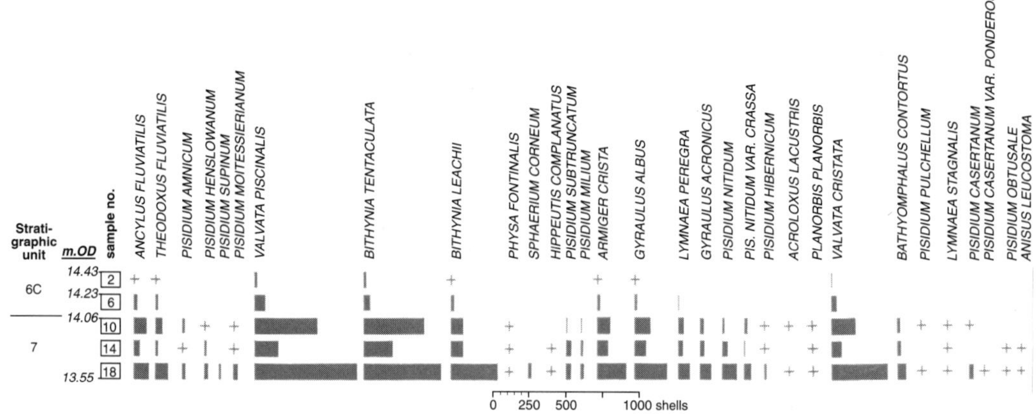

7.21   *Column 52, Area 27, aquatic molluscs.*

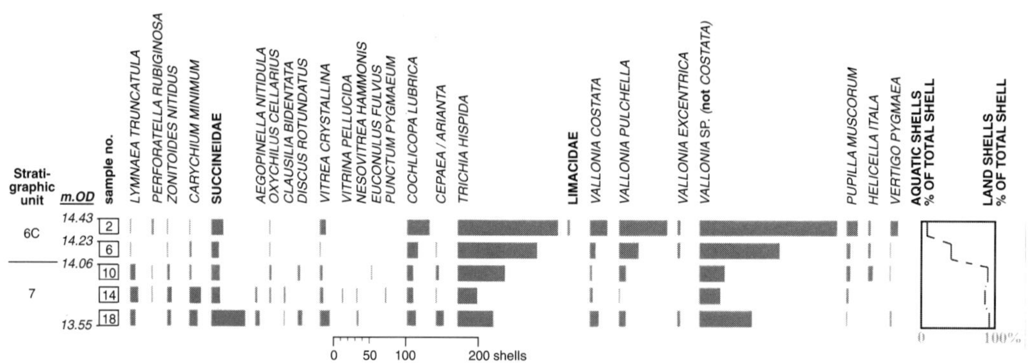

7.22   *Column 52, Area 27, land molluscs.*

The nature of the valley bottom is relevant to the interpretation of the Mollusca. As Robinson says (in Thomas *et al.* 1986), the present Thames is managed in the control of its levels by locks and weirs, by the control of its channel by revetments, and by the management of the land on either side for pasture, meadow and arable. In the recent (historical) past there was probably management of various other types of land such as fen, woodland and reed beds. In prehistory, however, although some of these practices may have been started – as seen at Runnymede itself in the LBA waterfront – there was probably a much greater diversity of land and water types, the divisions between which were less clear, and subject to much greater change, as in the lateral movement of the channel through time. What, for example, was the nature of the river channel? Did it have sharp, vertical edges, with a more or less abrupt change with land? Or were there one or more wetland or marginal aquatic habitats such as swamp, sand and gravel banks and mudflat?

This is discussed in Chapter 10, but the iterative and cyclical nature of interpretation require that some aspects be at least alluded to in discussion of the molluscan data.

Aquatic environments

The general nature of the river valley and its channel is given by the proportions of aquatic to land shells in the channel deposits (Sparks and West 1970). There was a clearly defined valley, of moderate gradient, and with a fast-flowing river. Slow-flowing rivers with practically no gradient and ill-defined valley sides, such as occur in the East Anglian fenland, have a very high proportion of freshwater to land shells and, among the land shells, a very high proportion of wetland species. This is not the case here.

The aquatic environment of the channel sediments was mostly river channel. The relatively coarse nature of the sediments (generally coarser than clays and fine silts), the high diversity of the fauna, and the presence of species characteristic of running water and rivers (groups 1 and 2) indicate flowing, highly aerated water, richly vegetated conditions and a diversity of habitats. The insect evidence is in general agreement with this.

A similar molluscan fauna, of the middle Holocene, is known from channel deposits near Staines (Preece and Robinson 1982), and others from later Holocene deposits

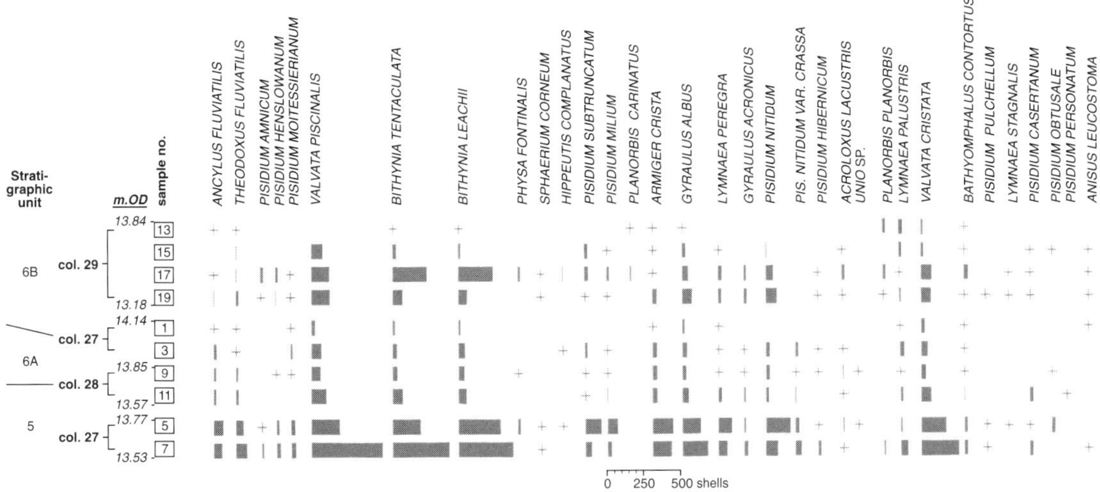

7.23 *Columns 27–29, Area 23, aquatic molluscs.*

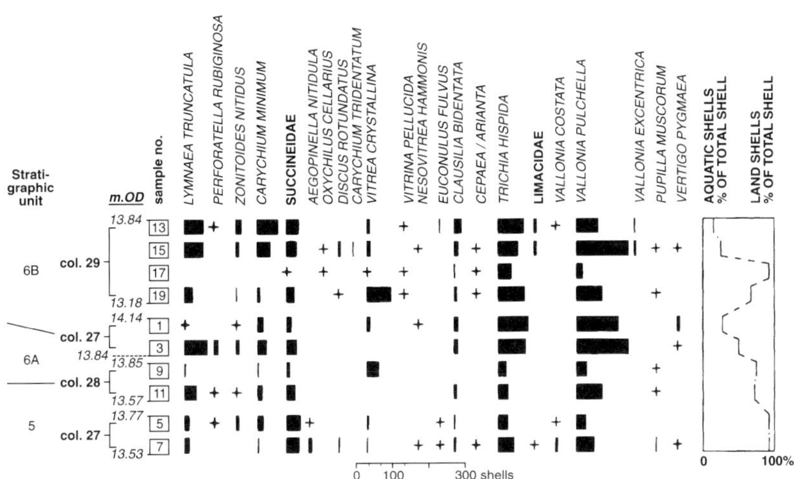

7.24 *Columns 27–29, Area 23, land molluscs.*

(roughly Bronze Age to Roman) from the River Kennet at Anslows Cottages near Reading (Evans 1992).

There is a hint of a drop in the diversity of river habitats, as seen in the decrease in diversity of the aquatic component of the floodplain faunas by comparison with those of the channels themselves. However, this may be a taphonomic effect, due to the small sample size and the fact that *Pisidium* valves are not transported onto the floodplain because, unlike gastropod shells, they do not float. If one looks at the channel assemblages themselves (parcels 1E, 2D, 2E, 4B, 5 and 7) there is no change in diversity and species composition from the late Boreal to the post-Late Bronze Age. The insect evidence shows exactly the same between parcels 1E through 1H to 4B (Chapter 8).

Other fully aquatic habitats were sparse. Reed swamp, sedge fen, and floodplain pools were present to a greater or lesser extent throughout, mostly as minor components of the valley floor and only widespread in parcel 1C, the early Boreal marl (column 31, Sn 31). This is shown by the general paucity of group 4 species (except for *Valvata cristata* and *Bathyomphalus contortus* which may often go with group 3 – see above).

Overbank flooding of the river with concomitant alluviation were normal features of the valley floor environment although sometimes infrequent or absent especially on the higher locations and where there was occupation. Almost all the molluscan assemblages have an aquatic component; where this is slight, one may suggest the shells to be residual; in a few samples, namely those of the Late Bronze Age occupation, it is totally absent.

There is not much evidence for water standing on the floodplain long enough for the establishment of breeding populations of aquatic species such as *L. truncatula* and *Anisus leucostoma*. The latter is exceedingly rare throughout

141

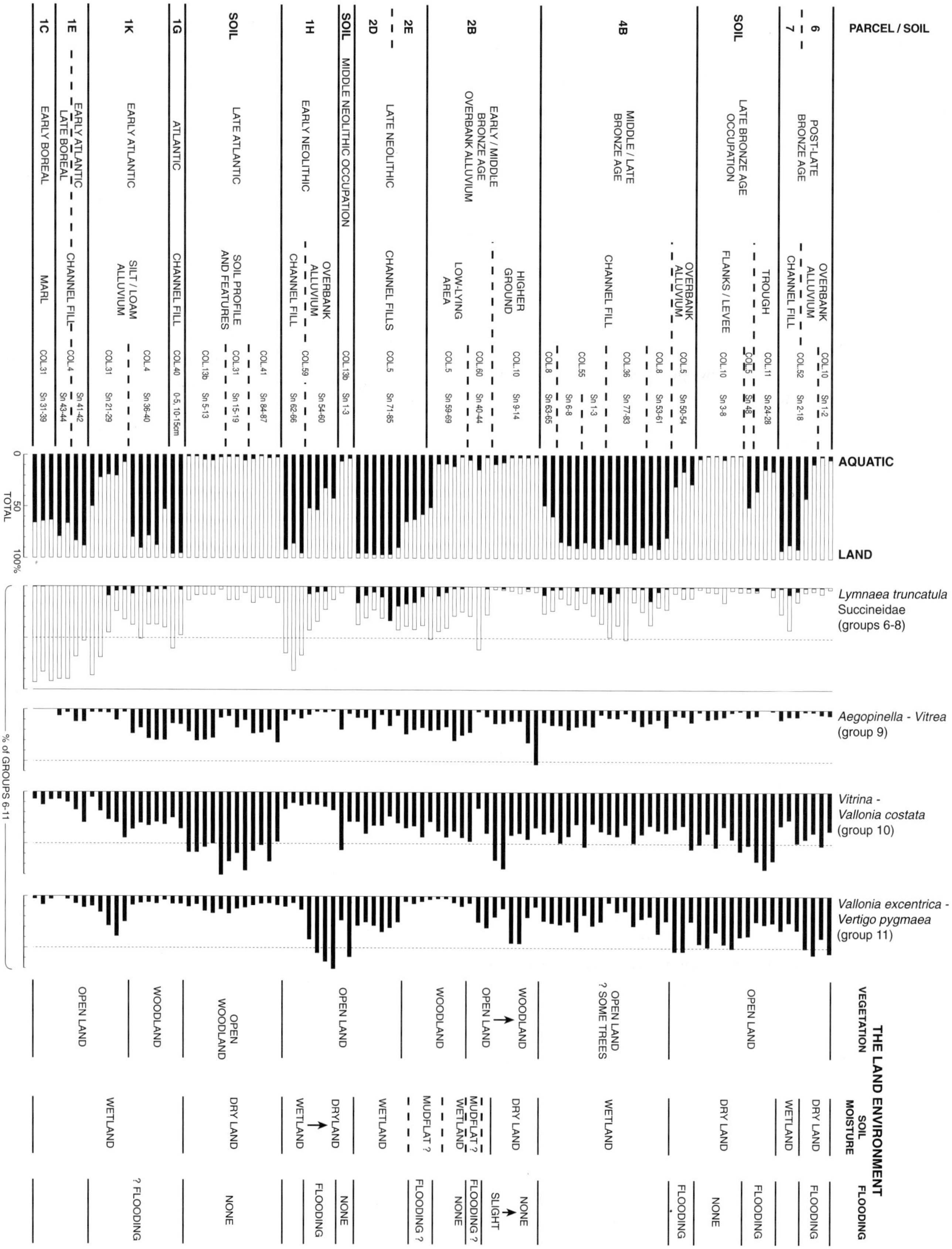

7.25 *Summary diagram of wetland and land mollusc groups. In groups 6–8,* L. truncatula *black, as a percentage of total groups 6–11.*

except in parcels 1C, 1E and 1K, where it probably reflects temporary pools in the drier areas of swamp or sedge fen. The *L. truncatula* peaks are seen rather as reflecting mudflat at the river edge or bare ground on the floodplain, particularly where associated with *Perforatella rubiginosa*, although they could alternatively be interpreted in terms of winter floodings (cf. Robinson 1988).

Land environments

The general nature of the land environment was almost always wetland with Mollusca of groups 7 and 8 present throughout. This does not mean constant waterlogging or permanently anoxic soil conditions, but at least relative humidities approaching 100% for the activity and breeding periods of the molluscs. The few species in question are very specific in their requirements and, except for *Carychium minimum* which can occur in drier woodland, are confined to wetland habitats. Where numbers of groups 7 and 8 are very low, it is likely that at least some of the shells are residual, so one can suggest that in these cases – Middle Neolithic occupation, upper part of MBA overbank alluvium (parcel 2B), and the flanks/levee contexts of the LBA occupation – the environment was dry. However, the soils associated with these horizons were heavily gleyed or at least gleyic (Chapter 5).

The wetland molluscan taxa (groups 7 and 8) and *V. pulchella* show a pattern in their behaviour through some of the overbank alluvial sequences. *Z. nitidus* drops off the most rapidly, followed by *C. minimum* and there is then an increase in *V. pulchella*. It is shown best in Figs 7.2 and 7.8, but the antipathetic behaviour of *Z. nitidus*/*C. minimum* with *V. pulchella* is clear in Figs 7.12, 7.13, 7.19 and 7.22. This is a feature of the previous sequences at Runnymede (Evans 1991a) as well as at other localities (Evans 1995) and in modern faunas (Bishop 1981). In the last, it was noted that *Z. nitidus* preferred shade, *V. pulchella* open habitats, and *C. minimum* wetness. At Runnymede, these changes reflect shaded vegetation giving way to open land with considerable wetness, followed by drier ground, probably grassland.

A feature of the Runnymede wetland assemblages is the absence of *Vertigo antivertigo*, *V. moulinsiana* and *V. angustior* and the rarity of *Pisidium obtusale*. They are often common in Pleistocene river valley deposits. These are species of very wet swamps and fens. It is unclear how these differ in their habitat requirements from the wetland land species of groups 7 and 8 on the one hand and the aquatic groups 4 and 5 on the other, but in terms of their water requirements they probably fall somewhere between the two, while at the same time being essentially land rather than aquatic species. Some may have rather narrow requirements, as *V. moulinsiana*, and whatever these were, they were absent from Runnymede.

Other land environments, in particular the nature of the vegetation, are discussed in the following section.

The environmental sequence

This section also draws on information from sediments, pollen, insects, macroscopic plant remains (macros), topography and archaeology. The general environment in parcel 1C was open wetland, of low structural diversity and quite extensive. A site environment of large sedge tussocks in permanent, gently flowing water is a possible scenario, with marl being deposited, probably by algae (including Characeae), under water; there may have been trees and bushes, but there was not heavy shade. Floodplain and channel are unrepresented and both, if present at all, were some distance from the locality of marl deposition. There was a decrease in open water and an increase in the swamp/pool element of the environment at the top. Openness of the valley bottom is indicated too by the pollen (Chapter 9, although the regional environment was woodland).

In parcel 1E, there was down-cutting through the parcel 1C marl by a river channel, and the infill deposits are river channel ones. On land, there was more structural diversity than in parcel 1C, although this may reflect the wider catchment of channel deposits by comparison with those of marl. Woody vegetation, if any, on the valley bottom was probably pine and willow.

In parcels 1C and 1E, land molluscs (groups 9–11) as a percentage of total land and wetland (groups 6–11) are less frequent than at any other time in the sequence (except for a few mudflat stages), suggesting that permanent wetland – initially reed swamp and open water, then sedge fen with willow – was extensive.

In parcel 1K, the site environment was probably overbank alluviating floodplain, the earliest evidence for this at Runnymede. The change from channel was less abrupt in column 4 than in column 31, perhaps reflecting the lower altitude of the former. The slightly higher deposits of column 31 reflect open grassland developing from fen in the lowest part, with at most sparse trees and bushes, while the slightly lower deposits of column 4 reflect woodland (alder according to the pollen, plant macros and insects); the evidence for woodland at the top of the column 4 sequence, with three species of tree snail, is strong. The difference in fauna between the two localities may be a function of catchment size since there is a greater percentage of aquatics in column 4 than in column 31, suggesting a greater degree of allochthony in the former and thus a wider catchment. On the other hand, it could reflect spatial variation of the vegetation between the sampling localities (the two columns are *c.* 30 m apart).

The molluscan evidence for openness in parcels 1E and 1K (column 31 part) is clear, but it is at odds with the insect evidence (Chapter 8). However, the following points should be noted:

1. Numbers of insects are low, so the absence of open land forms may be statistical;

2. the paucity of open land insects may be a function of the extensiveness of the sedge fen habitat and the paucity of insect species specific to this habitat;
3. the molluscan evidence is very local, as shown by the difference between woodland in column 4 and open land in column 31, whereas the insect evidence probably reflects a wider picture;
4. there are no insects from column 31, Sn 21–29, which show the clearest molluscan evidence for openness.

A trend through parcels 1C/E to 1K is the decrease in wetland species relative to true land species, a trend not subsequently reversed (apart from peaks of *Lymnaea truncatula*). This implies an increasing extensiveness of land habitats and shrinkage of the areas of sedge fen and may be due to build-up of sediment as overbank alluviation began.

The river channel of parcel 1G saw a similar aquatic environment to that of parcel 1E and a land environment of woodland or open woodland, as the lower locality of parcel 1K (column 4). The insect evidence is in agreement here at least.

The environment of the late Atlantic soil profile and features (Late Mesolithic–Early Neolithic), was drier than previous land environments, although there are still up to 13% wetland molluscs and the soils (Chapter 5) were heavily gleyed. Molluscs indicate a vegetation of open woodland, and this is supported by the pollen (from deposits before and after – columns 40 and 59) which indicates alder woodland with some openings.

The openness of some of the early environments at Runnymede – parcel 1C to late Atlantic soil profile – is a little unexpected in a river valley bottom. The evidence is based on the continuous presence of open-country shells; even in the most woodland of all the samples, column 4, Sn 36, there are a few. The main species is *Vallonia pulchella* which is probably a grassland species but which also occurs in unshaded fen (Bishop 1981). It is notable that the species occurs throughout many interglacial and Holocene river sequences (Sparks and Lambert 1961; Sparks and West 1970; Kerney 1971; Holyoak and Preece 1985).

However, the nature and extent of these open environments are uncertain and the causes of openness probably varied from one period to the next. Some environments may have been specialised wetland such as in parcel 1C and the distinctive open episode in parcel 1K (column 31, Sn 21–29), in which tree growth was inhibited by permanent water. This has been shown for large valleys in early Holocene Poland, for example (Alexandrowicz *et al.* 1984). In other cases, as in the succeeding late Atlantic woodland of the Mesolithic–Early Neolithic soil profile, where there is the additional pollen evidence of ruderals, a number of possible causes of openness are discussed by Scaife (Chapter 9). Floodplain instability caused by shifting river channels as well as alluviation itself (of whatever origin) may have been responsible for creating and maintaining localised openness.

Evidence for openness in river valleys during the middle Holocene has been presented for the Itchen valley in Hampshire (Waton 1982a), the Yorkshire Wolds (Bush 1988) and the Ouse in Sussex (Burrin and Scaife 1984), the last two possibly of Mesolithic origin, but there is only a little archaeological data against which the environments at Runnymede can be discussed (Chapter 3).

In the Early Neolithic, parcel 1H, channel cutting (laterally) and infilling gave way to overbank flooding and alluviation, with a trend from wetland to dry land; the general environment was grassland, especially later in the parcel when there was overbank alluviation. The Neolithic archaeology and the contrast with the preceding woodland suggest that humans were responsible for woodland recession. Insect and macro plant evidence indicate alder woodland, especially along the river edge, but this is from the lower, channel, deposits in column 59 where the molluscan evidence for grassland is less clear. Presumably, too, the top of parcel 1H (column 59, Sn 54 and 56) is contemporary with the late Atlantic Mesolithic–Early Neolithic soil and features on the high bank which are reflecting woodland (Fig. 2.2) so, as in parcel 1K, the molluscs are picking up local variation with woodland and open ground.

Open land continued into the Middle Neolithic occupation, with dryness and some flooding.

In parcels 2D and 2E, the Late Neolithic, there was further lateral channel cutting and aggradation. The land environment was wetland (perhaps a taphonomic effect of the wider catchment of the Mollusca as much as a change in the local hydrology) and grassland. The pollen indicates a mosaic of alder and sedge fen on the valley bottom, while the insects indicate reed swamp and more open land environments than previously.

The three sequences from parcel 2B, the Early/Middle Bronze Age overbank alluvium, show wetland and mudflat in low-lying areas and dry land on higher ground. If the *Lymnaea truncatula* peaks reflect floodplain pools rather than mudflat, then some flooding is indicated, but otherwise this was slight or absent. Presumably there was some flooding to account for the build-up of up to 1.5 m of deposit, tapering out over higher ground, but the molluscan evidence for it is often weak. There was variation in the degree of openness, with woodland and open land, but ultimately, at the top of the autochthonous deposits of column 10, Sn 9–14, there was woodland (with the highest percentage of the structure-loving molluscs, group 9, in the entire sequence). However, the top of column 10 (Sn 9) is probably a soil, and as such would have been active after alluviation in parcel 2B had ceased and well into the M/LBA of parcel 4B, with concomitant conflation of the sequence. The precise age of the woodland, therefore, is difficult to assess. Even so, by comparison with the Late Neolithic, there was woodland regeneration on the valley floor late in the second millennium BC.

In parcel 4B, the Middle/Late Bronze Age, there was

river channel erosion cutting through the previous parcel 2B overbank alluvium. There was wetland throughout, although the insects indicate relatively well-drained conditions (Chapter 8). The vegetation was either open woodland or, more likely, a mixture of open and wooded habitats in the catchment from which the shells derived; pollen indicates the latter (Chapter 9). There was possible mudflat (column 36, Sn 81–83) in which *L. truncatula* and *Perforatella rubiginosa* are represented and in which the sediments are clay and shelly lenses, the latter possibly rafts of shell washed onto the mudflat.

During the parcel 4B overbank alluviation which succeeded the channel deposits, there was a change to dry land, but with flooding (except in the lowest sample); soils were gleyic (Chapter 5). The land environment was open, and probably quite free of woody vegetation locally. Xerophile species, *Pupilla muscorum* and *Helicella itala*, indicate dry, calcareous grassland.

Whatever our interpretation of the Middle/Late Bronze Age channel fill deposits of parcel 4B (local open woodland or separate open and woodland habitats in a wider catchment), there was a significant reduction in tree cover between the Early/Middle Bronze Age woodland (overbank alluvial soil – parcel 2B, column 10, Sn 9; column 5, Sn 59) and the Middle/Late Bronze Age grassland (overbank alluvium – parcel 4B, column 5, Sn 50–54). The evidence from both is comparable, since both contexts are dry land and represented by at least some autochthonous materials in which there was little or no flooding. Robinson suggests substantial woodland clearance in the Early Bronze Age or 'first part' of the Middle Bronze Age (Chapter 8), but the data require clearance to have taken place after the Early/Middle Bronze Age (parcel 2B) and before the Middle/Late Bronze Age (parcel 4B). Woodland clearance at this time, given the slightly later intensive on-site human settlement and the waterfront construction, was almost certainly brought about by man.

This open, dry land environment (although without the xerophiles) continued into and through the Late Bronze Age occupation and soil. The site-specific nature of the environment here is clear, since there is a virtual absence of aquatic shells and the deposits are thus autochthonous, at any rate on the flanks and levee; in the trough there was continued flooding.

The post-Bronze Age activity, parcels 5, 6 and 7, saw a phase of wetland (in channel deposits) followed by dry land in alluvium. Both are in open land, and not significantly different from the situation in the Late Bronze Age.

Summary points

1. There was open sedge fen and grassland in the Boreal and early Atlantic periods. This could be a feature of broad lowland rivers generally at this time, the open land being maintained by high water tables. It is relevant to the exploitation of the resources of these environments by animals and humans, especially for winter feeding.
2. There was woodland in the Atlantic period, with some openings.
3. Woodland recession took place in the Early Neolithic and open conditions continued until the Early/Middle Bronze Age.
4. There was woodland regeneration in the later part of the Early/Middle Bronze Age sequence, recorded by the molluscs alone. It would be interesting to know how widespread this was and how it related to the archaeology.
5. Woodland recession ensued before the Late Bronze Age occupation.
6. The earliest overbank alluviation occurred in the early Atlantic period (parcel 1K). It would be interesting to know its cause, and the reasons for the change from marl deposition.
7. Except in parcel 1C, the early Boreal marl, there was generally a clear distinction between channel and overbank floodplain environments. The latter, although wetland, did not have standing water long enough for molluscs of aquatic habitats to breed.

CHAPTER 8

Middle Mesolithic to Late Bronze Age insect assemblages and an Early Neolithic assemblage of waterlogged macroscopic plant remains

MARK ROBINSON

Introduction

The 1978 excavation of the Late Bronze Age waterfront site at Runnymede Bridge yielded important Middle Neolithic and Late Bronze Age insect assemblages from waterlogged channel sediments (Robinson 1991). Further excavations on the site have greatly extended the time-span covered by palaeochannel sediments and provided the opportunity to study the development of its insect fauna leading up to the Middle Neolithic occupation and also to fill in part of the gap in the sequence prior to the construction of the Late Bronze Age waterfront.

The samples

The samples were taken in the form of columns through the organic sediments, mostly divided at 5 cm or 10 cm intervals (Table 1.2). Following analysis, the units within a column were combined when necessary to give sufficient numbers of insects for detailed interpretation.

Column 31: Boreal marl. Insect remains absent (parcel 1C; 13.18–12.88 m OD).

Column 4: late Boreal riverine sediments. Insects present in the bottom two samples, I4 and I5 (parcel 1E; 12.91–12.71 m OD).

Column 39: Boreal/Atlantic transition riverine sediments. 20 cm column divided at 10 cm; insect remains present throughout (parcel 1D; 13.01–12.81 m OD).

Column 40: Atlantic riverine sediments. Column divided at 5 cm intervals; insect remains present from 5–20 cm depth (parcel 1F–1G; 12.97–12.82 m OD).

Sample A32.040 (equivalent to lower part of column 59, parcel 1H; 13.00–12.95 m OD): Early Neolithic riverine sediments pre-dating the Neolithic occupation of the site.

Sample A14 W1, grid 79/30 (equivalent to mid–low column 6, contexts 14.118 and 14.120; parcel 2E; *c.* 12.80–12.60 m OD): final Neolithic/Beaker riverine sediments.

Column 55: Middle/Late Bronze Age riverine sediments.

Column divided at 5 cm intervals; insect remains present from 0–40 cm depth (parcel 4B; 13.01–12.61 m OD).

Laboratory procedure

The samples for insects were gently disaggregated in warm water and washed onto a 0.2 mm aperture sieve. The fraction retained by the sieve was subjected to paraffin flotation and then sorted to recover the insect remains as described in Robinson (1991, 277). The insect fragments were identified with reference to the Hope Entomological Collections of the University Museum, Oxford.

Macroscopic plant remains were also recovered from sample A32.040. A sub-sample was wet-sieved down to 0.2 mm and sorted under water using a binocular microscope. Only a tenth of the size fraction between 0.5 and 0.2 mm was sorted and the results have been multiplied up accordingly. Seeds and other macroscopic remains were identified with reference to the collections of the author.

Total sample weights for insects are given in Table 8.3 and for macroscopic plant remains in Table 8.1.

Results

The results have been listed in Tables 8.1–4, giving the minimum number of individuals represented by the fragments of each species for each sample, or recording species present but unquantified (+). The nomenclature follows the Royal Entomological Society's revised checklists of British insects (Kloet and Hincks 1964, 1977, 1978) and for plants, Clapham *et al.* (1987).

Along with the identifications is given a short description of the habitat or food of each species. The abbreviations used for the insects are: A: aquatic; B: bankside/water's edge; C: carrion; D: disturbed/bare ground; F: dung; G: grassland; M: marsh; P: pest of stored farinaceous foods; T: terrestrial and occurring in several habitats; V: decaying plant remains; W: woodland or scrub. Less normal habitats are given in brackets.

The abbreviations used for the plant habitats are: A: aquatic; B: bankside; C: cultivated crop; D: disturbed ground; Da: disturbed ground including arable; G: grassland; M: marsh and fen; S: scrub; W: woodland. Less usual habitats are given in brackets.

A wide range of sources has been used for ecological information about the insects. The main references are given in Robinson (1991, 277–8), with the important addition of Koch (1989a, 1989b, 1992). Habitat details about the plants have been derived particularly from Clapham *et al.* (1987) and the 'Biological Flora of the British Isles' (*Journal of Ecology*, various years).

Notes on identifications

Cossoninae: indeterminate elytral fragments of a species of weevil belonging to the subfamily Cossoninae, which could not be matched with any British species, were found in column 40 and sample A32.040. Attempts to match them against non-British European species of Cossoninae were also unsuccessful, but there are many gaps in the European Cossoninae in the Hope Collections.

Apis mellifera. The flattened part of the hind tibia of a worker of *A. mellifera* (honey bee) from column 55, 35–40 cm, was identified using the distinctive pattern of rows of setae. Since their record is of major significance, the fragment (popularly known as the pollen basket) was compared with the tibiae of a wide range of bees in order to confirm it.

Analysis of the insect data

The same considerations outlined in Robinson (1991, 278–81) are applicable to the interpretation of the insects from the present excavations. Analysis has again concentrated on the Coleoptera (beetles) because they were abundant in the samples, could mostly be identified quite closely and had been derived from a wide range of terrestrial as well as aquatic habitats. The Coleoptera have been divided into the same species groups as before and are expressed in Fig. 8.1 as percentages of the minimum number of individuals of terrestrial Coleoptera in each assemblage. The members of the groups from the samples are as follows:

1. AQUATIC

Beetles from the families listed below can spend much of their adult life under water (A in Table 8.3) plus *Macroplea appendiculata*.

Haliplidae	Hydrophilidae	Elmidae
Dytiscidae	Hydraenidae	
Gyrinidae	Dryopidae	

Members of this group have been excluded from the coleopteran sum on which the percentages are calculated because the assemblages accumulated under water and it enables some of the differences due to the environment of the deposit itself to be eliminated.

2. PASTURE/DUNG

Scarabaeoid dung beetles of the genera *Geotrupes*, *Aphodius* and *Onthophagus*.

3. ?MEADOWLAND

Weevils of the genera *Apion* and *Sitona*, which mostly feed on vetches, clovers and other grassland trefoils (*A. urticarium* excluded).

4. WOOD AND TREES

Coleoptera which feed on wood in various stages of decay, leaves, fruits, bark and live wood of trees and shrubs; plus fungal feeders and predators which are strictly associated with wood. The following families have some members in species group 4 from the samples:

Lucanidae	Lathridiidae	Chrysomelidae
Elateridae	Mycetophagidae	Curculionidae
Eucnemidae	Tenebrionidae	Scolytidae
Anobiidae	Oedemeridae	Platypodidae
Melyridae	Cerambycidae	

Anobium punctatum has been placed in a separate category.

5. MARSH/AQUATIC PLANTS

Species of Chrysomelidae and Curculionidae which feed exclusively on marsh and aquatic plants.

6A. GENERAL DISTURBED GROUND/ARABLE

Harpalus rufipes.

6B. SANDY/DRY, DISTURBED GROUND/ARABLE

Amara tibialis.

7A. DUNG/FOUL ORGANIC MATERIAL

Various Hydrophilidae and Staphylinidae of foul organic material: *Cercyon* spp., *Megasternum obscurum*, *Platystethus arenarius*, *Anotylus rugosus* and *A. sculpturatus* gp.

8. LATHRIDIIDAE

Lathridiidae excluding *Enicmus fungicola* or *rugosus*.

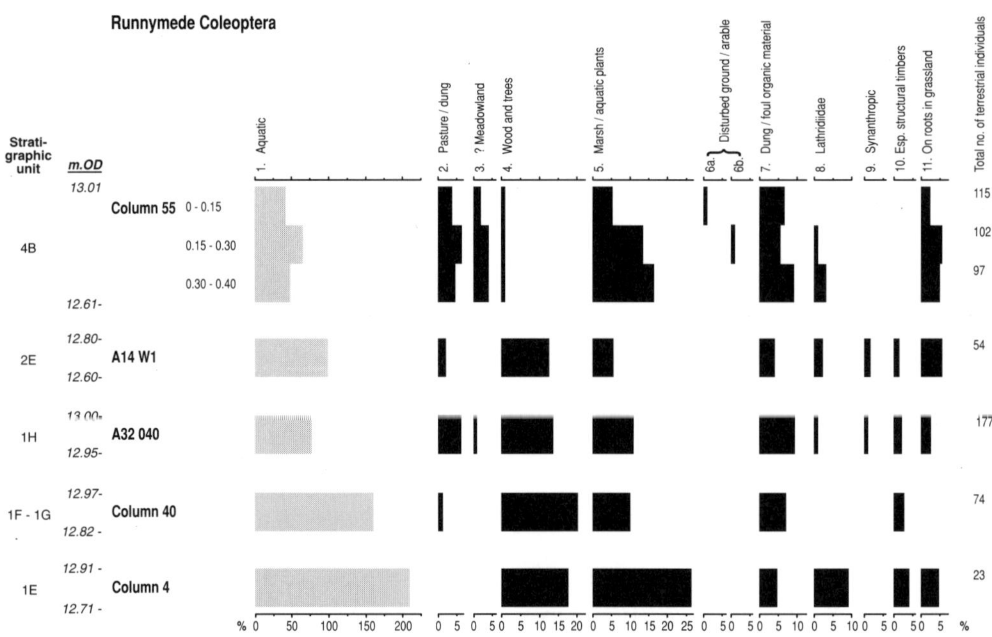

8.1   *Species groups of Coleoptera from Runnymede. Species groups 2–11 are expressed as a percentage of the total terrestrial Coleoptera (i.e. excluding aquatics). Not all the terrestrial Coleoptera have been classified into groups.*

9. SYNANTHROPIC

Ptinus fur.

10. ESPECIALLY STRUCTURAL TIMBERS

Anobium punctatum (the woodworm beetle).

11. ON ROOTS IN GRASSLAND

Various Scarabaeidae and Elateridae with larvae which live on the roots of herbs in permanent grassland: *Hoplia philanthus, Phyllopertha horticola, Agrypnus murinus, Athous* spp, *Agriotes* spp.

Interpretation of the insect assemblages: their taphonomy

The insect assemblages from these Runnymede samples all accumulated under water as a result of natural agencies in riverine sediments. They had been derived from the faunas of both the river and the surrounding landscape. It was argued previously that in the order of 50% of the terrestrial Coleoptera had their origins in a strip 50 m wide on either side of the channel but extending 0.5 km upstream (Robinson 1991, 316) and this assumption remains a useful aid to interpretation. Unlike some of the deposits examined earlier, no midden material had been incorporated into the sediments. None of the samples showed evidence of size bias of insect remains due to water sorting.

The Mesolithic woodland succession

Columns 4, 39 and 40 provided a useful sequence spanning the Boreal to the Atlantic periods of the Mesolithic. Unfortunately, the results from the individual samples in each column had to be combined to give sufficient numbers of insects for full interpretation and even so, there were insufficient terrestrial Coleoptera from column 39 for its inclusion in Fig. 8.1. Although differences could readily be detected in the insect assemblages between columns, no significant variation was noted within each column.

The aquatic and waterside environment

A very high proportion of the insects from the three columns were aquatic species, for example almost 70% of the Coleoptera from column 4 were water beetles (i.e., they were at a level of over 200% of the total terrestrial Coleoptera, Fig. 8.1). The most abundant group of terrestrial Coleoptera from the columns comprised species which feed on marsh or aquatic plants.

The aquatic insects from all three columns were the by now familiar fauna of the early to mid-Holocene River Thames, being dominated by species of clean flowing water over a stony bed, particularly the caddis *Ithytrichia* sp. and beetles from the family Elmidae (Robinson 1991, 316–17). The elmids included two species, *Macronychus quadrituberculatus* and *Stenelmis canaliculata*, that are so fastidious in their requirements of well-oxygenated, unpolluted water that they have long since disappeared from the major English lowland river systems, including the Thames, and now have very restricted distributions.

The Chrysomelidae suggest a rich aquatic and marginal flora with, for example, *Donacia versicolorea*, which feeds on the floating-leaved species of *Potamogeton* (pond weed), *D. dentata*, which feeds on *Sagittaria sagittifolia* (arrowhead) and *Alisma* spp. (water plantain), and *D. clavipes*, whose host plant is *Phragmites australis* (reed). *Prasocuris phellandrii*, which feeds on aquatic Umbelliferae (probably *Oenanthe aquatica* gp. (water dropwort)), was also recorded from two of the columns. The Carabidae included various species of bankside habitats such as *Bembidion semipunctatum*, which occurs on wet sand. There were also amphibious members of the Hydraenidae and Dryopidae which readily leave water to crawl over wet mud. The insects from these three columns do not, however, suggest that there were extensive areas of fen or marsh.

The overall picture given by the insects for the Mesolithic river at Runnymede is somewhat similar to that obtained earlier for the Neolithic river (Robinson 1991, 316–18): a channel with lengths of hard bed and relatively rapidly flowing water, but also with lengths rich in aquatic vegetation. The transition from channel with reed swamp vegetation to dry land was probably abrupt.

Although there were differences between the aquatic insect faunas from columns 4 and 39 – for example, *Ochthebius bicolon* was absent from column 4 but quite well represented in column 39 – these differences cannot obviously be related to changing conditions in the river.

Woodland and scrub

There were significant differences between the woodland Coleoptera from the columns which reflected woodland succession on the river bank. Wood and tree-dependent Coleoptera comprised approximately 20% of the terrestrial Coleoptera from column 4. The sample was too small to use this figure to give an estimate of the degree of tree cover but there was clearly a significant presence of trees, as would have been expected during the Mesolithic. The host-specific of these beetles, *Phyllodecta* sp. and *Chalcoides* sp., both feed on *Salix* spp. and *Populus* spp. Leaf fragments of *Salix* subgenus *Caprisalix* sp. (sallows), identified from the glands on the leaf margin, occurred in some of the samples and this tree probably grew along the edge of the channel. *Populus nigra* (black poplar) would also be a plausible tree to have grown on the floodplain, especially if there was some channel instability, although the occurrence of this tree cannot be confirmed. The insect assemblage was not large enough to give an indication of the type of woodland which was growing on the higher ground beyond the floodplain.

The number of terrestrial Coleoptera from column 39 was even smaller than that from column 4, and there was only a single specimen of a tree-feeding beetle, *Phyllodecta* sp. However, the macroscopic plant remains included fruits

and cones of *Alnus glutinosa* (alder), which were absent from column 4. The great expansion of *Alnus* in the British Isles tended to occur at the end of the Boreal period and pollen evidence suggests that it was the predominant wetland tree during the Atlantic over much of England. However, *A. glutinosa* recolonised the British Isles in the early post-glacial, the population remaining low throughout most of the Boreal (Brown 1988). It is possible that *A. glutinosa* became established at Runnymede at an early date and that the sediments of column 39 were indeed Early Mesolithic (see Chapter 10).

The larger assemblage of insects from column 40 resulted in the occurrence of a wider range of woodland insects. Members of species group 4 (Fig. 8.1) comprised just over 20% of the terrestrial Coleoptera which, it has been argued, is consistent with old closed woodland overhanging the deposit (Robinson 1991, 280). The samples from this column again contained fruits and cones of *Alnus glutinosa*, in this case in association with *Agelastica alni*, the alder leaf beetle. Dense alder woodland is likely to have lined the river banks. The other host-specific tree- and shrub-feeding beetles from column 40 included two bark beetles, *Scolytus scolytus* and *Hylesinus oleiperda*. The former feeds mostly on *Ulmus* spp. (elm) and the latter occurs on *Fraxinus excelsior* (ash). There was also a possible example of the anobiid *Ochina ptinoides* which bores into dead stems of *Hedera* spp. (ivy). Mixed deciduous woodland including elm and ash probably grew on the drier ground. It is also likely by analogy with Neolithic insect assemblages from Runnymede (below and Robinson 1991, 318) that oak was a major component of the Late Mesolithic woodland on the drier ground, but the sample was not large enough to include beetles from the full range of trees.

S. scolytus is now notorious as the vector of *Ceratocystis ulmi*, the Dutch elm disease fungus. Great interest was shown when remains of *S. scolytus* were first discovered in an archaeological context, just below the elm decline horizon at Hampstead Heath (Girling 1988). This encouraged reasonable speculation that the elm decline was due to Dutch elm disease. However, further speculation (not by Girling) implied that the beetle always carries the disease and that its introduction at the start of the Neolithic period caused the elm decline. The belief that *S. scolytus* always carries fungus with it is probably the result of bark beetles being confused with ambrosia beetles (both are members of the family Scolytidae). Most British trees have Scolytidae associated with them and, given the present pan-European distribution of *S. scolytus*, its occurrence in a Late Mesolithic deposit well prior to the elm decline at Runnymede should be seen as unexceptional.

The Coleoptera included an 'old woodland' element of beetles which are now associated with over-mature trees or woodland of great age, as would be expected for an assemblage of this date (Robinson 1991, 319–20). The three species of weevils from the subfamily Cossoninae illustrate this aspect well. They all feed on damp, decaying wood. *Rhyncolus truncorum* is a rare beetle which has been recorded from several old woodlands in southern England including Windsor Forest (Fowler 1891, 394; Donisthorpe 1939, 111). It had previously been found in a Neolithic context at Runnymede (Robinson 1991, 319). *Dryophthorus corticalis* was only added to the British list in 1925 when it was found at Windsor Forest (Donisthorpe 1939, 111). Although well established at Windsor, it remains unique in Britain to that locality. It has a widespread distribution in Europe, occurring in a variety of dead wood (e.g., Hoffmann 1954, 746–7). In Windsor Forest, it tends to occur in damp, tough wood inside old oak trees infested by the ant *Lasius brunneus*. It is not, however, dependent on the ant. *D. corticalis* is on the edge of its ecological range in England, which is probably why it is more fastidious about the wood in which it occurs here than in parts of continental Europe. It is also likely to have better colonising powers on the European mainland. Nevertheless, when conditions in England are suitable, it is able to flourish and *D. corticalis* seems to have been widely distributed prior to the loss of its habitat by large-scale clearance and intensive forest management. It has also been identified from an Early Neolithic woodland assemblage at Shustoke, Warwickshire (Kelly and Osborne 1963–64), Late Bronze Age woodland peat at Thorne Moors, Humberside (Buckland 1979, 109), and from a Bronze Age pit set against a background of woodland at Northwold, Norfolk (Robinson, unpublished). The final member of this subfamily could not be identified any further and was apparently a species which no longer occurs in Britain (see above).

The insects from the three columns clearly showed the succession from Early Mesolithic (Boreal) willow woodland to Late Mesolithic (Atlantic) alder woodland lining the river bank. The appearance of an 'old woodland' element to the fauna in column 40 might, in some small part, have been due to habitat maturation but the larger assemblage of terrestrial Coleoptera from this column in comparison with the earlier columns would have been the main factor. Inadequate sample size was also the reason that it was not possible to trace the ecological succession of the mixed deciduous woodland on the drier ground some distance beyond the river.

The open environment

There was little evidence from any of the Mesolithic columns for open conditions or even woodland glades. There were single examples of *Athous hirtus*, from columns 4 and 39, an elaterid whose larva feeds on the roots of herbs, especially in grassland, and is a member of species group 11 (Fig. 8.1). However, the adults do sometimes occur on trees (Fowler 1890, 99; Koch 1989b, 77) and it is possible that there were suitable places on the river bank for the larvae, with well-drained soil and herbaceous vegetation. Weevils

which feed on grassland clovers and vetches (species group 3) were absent and there was only a single scarabaeoid dung beetle from species group 2, a specimen of *Geotrupes* sp. from column 40. Other insects which might suggest open conditions were absent; all the phytophagous species which feed on herbaceous vegetation, for example, can occur on aquatic, bankside or woodland plants.

Other habitats

The insects provided no evidence for human habitation or other human activities in the area. Some examples of *Anobium punctatum*, the woodworm beetle (species group 7), could easily have been derived from dead wood in the woodland. Beetles of decaying organic material including species group 7 are present, but no more abundantly than would be expected in plant detritus along a river bank and in woodland.

The earliest Neolithic environment

Sample A32.040 was dated to the start of the fourth millennium cal. BC and thus around the time of the elm decline. It was decided that it would be particularly useful to examine the macroscopic remains in detail, as well as the insects, to look for any evidence of human activity.

The aquatic and waterside environment

While the proportion of aquatic Coleoptera had fallen in comparison with the earlier columns to about 45% of the total (i.e., 77% of the total terrestrial Coleoptera), riverine conditions did not seem to have changed. The same groups of aquatic species were present as were recorded from the earlier columns, including the now rare Elmidae. A similar range of phytophagous Coleoptera of aquatic and marginal plants was also present, with the addition of the now rare *Macroplea appendiculata*, which feeds on *Potamogeton* (pondweed) and *Myriophyllum* (water milfoil) spp. Seeds of aquatic and marginal plants were not particularly abundant, which was possibly due to the shading effect of trees along the river bank. However, they included *Alisma* sp. (water plantain), the host of *Donacia dentata*. *Nuphar lutea* (yellow water lily) and *Oenanthe aquatica* gp. (water dropwort) were also present. Both plant and insect remains were consistent with a relatively sharp transition from water to land.

Woodland and scrub

Both the macroscopic plant and insect remains from this sample showed that the site retained a strong presence of woodland. About 14% of the terrestrial Coleoptera belonged to species group 4, wood and tree-dependent species (Fig. 8.1). Although the figure is lower than the values for species group 4 from the earlier columns and the proportion of Coleoptera from this group can be as high as 20% under conditions of closed woodland (Robinson 1991, 280), this does not by itself imply an opening of the tree canopy. Although it is argued below that there had been slight clearance, the woodland Coleoptera from the site would suggest it to have been small or remote.

The macroscopic plant remains were dominated by *Alnus glutinosa* (alder), especially seeds but also bud scales, female catkins and wood. Dense alder woodland probably grew along the river bank. This impression was further supported by the occurrence of several examples of the alder leaf beetle, *Agelastica alni*. The macroscopic plant remains did not include any other trees, but the Coleoptera, which would have been derived from a larger catchment area, included *Leperisinus varius*, a bark beetle which occurs mainly on *Fraxinus excelsior* (ash) and *Rhynchaenus quercus*, a leaf weevil of *Quercus* sp. (oak). These two tree species probably grew in mixed deciduous woodland on drier ground.

The following shrubs were represented by their seeds:

Rhamnus catharticus (purging buckthorn)
Rubus fruticosus agg. (blackberry)
Crataegus cf. *monogyna* (hawthorn)
Cornus sanguinea (dogwood)
Sambucus nigra (elder)
Viburnum opulus (guelder rose)

All would have been able to grow as undershrubs in the woodland on the floodplain. Some show a preference for calcareous soils and indeed most of the Neolithic sediments of the site are highly calcareous. One of the shrubs, *R. catharticus*, does not tolerate such deep shade as the others and can only occur as an undershrub when the woodland is open (Godwin 1943, 70). The river itself would have provided a break in the tree canopy and shrubs would be favoured by gaps due to fallen trees. However, any clearance, especially if combined with grazing, for *R. catharticus* is a thorny shrub, would encourage this species to spread.

The seeds of terrestrial herbaceous species mostly comprised a flora of damp woodland. The most abundant were from *Urtica dioica* (stinging nettle) and *Eupatorium cannabinum* (hemp agrimony), which often dominate the ground vegetation of alder woodland.

The Coleoptera included an element from a wide range of woodland habitats, for example *Colenis immunda*, which occurs among mouldy leaf litter; *Silpha atrata*, a predator under bark or in rotten wood; species of *Rhynchaenus* which feed on tree leaves; and species of *Curculio* which feed on acorns or nuts, as well as beetles which feed on wood in various stages of decay. As might be expected, the latter group included an 'old woodland' element, with *Rhyncolus truncorum* and the unidentified species of Cossoninae present again. There were also more widespread species, for example *Phymatodes alni*, which occurs on recently dead

hardwood with the bark on; *Tetrops praeusta*, in small dry dead hardwood branches; and *Ischnomera* cf. *caerulea*, in rotten wood.

Clearings and grazing

The waterlogged seeds included a very few from light-demanding species. *Stellaria media* (chickweed), *Chenopodium polyspermum* (all-seed) and *Chenopodium album* (fat hen) are all annual weeds which would be unable to grow under conditions of closed woodland. Each was represented by a single seed. They would certainly have been able to grow in fresh clearings, but they would also be potential colonists of temporary mud banks created by the river, in company with such plants as *Myosoton aquaticum* (water chickweed). Apart from *Rhamnus catharticus* mentioned above, the macroscopic plant remains did not provide any evidence for woodland disturbance.

The Coleoptera gave a little evidence for grassland. Elateridae with larvae which feed on roots in grassland, including *Agrypnus murinus*, comprised 2% of the terrestrial Coleoptera. However, very few of the Chrysomelidae, Apionidae or Curculionidae can be tied down as feeding on light-demanding herbs, other than those which could have been growing in marginal aquatic habitats. For example, there were at least three individuals of two species of *Phyllotreta*, but they could have fed on waterside species of *Rorippa*, and seeds of *Rorippa* cf. *amphibia* (great yellow-cress) were present. This just leaves single specimens of *Apion* sp. and *Longitarsus* sp., neither of which can be attributed to host plants. By themselves, these two beetles need not imply open conditions on the site. In contrast, the admittedly much larger assemblage from Mid- to Late Neolithic sediments which were investigated earlier included eight host-specific phytophagous grassland taxa of Coleoptera from these families in addition to less host-specific species (Robinson 1991, 320). The Mid- to Late Neolithic catchment for these insects was interpreted as having between one-third and two-thirds of woodland cover.

Surprisingly, the proportion of scarabaeoid dung beetles of species group 2 was as high as from the Mid- to Late Neolithic sediments, at 6% of the terrestrial Coleoptera. *Aphodius* cf. *sphacelatus* was similarly the most numerous, but there were also other species of *Aphodius*, *Onthophagus* spp. and *Geotrupes* sp. There was clearly a concentration of large herbivorous mammals in the area above the usual level for wild species. Unfortunately it is not possible to determine from the dung beetles whether the herbivores were wild or domestic. The two most likely domestic animals to be grazed under woodland conditions, cattle and pig, would have had wild native counterparts of the same species, aurochs and wild pig. If there had been an open space on the river bank where wild mammals congregated to drink, some evidence would have been expected from the macroscopic plant remains. The other possibility is that domestic animals were being herded over a greater area of woodland in which the tree canopy was still largely intact. The *Rhamnus catharticus* (purging buckthorn) would perhaps have been favoured by some reduction in tree cover and pressure of browsing.

The various members of the Hydrophilidae, particularly *Megasternum obscurum*, and Staphylinidae, such as *Anotylus rugosus*, which make up species group 7 and live in a range of foul organic materials including dung, had risen in comparison with the Mesolithic columns. However, at 7% of the terrestrial Coleoptera, they had not reached the level seen for this group in the Mid- to Late Neolithic deposits. They would have found suitable habitats in both the animal droppings and in naturally occurring accumulations of decaying vegetation along the river bank in company with beetles such as *Corylophus cassidoides* in the latter habitat.

Other habitats

All the remaining insects could have been members of the woodland, marginal and aquatic communities described above. It is possible that a few of the water beetles which are usually associated with stagnant water – for example, *Colymbetes fuscus* and the now very rare *Hydrophilus piceus* (great silver diving beetle) – were living in cut-off lengths of palaeochannel rather than slowly moving reaches of the main river.

There was no reliable evidence for human settlement. As argued for the Mesolithic columns, *Anobium punctatum* (species group 7) could have been living in naturally occurring dead dry wood. A single synanthropic beetle (species group 9), *Ptinus fur*, was present. It is now best known as an indoor omnivorous species, feeding on rather dry decaying animal and vegetable matter (Hinton 1940–1), but its natural habitat includes birds' nests (Fowler 1890, 181). There is no reason why this specimen could not have been derived from a nest.

The Beaker period partly cleared landscape

Sample A14 W1 was dated slightly later than the top part of the SS1 column from the earlier excavation, the Late Neolithic (Robinson 1991, 277). The terrestrial insect fauna from the sample was small and similar to that from the top of the SS1 column, so need not be considered in detail. The proportion of aquatic Coleoptera was much higher than towards the top of the SS1 column but otherwise the fauna was similar.

The aquatic and waterside environment

Half the Coleoptera from the sample were water beetles. The now rare Elmidae were again present. The occurrence of *Donacia impressa* suggests that *Schoenoplectus lacustris*

(true bulrush) grew in reed swamp alongside the channel while the usual bankside Carabidae such as *Bembidion assimile* were present.

The terrestrial environment

Wood and tree-dependent Coleoptera comprised 13% of the terrestrial Coleoptera (Fig. 8.1, species group 4). This was a higher value than for the SS1 column and may continue the trend seen there with this proportion increasing towards the top (Robinson 1991, Fig. 125). *Agelastica alni* confirmed the continuing presence of *Alnus glutinosa* (alder) woodland.

The grassland Elateridae of species group 11 were, at 6% of the terrestrial Coleoptera, better represented than in the SS1 column, whereas the scarabaeoid dung beetles of species group 2 were, at 2% of the total, less abundant than in that column. Phytophagous Coleoptera which feed on herbs were few, but included *Chaetocnema concinna* which feeds on Polygonaceae, especially *Polygonum aviculare* (knotgrass). The overall picture is of a partly cleared landscape, but the insect assemblage was not large enough to give the degree of detail obtained from the SS1 column.

The late second millennium BC open landscape

Column 55 essentially pre-dated the WF1b column of the earlier excavation, which spanned part of the Late Bronze Age (Robinson 1991, 277), although it is possible that there was some limited overlap between the columns.

The aquatic and waterside environment

About a third of the Coleoptera from column 55 were aquatic species (i.e., they were at a value of 54% of the total terrestrial Coleoptera). This was a somewhat higher value than for the WF1b column. The Elmidae of clean flowing water and stony beds retained their importance from the Neolithic which indeed continued through the Late Bronze Age. A decline to only a single example of *Macronychus quadrituberculatus* was perhaps a reflection of less decaying wood in the water. Submerged to floating leaved vegetation of *Potamogeton* (pondweed) or *Myriophyllum* spp. (water milfoil), the food plant of *Macroplea appendiculata* evidently continued to flourish. The donaciine chrysomelids suggested that there was much emergent vegetation and it is possible that the marginal reed swamp had become more dense or extensive compared with the Mesolithic and Early Neolithic as a result of the reduction in shade caused by clearance. *Donacia clavipes*, which feeds on *Phragmites australis* (reed), was the most numerous of the beetles, as it was from the WF1b column. *D. impressa*, which feeds on *Schoenoplectus lacustris* (true bulrush) was also present. As before, there was a distinctive bankside fauna of Carabidae, with various species of *Bembidion*, such as *B. semipunctatum*, *B. gilvipes* and *B. assimile*, but no evidence for extensive marshland. The phytophagous Coleoptera which feed on marsh and aquatic plants declined from 17% to 5% of the total terrestrial Coleoptera from the bottom to the top of the column (Fig. 8.1, species group 5). This possibly represented a decline in reed swamp, but there was no other evidence for a changing aquatic environment.

Woodland and scrub

Wood and tree-dependent beetles (species group 4) comprised 1% of the terrestrial Coleoptera. Substantial clearance must have occurred in the catchment since the Beaker period. It is possible that a few trees survived from the woodland because *Scolytus* cf. *intricatus*, which feeds mainly on *Quercus* spp. (oak) and *Acrantus vittatus*, which feeds mainly on decaying *Ulmus* spp. (elm), were present. However, any 'old woodland' element was missing. No fruits or catkins of *Alnus glutinosa* (alder) were observed during the processing of the samples, whereas they were very much in evidence during the sieving of the later Mesolithic and Neolithic samples. The landscape towards the end of the Middle Bronze Age was at least as open as during the Late Bronze Age. This proportion of wood and tree-dependent Coleoptera was lower than for the WF1b column as a whole, although similar to the first half of the column.

Grassland and the open landscape

The insects suggest that the late second millennium BC landscape was mostly grassland. *Phyllopertha horticola*, *Agrypnus murinus* and the other beetles of species group 11, whose larvae feed on the roots of grassland herbs, made up 4.5% of the terrestrial Coleoptera, a similar result to that from the WF1b column. The occurrence of *A. murinus* likewise suggests relatively well-drained conditions. Grazing does not seem to have been heavy; the scarabaeoid dung beetles of species group 2 only comprised 5.4% of the terrestrial Coleoptera, little more than half their occurrence in the WF1b column. The two main species were *Aphodius* cf. *sphacelatus* and *Onthophagus joannae*. None of the exotic Scarabaeidae recorded during the earlier work at Runnymede were found.

Clover and vetch-feeding weevils of the genera *Apion* and *Sitona* (species group 2) comprised 3.2% of the terrestrial Coleoptera, sufficient to show the presence of these plants in the grassland, but too low to indicate hay meadow. Another grassland herb, *Plantago lanceolata* (ribwort plantain) was indicated by the weevils *Ceutorhynchidius troglodytes*, *Mecinus pyraster* and *Gymnetron pascuorum*.

The majority of the other terrestrial Coleoptera were appropriate to a grassland fauna, including most of the Carabidae and many of the Staphylinidae. A high proportion of the less host-specific phytophagous Coleoptera from the

Chrysomelidae and Curculionidae can feed on grassland plants. The ant *Lasius flavus*, which makes mounds in grassland, was present. The dung of domestic animals on the grassland would have provided a suitable habitat for the various Hydrophilidae and Staphylinidae of foul organic material, including the members of species group 7.

Not all the phytophagous Coleoptera feed on grassland plants. There was a group which fed on plants of waste places or incipient scrub:

Batophila rubi	on	*Rubus* spp. (blackberry)
Epitrix pubescens	on	Solanaceae (nightshades etc.)
Apion urticarium	on	*Urtica* spp. (nettles)
Cidnorhinus quadrimaculatus	on	*Urtica* spp. (nettles)

This vegetation could have grown on parts of the river bank or along the edge of any hedges or scrub.

There was no conclusive evidence for cultivated land. There were only a couple of examples of the Carabidae, which tend to favour weedy, disturbed ground. However, members of the genus *Phyllotreta*, particularly *P. vittula* which feed on Cruciferae, and *Reseda* spp. were abundant and two species of *Ceutorhynchus* which feed on Cruciferae were present. While it is possible that they had all been derived from bankside Cruciferae, they can be very numerous on some weeds of disturbed ground and arable.

The insects gave no certain evidence for changing conditions of the open landscape during the period of deposition of the column. There was a decline in both species group 3, the meadowland weevils, and species group 11, beetles which feed on the roots of grassland plants, at the top of the column (Fig. 8.1), which might be interpreted as a decrease in the area of grassland, but the sequence does not continue for long enough to be certain.

Other aspects

There was no evidence from the insects in column 55 for human habitation on the site. However, one very interesting discovery was part of the hind tibia (the 'pollen basket') of a worker honey bee (*Apis mellifera*) from the bottom of the column (35–40 cm). Although there have now been several Iron Age finds of honey bee (e.g., Allen and Robinson 1993, 139), this is at present the only pre-Iron Age record from Britain. While the Runnymede bee need not imply Middle Bronze Age bee keeping, it does show that bee products would have been available for exploitation. Earlier evidence for the exploitation of bees at Runnymede was provided by the detection of a honey residue by chemical analysis of a Neolithic sherd (Needham and Evans 1987).

As in the Early Neolithic, there was a stagnant water element to the aquatic insect fauna that was possibly living in cut-off branches of palaeochannel or other isolated bodies of water rather than in one of the flowing channels. In this instance they were small water beetles of the *Helophorus brevipalpis* group.

The prehistoric environmental sequence at Runnymede

The results from the 1978 excavation and the current excavations have given one of the longest Holocene insect sequences from Britain, spanning much of at least 6000 calendar years. The only serious major gap was for the Early–Middle Bronze Age (*c.* 2000–1200 BC), regrettably a period when very interesting changes were occurring at Runnymede.

The Mesolithic part of the sequence, represented by columns 4, 39 and 40 covered the transition of the Boreal willow woodland on the floodplain to alder woodland. It also saw the rise of the 'old woodland' insect fauna characteristic of over-mature trees and dead wood. These developments were not unexpected.

The Early Neolithic stage of the sequence, represented by sample A32.040, gave very interesting results, with domestic animals apparently being grazed in woodland that had only experienced a slight opening of the tree canopy, at a date around that of the elm decline (perhaps just before it – Scaife, Chapter 9).

The Middle–Late Neolithic was represented by the SS1 and A4 columns from the 1978 excavation (Robinson 1991) and sample A14 W1 from the current work. They showed a landscape with a mosaic of clearings and woodland. Grazing was occurring, but the areas of alder woodland retained their 'old woodland' beetle fauna. A few of these beetles are now extinct in Britain and others are very rare, some only being found in the old trees of Windsor Great Park, not far from Runnymede.

Substantial clearance must have occurred in the Early to Middle Bronze Age. The Middle–Late Bronze Age transition was covered by column 55. There was no evidence for alder woodland on the floodplain; it had been replaced by a landscape of lightly grazed grassland.

The early first millennium BC was represented by column WF1b from the 1978 excavation (Robinson 1991). This showed an increase in the intensity of grazing compared with the late second millennium BC and also some evidence for arable beyond the floodplain. A little scrub was present, but willow had replaced alder along the water courses. Waterlogged material from the Late Bronze Age settlement was represented by the midden samples (Robinson 1991), which contained a fauna of flies etc. appropriate to foul organic material, but they gave surprisingly little evidence for structures or indoor habitats.

Throughout the Runnymede sequence, the river supported a rich fauna of elmid beetles, which are fastidious in their requirements for clean, well-oxygenated water and are now extinct in the Thames. Some of them now have an extremely restricted distribution in Britain. The Neolithic and Bronze Age samples also showed the development of the open country fauna following clearance. One interesting aspect was the occurrence of scarabaeoid dung beetles which are now extinct in Britain, although the reason for their disappearance is uncertain (Robinson 1991, 326).

The Neolithic and Late Bronze Age settlements at Runnymede are of considerable archaeological importance and certain aspects of the Late Bronze Age site are unusual. The environmental sequence, however, is probably typical for the middle Thames. What was fortuitous was that such a range of waterlogged sediments survived on one site, allowing their sampling for palaeoecological study.

Table 8.1 *Waterlogged seeds from Early Neolithic sediments in Area 32 (for abbreviations see text).*

Species	Common name	No. of seeds Sample A32.040 1kg sample	Habitat
RANUNCULACEAE			
Ranunculus cf. *repens* L.	buttercup	3	damp G and W, D(a)
R. S. Batrachium sp.	water crowfoot	1	A
NYMPHAEACEAE			
Nuphar lutea (L.)Sm.	yellow water-lily	3	A-0.5-3m deep
CRUCIFERAE			
Rorippa cf. *amphibia* (L.)Bes.	great yellow-cress	2	B, seasonally exposed mud
CARYOPHYLLACEAE			
Myosoton aquaticum (L.)Moen.	water chickweed	3	W-damp MB
Stellaria media gp.	chickweed	1	Da
S. cf. *palustris* Retz.	marsh stitchwort	1	M
CHENOPODIACEAE			
Chenopodium polyspermum L.	all-seed	1	Da-esp. nitrogen-rich soils
C. album L.	fat hen	1	Da-esp. nitrogen-rich soils
Atriplex sp.	orache	1	Da
RHAMNACEAE			
Rhamnus catharticus L.	buckthorn	8	SW esp. on calcareous soils
ROSACEAE			
Rubus fruticosus agg	blackberry	15	WSD
Crataegus cf. *monogyna* Jacq.	hawthorn	1	SW
ONAGRACEAE			
Epilobium sp.	willow herb	2	MBWSDa
CORNACEAE			
Cornus sanguinea L.	dogwood	1	SW on calcareous soil
UMBELLIFERAE			
Oenanthe aquatica gp.	water-dropwort	3	A
POLYGONACEAE			
Rumex sp.	dock	1	DaGMSW
URTICACEAE			
Urtica dioica L.	stinging nettle	58	DWS&B - often nitrogen & phosphorus-rich soils
BETULACEAE			
Alnus glutinosa (L.) Gaertn.	alder	126	SW-esp. damp
SOLANACEAE			
Solanum dulcamara L.	woody nightshade	7	DWSB(A)
SCROPHULARIACEAE			
Scrophularia sp.	figwort	2	SW&G-wet, B

Species	Common name	No. of seeds Sample A32.040 1kg sample	Habitat
LABIATAE			
Mentha sp.	mint	5	G&W-wet, Da MA
Lycopus europaeus L.	gipsy-wort	11	MB
CAPRIFOLIACEAE			
Sambucus nigra L.	elder	24	SW&D-esp. base and nitrogen-rich soils
Viburnum opulus L.	guelder rose	1	S&W-esp. damp
COMPOSITAE			
Senecio cf. *aquaticus* Hill	ragwort	1	G-wet, MB
Eupatorium cannabinum L.	hemp agrimony	43	W-moist, MB
ALISMATACEAE			
Alisma sp.	water-plantain	5	AB
IRIDACEAE			
Iris pseudacorus L.	yellow flag	1	AMB
SPARGANIACEAE			
Sparganium sp.	bur-reed	1	AM-ungrazed
TYPHACEAE			
Typha sp.	reedmace	10	reedswamps
CYPERACEAE			
Schoenoplectus lacustris(L.) Pal.	bulrush	1	A
GRAMINEAE			
Gramineae indet.	grass	1	
Varia		1	
Total		346	

Identification	Anatomy	No. of items Sample A32.040	Habitat
Alnus glutinosa (L.) Gaertn. (alder)	female catkin	15	SW - esp. damp
A. glutinosa (L.) Gaertn. (alder)	bud scale	16	SW - esp. damp
A. glutinosa (L.) Gaertn. (alder)	wood	+	SW - esp. damp
Bryophyta indet. (moss)	stem with leaves	+	
Bud scales indet.		+	SW
Deciduous tree leaves indet.		+	SW
Leaf abscission pads indet.		+	SW
Rosa or *Rubus* sp. (rose or blackberry)	prickle	1	SW

Table 8.2 *Other waterlogged plant remains from Early Neolithic sediments in Area 32 (for abbreviations see text).*

Table 8.3 *Coleoptera (for abbreviations see text).*

Species	Minimum numbers of individuals									Habitat or food
	Column 4	Column 39	Column 40	A32 040	A14 W1	Column 55				
Depth (cms): to:	40 60	0 20	5 20			30 40	15 30	0 15	Total	
Sample weight (kg):	2	3	4.5	5	3	2	3	3	8	
CARABIDAE										
Carabus granulatus L.	-	-	1	1	1	-	-	-	-	T - often near water, in rotten trees
Carabus sp.	-	-	-	-	-	-	-	1	1	T
Elaphrus cupreus Duft.	-	-	-	1	1	-	-	-	-	MB - at edge of water
Loricera pilicornis (F.)	-	-	-	2	1	-	1	-	1	T - usually moist
Dyschirius globosus (Hbst.)	-	-	1	1	-	3	-	1	4	T - moist ground, M
Clivina collaris (Hbst.) or *fossor* (L.)	-	-	-	2	-	-	1	2	3	moist T, often under dung
Trechus obtusus Er. or *quadristriatus* (Schr.)	-	-	-	1	-	-	1	2	3	T
T. secalis (Pk.)	-	-	-	1	-	-	-	4	4	moist G and W, B
Asaphidion flavipes (L.)	-	-	-	1	-	-	-	-	-	G - damp BMW
Bembidion lampros (Hbst.) or *properans* Step.	-	-	-	-	-	1	-	-	1	T
B. dentellum (Thun.)	-	-	-	-	-	-	1	-	1	M and (G) - well vegetated
B. semipunctatum Don	2	1	-	-	-	-	-	-	-	B - on sand
B. tetracolum Say	-	-	-	-	1	-	-	2	2	TM
B. gilvipes Sturm	-	-	-	-	-	-	1	1	2	(W) B also wet meadowland
B. assimile Gyl.	-	-	-	-	2	-	-	1	1	BM - well vegetated and close to water
B. assimile Gyl. or *clarki* Daw.	1	-	-	-	-	-	-	-	-	BMW - well vegetated and close to water
B. doris (Pz.)	-	-	2	1	-	-	-	-	-	BM
B. biguttatum (F.)	-	-	1	-	-	-	-	1	1	BG and W, usually near water
B. guttula (F.)	-	-	1	-	-	-	1	-	1	MG & W - moist
B. guttula (F.) or *mannerheimi* Sahl.	1	-	-	-	2	-	-	-	-	MG & W - moist (in manure heaps)
Bembidion spp.	-	-	1	3	-	-	-	-	-	mostly in wet or marshy places
Pterostichus anthracinus (Pz.)	-	-	-	1	-	-	-	-	-	M & B - shaded
P. cf. *gracilis* (Dej.)	-	1	-	-	-	-	-	1	1	G & W - wet, M
P. melanarius (Ill.)	-	-	-	-	-	-	-	1	1	DG (W)
P. minor (Gyl.)	-	1	-	-	1	1	-	1	2	M & B - both wooded and open
P. nigrita (Pk.)	-	-	-	1	-	1	-	-	1	MB
P. oblongopunctatus (F.)	-	-	1	-	-	-	-	-	-	W
P. strenuus (Pz.)	-	-	1	3	-	1	-	-	1	T - often near water
P. cf. *strenuus* (Pz.)	-	-	1	-	-	-	-	-	-	T - often near water
P. cupreus (L.) or *versicolor* (Sturm)	-	-	-	-	-	1	-	-	1	G (DW)

Species	Minimum numbers of individuals									Habitat or food
	Column 4	Column 39	Column 40	A32 040	A14 W1	Column 55				
Depth (cms): to:	40 60	0 20	5 20			30 40	15 30	0 15	Total	
Sample weight (kg):	2	3	4.5	5	3	2	3	3	8	
Calathus melanocephalus (L.)	-	-	-	-	1	-	-	-	-	GD (W)
Synuchus nivalis (Pz.)	-	-	-	-	-	-	-	2	2	G & D - often in sandy or gravelly places
Agonum obscurum (Hbst.)	-	-	1	-	-	-	-	-	-	W - wet. M - well vegetated
Agonum sp.	-	-	1	-	1	-	-	-	-	mostly wet habitats
Amara aulica (Pz.)	-	-	-	-	-	-	-	1	1	G & D - often feeding on Compositae seeds
A. tibialis (Pk.)	-	-	-	-	-	-	1	-	1	D & (G) usually sandy & open
Amara sp. (not *aulica*)	-	-	-	2	-	-	-	1	1	T
Harpalus rufipes (Deg.)	-	-	-	-	-	-	-	1	1	D - often cultivated (G)
Bradycellus sp.	-	-	-	-	-	-	-	1	1	T
Badister bipustulatus (F.)	-	-	-	-	-	-	-	1	1	mostly wet places
Dromius melanocephalus Dej.	-	-	-	-	-	-	-	1	1	G
D. quadrimaculatus (L.)	-	-	-	-	-	-	-	1	1	W - often on trees
Metabletus cf. *truncatellus* (L.)	-	-	-	-	-	-	-	2	2	G (D)
HALIPLIDAE										
Haliplus sp.	-	-	-	-	-	2	2	3	7	A
DYTISCIDAE										
Hydroporus sp.	1	-	-	-	1	1	-	1	2	A
Potamonectes depressus (F.)	-	-	-	-	-	-	1	-	1	A
Stictotarsus duodecimpustulatus (F.)	-	1	-	-	1	1	-	-	1	A - running water and lakes
Platambus maculatus (L.)	-	-	-	-	-	-	1	-	1	A - running water and lakes
Agabus bipustulatus (L.)	-	-	1	-	-	-	1	-	1	A - still water, ponds, puddles and ditches
Agabus sp. (not *bipustulatus*)	-	-	-	6	-	-	1	-	1	A
Colymbetes fuscus (L.)	-	-	-	4	-	-	-	-	-	A - stagnant water, ponds and ditches
Dytiscus sp.	-	-	-	-	1	-	-	-	-	A - mostly stagnant
GYRINIDAE										
Gyrinus sp.	-	-	1	-	2	-	-	-	-	A
Orectochilus villosus (Müll.)	-	-	2	2	2	1	1	1	3	A - running water
HYDROPHILIDAE										
Hydrochus sp.	1	-	1	-	-	-	-	-	-	A - mostly stagnant
Helophorus aquaticus (L.)	-	-	-	-	-	-	1	1	2	A - puddles, ponds, rarely flowing water

Species	Column 4	Column 39	Column 40	A32 040	A14 W1	Column 55 30–40	Column 55 15–30	Column 55 0–15	Total	Habitat or food
Depth (cms): to:	40 60	0 20	5 20			30 40	15 30	0 15		
Sample weight (kg):	2	3	4.5	5	3	2	3	3	8	
H. arvernicus Muls.	-	-	-	1	1	-	-	-	-	A - clean, flowing
H. grandis Ill.	-	-	-	-	-	1	-	-	1	A - puddles, ponds, rarely flowing water
H. aquaticus (L.) or *grandis* Ill.	-	-	-	1	-	-	2	1	3	A - puddles, ponds, rarely flowing water
Helophorus spp. (*brevipalpis* size)	1	-	-	1	1	4	10	7	21	A - but readily leave water
Cercyon haemorrhoidalis (F.)	-	-	-	-	-	1	-	-	1	FV
C. melanocephalus (L.)	-	-	-	-	-	-	1	-	1	mostly F
C. pygmaeus (Ill.)	-	-	-	-	-	-	1	-	1	mostly F
C. sternalis Sharp.	-	-	-	4	-	-	-	-	-	MB - on deep ground
C. cf. *tristis* (Ill.)	-	-	-	-	1	-	-	-	-	under plant detritus
Cercyon spp.	1	-	2	-	-	-	-	1	1	FVC, some species on wet mud
Megasternum obscurum (Marsh.)	-	-	1	10	-	5	3	3	11	FVC
Hydrobius fuscipes (L.)	-	-	-	1	-	2	1	1	4	A - stagnant water often with detritus bottom
Anacaena bipustulata (Marsh.) or *limbata* (F.)	-	-	1	-	1	2	2	1	5	G & W in wet places, VA
Laccobius sp.	-	-	-	1	1	1	1	1	3	A
Hydrophilus piceus (L.)	-	-	-	1	-	-	-	-	-	A - well vegetated ponds and marsh dykes
HISTERIDAE										
Abraeus globosus (Hoff.)	-	-	-	1	-	-	-	-	-	rotten wood
HYDRAENIDAE										
Ochthebius bicolon Germ.	-	-	1	-	-	1	-	-	1	B - mud and decaying vegetation at water's edge. A
O. cf. *bicolon* Germ.	-	-	10	5	2	-	1	2	3	B - as above. A
O. minimus (F.)	3	-	10	4	1	-	2	1	3	B - as above, A
O. cf. *minimus* (F.)	-	1	-	11	1	3	-	1	4	B - as above, A
Hydraena minutissima Step.	-	-	1	7	-	-	1	-	1	A - flowing
H. pulchella Germ.	-	-	3	-	-	-	-	-	-	A - flowing
H. riparia Kug.	4	1	9	9	5	2	2	1	5	A
H. testacea Curt.	-	1	3	2	1	-	1	-	1	A - usually stagnant
Limnebius aluata (Bed.)	-	-	-	-	-	1	2	-	3	B - mud at water's edge, M(A)
L. papposus Muls.	-	-	-	-	-	1	1	1	3	A B - mud at water's edge
L. truncatellus (Thun.)	-	-	-	1	-	-	-	-	-	A
PTILIIDAE										
Ptenidium sp.	-	1	2	4	1	-	1	-	1	dung heaps, rotten wood, VM

Species	Minimum numbers of individuals									Habitat or food
	Column 4	Column 39	Column 40	A32 040	A14 W1	Column 55				
Depth (cms): to:	40 60	0 20	5 20			30 40	15 30	0 15	Total	
Sample weight (kg):	2	3	4.5	5	3	2	3	3	8	
Ptiliidae indet. (not *Ptenidium*)	-	-	2	4	1	-	-	2	2	VM (TF)
LEIODIDAE										
Colenis immunda (Sturm.)	-	-	-	1	-	-	-	-	-	mouldy leaves and soil fungi
SILPHIDAE										
Silpha atrata L.	-	-	-	2	1	-	1	1	2	mostly under bark or in rotten wood (GDV)
S. tristis Ill.	-	-	-	-	1	-	-	-	-	esp. C also DGW
SCYDMAENIDAE										
Scydmaenidae indet.	-	-	1	-	1	-	-	-	-	TV
STAPHYLINIDAE										
Metopsia retusa (Step.)	-	-	1	-	-	-	-	-	-	GV
Olophrum fuscum (Grav.)	-	-	-	-	-	-	-	1	1	M - often under dead vegetation
Lesteva longoelytrata (Gz.)	-	-	-	-	-	-	2	1	3	B - often at water's edge, M
Carpelimus bilineatus Step.	-	-	-	3	1	-	1	-	1	B - on wet mud (GV & F on wet soils)
C. cf. *corticinus* (Grav.)	-	-	3	1	2	-	1	-	1	B - on wet mud and sand M
C. rivularis (Mots.)	-	-	-	-	1	-	-	-	-	B & M on mud or in V
Platystethus arenarius (Fouc.)	-	-	-	-	1	-	-	-	-	FV
P. cornutus gp.	-	-	-	-	-	-	-	1	1	M & B - often on mud (VF)
P. nodifrons (Man.)	-	-	3	1	-	-	-	-	-	V
Anotylus nitidulus (Grav.)	-	-	-	-	-	-	1	1	2	VFC(M)
A. rugosus (F.)	-	1	2	2	-	2	1	2	5	FV(C)
A. sculpturatus gp.	-	-	-	1	-	1	-	2	3	FVC(also T)
Oxytelus sculptus Grav.	-	-	-	-	-	1	-	-	1	FV(C)
Stenus spp.	1	1	3	5	5	-	2	4	6	TM
Paederus littoralis Grav.	-	-	-	-	-	-	-	1	1	G&D - mostly dry
Lathrobium longulum Grav.	-	-	-	3	-	-	-	-	-	T
Lathrobium sp. (not *longulum*)	-	-	3	2	-	1	1	1	3	TV(C)
Sunius sp.	-	-	-	-	-	1	-	-	1	V
Rugilus cf. *erichsoni* (Fauv.)	-	-	-	1	-	-	-	-	-	V(G)
R. geniculatus (Er.)	-	-	-	-	-	-	-	1	1	V(G)
Gyrohypnus angustatus Step.	-	1	-	-	-	1	-	-	1	V - sometimes at water's edge
Xantholinus linearis (Ol.) or *longiventris* Heer	-	-	-	-	-	1	1	-	2	WGV(FC)
Philonthus spp.	-	-	-	3	-	1	-	4	5	FVC(T)
Gabrius sp.	-	-	-	1	-	-	-	-	-	WGFVC

Species	Minimum numbers of individuals									Habitat or food
	Column 4	Column 39	Column 40	A32 040	A14 W1	Column 55				
Depth (cms): to:	40 60	0 20	5 20			30 40	15 30	0 15	Total	
Sample weight (kg):	2	3	4.5	5	3	2	3	3	8	
Tachyporus sp.	-	-	-	1	-	1	-	-	1	T
Tachinus sp.	-	-	1	1	1	-	-	-	-	T
Aleocharinae indet.	1	-	2	6	2	4	4	4	12	TFCV
PSELAPHIDAE										
Pselaphidae indet.	1	-	2	2	2	-	2	-	2	V(GWM)
LUCANIDAE										
Dorcus parallelipipedus (L.)	-	-	1	1	1	-	-	-	-	rotten hardwood
GEOTRUPIDAE										
Geotrupes sp.	-	-	1	1	1	-	-	-	-	F
SCARABAEIDAE										
Aphodius ater (Deg.)	-	-	-	1	-	-	-	-	-	FV
A. depressus (Kug.)	-	-	-	1	-	-	-	-	-	F
A. foetens (F.)	-	-	-	1	-	-	-	-	-	F
A. pusillus (Hbst.)	-	-	-	1	-	1	1	2	4	FV
A. cf. *sphacelatus* (Pz.)	-	-	-	4	-	2	3	2	7	FVC
Aphodius sp.	-	-	-	1	-	-	1	-	1	mostly F
Onthophagus joannae Golj. (*ovatus*)	-	-	-	-	-	2	2	1	5	FCV
Onthophagus sp. (not *joannae*)	-	-	-	1	-	-	-	-	-	F(C)
Hoplia philanthus (Fues.)	-	-	-	-	-	1	1	-	2	larvae on roots in permanent grassland
Phyllopertha horticola (L.)	-	-	-	-	-	-	3	1	4	larvae on roots in permanent grassland
DASCILLIDAE										
Dascillus cervinus (L.)	-	-	-	1	-	-	-	-	-	on flowers and bushes
SCIRTIDAE										
cf. *Cyphon* sp.	-	-	2	1	-	2	1	1	4	larvae A, adults T, but close to water and M
DRYOPIDAE										
Helichus substriatus (Müll.)	1	-	-	2	1	1	-	1	2	A & water's edge
Dryops sp.	3	-	5	4	1	-	-	1	1	BA &M in or close to water (V)
ELMIDAE										
Elmis aenea (Müll.)	-	-	1	1	1	-	-	-	-	A - clean flowing water, clinging to stones and aquatic plants
Esolus parallelepipedus (Müll.)	1	-	6	8	3	-	-	2	2	A - as above
Limnius volckmari (Pz.)	1	-	6	7	1	-	1	-	1	A - as above

Species	Minimum numbers of individuals									Habitat or food
	Column 4	Column 39	Column 40	A32 040	A14 W1	Column 55				
Depth (cms): to:	40 60	0 20	5 20			30 40	15 30	0 15	Total	
Sample weight (kg):	2	3	4.5	5	3	2	3	3	8	
Macronychus quadrituberculatus (Műll.)	3	-	4	11	4	1	-	-	1	A - on submerged decaying wood
Normandia nitens (Műll.)	18	3	23	28	14	5	7	8	20	A - clean flowing water clinging to stones and aquatic plants
Oulimnius sp.	11	2	24	13	8	18	25	13	56	A - mostly clean, flowing
Stenelmis canaliculata (Gyl.)	-	2	6	5	1	1	2	3	6	A - clean flowing, also large lakes
ELATERIDAE										
Agrypnus murinus (L.)	-	-	-	1	1	1	1	1	3	G
Melanotus erythropus (Gm.)	-	-	-	1	-	-	-	-	-	rotten wood
Athous hirtus (Hbst.)	1	1	-	1	1	-	-	-	-	WG - esp. meadowland, larvae esp. on the roots of grasses, also trees and shrubs
Selatosomus nigricornis (Pz.)	-	-	-	2	-	-	-	-	-	G - at woodland edge, W
Agriotes obscurus (L.)	-	1	-	1	-	-	-	-	-	larvae mostly on roots of grassland plants
A. sputator (L.)	-	-	-	-	-	3	-	-	3	as above
Agriotes sp.	-	-	-	1	1	-	1	1	2	as above
Synaptus filiformis (F.)	-	-	-	1	-	-	-	-	-	G - damp. M
THROSCIDAE										
Trixagus dermestoides (L.)	-	-	-	1	-	-	-	-	-	rotten wood, leaf litter
T. elateroides (Heer)	-	-	-	1	-	-	-	-	-	grassy places at the edge of woodland
Trixagus sp.	-	1	-	-	-	-	-	-	-	G W
EUCNEMIDAE										
Melasis buprestoides (L.)	-	-	-	1	-	-	-	-	-	in rotten hardwood
CANTHARIDAE										
Cantharis sp.	-	-	-	1	-	-	-	-	-	adults on flowers of herbs and shrubs
Rhagonycha sp.	-	-	-	-	1	-	-	-	-	as above
Cantharis or *Rhagonycha* sp.	-	-	-	1	-	1	-	-	1	as above
ANOBIIDAE										
Ptinomorphus imperialis (L.)	-	-	-	1	-	-	-	-	-	dead hardwood twigs
Grynobius planus (F.)	-	-	3	2	1	-	-	-	-	dead hardwood
cf. *Ochina ptinoides* (Marsh.)	-	-	1	-	-	-	-	-	-	dead dry *Hedera* wood
Anobium punctatum (Deg.)	1	-	2	4	1	-	-	-	-	dead wood
PTINIDAE										

Species	Minimum numbers of individuals									Habitat or food
	Column 4	Column 39	Column 40	A32 040	A14 W1	Column 55				
Depth (cms): to:	40 60	0 20	5 20			30 40	15 30	0 15	Total	
Sample weight (kg):	2	3	4.5	5	3	2	3	3	8	
Ptinus fur (L.)	-	-	-	1	1	-	-	-	-	straw and birds' nests etc. P - grain (C, old wood)
MELYRIDAE										
Aplocnemus nigricornis (F.) or *pini* Redt.	-	-	1	-	1	-	-	-	-	larvae in decaying wood
NITIDULIDAE										
Brachypterus urticae (F.)	-	-	2	-	-	-	-	-	-	*Urtica* sp.
Meligethes sp.	-	-	-	-	-	-	1	1	2	herbs and trees - mostly on flowers
Epuraea sp.	-	-	-	1	-	-	1	-	1	in fungi, under bark, at sap, on flowers
CRYPTOPHAGIDAE										
Cryptophagidae indet. (not *Atomaria*)	-	-	-	1	-	-	-	-	-	V - of all sorts, T
Atomaria sp.	-	-	-	1	-	-	1	-	1	VT(F)
PHALACRIDAE										
Stilbus sp.	-	-	-	-	-	-	-	1	1	in dry grass, hay and on *Typha* sp.
CORYLOPHIDAE										
Corylophus cassidoides (Marsh.)	-	-	1	3	-	1	1	1	3	V - esp. decaying reeds
Orthoperus sp.	-	-	-	-	-	2	1	3	6	V
COCCINELLIDAE										
Platynaspis luteorubra (Gz.)	-	-	-	-	-	-	-	1	1	on trees and in dry grassland
Propylea quattuordecimpunctata (L.)	-	-	-	-	-	-	1	-	1	T
LATHRIDIIDAE										
Enicmus fungicola Thom. or *rugosus* (Hbst.)	-	-	1	-	-	-	-	-	-	fungi on trees and under bark
E. transversus (Ol.)	-	-	-	-	-	1	-	-	1	V(GW)
Corticariinae indet.	2	-	-	1	1	2	1	-	3	mostly V
MYCETOPHAGIDAE										
Mycetophagus cf. *multipunctatus* F.	-	-	-	1	-	-	-	-	-	in tree fungi and mouldy wood
TENEBRIONIDAE										
Corticeus unicolor P. & M.	-	-	-	1	-	-	-	-	-	under rotten bark esp. of *Quercus*
MELANDRYIDAE										
Anaspis sp.	-	-	-	-	-	-	1	-	1	adults on blossom esp. shrubs
OEDEMERIDAE										
Ischnomera cf. *caerulea* (L.)	-	-	-	-	1	-	-	-	-	rotten wood esp. *Quercus*
CERAMBYCIDAE										

Species	Column 4	Column 39	Column 40	A32 040	A14 W1	Column 55				Habitat or food
Depth (cms): to:	40 60	0 20	5 20			30 40	15 30	0 15	Total	
Sample weight (kg):	2	3	4.5	5	3	2	3	3	8	
Grammoptera sp.	-	-	2	-	-	-	-	-	-	larvae on twigs of deciduous trees
Phymatodes alni (L.)	-	-	-	1	-	-	-	-	-	recently dead hardwood with bark on esp. *Quercus*
P. testaceus (L.)	-	-	-	-	1	-	-	-	-	dead hardwood with bark on
Tetrops praeusta (L.)	-	-	-	1	-	-	-	-	-	small dry dead hardwood branches
BRUCHIDAE										
Bruchus or *Bruchidius* sp.	-	-	-	-	-	-	1	-	1	on Leguminosae
CHRYSOMELIDAE										
Macroplea appendiculata (Pz.)	-	-	-	1	-	2	1	2	5	*Potamogeton* and *Myriophyllum* spp.
Donacia cinerea Hbst.	-	-	-	-	-	-	1	-	1	various reedswamp monocotyledons
D. clavipes F.	2	-	-	-	-	7	7	2	16	*Phragmites australis* (? & *Phalaris* sp.)
D. crassipes F.	-	-	-	-	-	1	1	-	2	*Nymphaea alba* L. & *Nuphar lutea* (L.) Sm.
D. dentata Hoppe	1	-	2	3	1	2	-	1	3	*Sagittaria sagittifolia* L. & *Alisma* spp.
D. impressa Pk.	-	-	-	-	1	2	2	-	4	*Schoenoplectus lacustris* (L.) Pal.
D. simplex F.	-	-	1	2	-	-	-	-	-	various reedswamp monocotyledons
D. versicolorea (Brahm)	-	-	1	-	-	-	-	-	-	floating-leaved species of *Potamogeton*
Donacia sp.	-	-	-	1	-	-	-	-	-	various reedswamp monocotyledons
Donacia or *Plateumaris* sp.	1	1	-	1	-	-	-	1	1	various aquatic plants
Timarcha tenebricosa (F.)	-	-	-	-	-	-	-	1	1	esp. *Galium* spp. in grassy places
Chrysolina polita (L.)	-	-	-	1	-	-	1	-	1	Labiatae - often in marshes
Gastrophysa polygoni (L.)	-	-	1	-	-	-	-	-	-	*Rumex* and *Polygonum* spp.
Phaedon sp. (not *tumidulus*)	-	-	-	1	-	-	-	-	-	various herbs
Prasocuris phellandrii (L.)	1	-	1	-	-	1	-	-	1	aquatic Umbelliferae
Phyllodecta sp.	3	1	-	-	-	-	-	-	-	*Salix* and *Populus* spp.
Galerucella sp.	-	-	-	2	-	-	-	-	-	various herbs
Agelastica alni (L.)	-	-	1	4	1	-	-	-	-	*Alnus glutinosa* (L.) Gaert.
Phyllotreta cf. *consobrina* (Curt.)	-	-	-	-	-	1	-	2	3	Cruciferae and *Reseda* spp.
P. nigripes (F.)	-	-	-	-	-	2	-	1	3	Cruciferae and *Reseda* spp.
P. nemorum (L.) or *undulata* Kuts.	-	-	-	2	-	1	2	-	3	Cruciferae and *Reseda* spp.
P. vittula Redt.	-	-	-	1	-	4	5	5	14	Cruciferae and *Reseda* spp.

Species	Minimum numbers of individuals									Habitat or food
	Column 4	Column 39	Column 40	A32 040	A14 W1	Column 55				
Depth (cms): to:	40 60	0 20	5 20			30 40	15 30	0 15	Total	
Sample weight (kg):	2	3	4.5	5	3	2	3	3	8	
Aphthona nonstriata (Gz.)	-	-	-	-	-	1	-	-	1	*Iris pseudacorus* L.
Longitarsus spp.	-	-	-	1	-	8	4	3	15	various herbs
Altica sp.	-	-	-	-	-	1	2	3	6	includes *Corylus, Salix, Rumex* & *Epilobium* spp.
Batophila rubi (Pk.)	-	-	-	-	-	-	-	1	1	*Rubus* spp. (*Fragaria* sp.)
Crepidodera ferruginea (Scop.)	-	-	-	-	-	1	-	2	3	various herbs
Chalcoides sp.	1	-	-	-	-	-	-	-	-	*Salix* and *Populus* spp.
Epitrix pubescens (Koch)	1	-	-	1	1	-	2	-	2	*Solanum* spp., *Hyoscyamus* sp.
Chaetocnema concinna (Marsh.)	-	-	-	-	1	1	3	2	6	Polygonaceae esp. *P. aviculare* agg.
Chaetocnema sp. (not *concinna*)	-	-	-	-	-	-	-	1	1	various herbs
Psylliodes sp.	-	-	-	3	1	-	-	1	1	various herbs
APIONIDAE										
Apion urticarium (Hbst.)	-	-	-	-	-	-	-	1	1	*Urtica dioica* L. & *U. urens* L.
Apion spp. (not above)	-	-	-	1	-	2	2	-	4	various herbs
CURCULIONIDAE										
Phyllobius roboretanus Gred. or *viridiaeris* (Laich.)	-	-	-	-	-	-	1	-	1	trees, shrubs and Compositae
Phyllobius sp.	-	-	-	-	-	2	-	1	3	trees, shrubs and herbs
Polydrusus mollis (Strőm.)	-	-	-	1	-	-	-	-	-	deciduous shrubs and trees
Phyllobius or *Polydrusus* sp.	-	-	1	1	-	-	-	-	-	trees, shrubs and herbs
Barypeithes araneiformis (Schr.)	-	-	-	-	-	-	1	-	1	T
Brachysomus echinatus (Bons.)	-	-	-	-	-	-	-	1	1	various plants
Liophloeus tessulatus (Műll.)	-	-	-	-	-	-	1	-	1	larvae esp. on *Heracleum sphondylium* L.
Barynotus obscurus (F.)	-	-	-	-	-	1	2	1	4	T
Sitona hispidulus (F.)	-	-	-	-	-	-	1	-	1	Leguminosae esp. *Medicago* and *Trifolium* spp.
Sitona sp.	-	-	-	-	-	2	1	2	5	Leguminosae esp. *Trifolium* spp.
Hypera sp. (not *punctata*)	-	-	-	-	-	-	-	1	1	various herbs
Alophus triguttatus (F.)	-	-	-	-	-	1	3	-	4	various herbs
Rhyncolus truncorum (Germ.)	-	-	1	1	-	-	-	-	-	sapwood of various hardwoods mostly in large woods
Dryophthorus corticalis (Pk.)	-	-	1	-	-	-	-	-	-	wood
Cossoninae indet.	-	-	1	1	-	-	-	-	-	wood
Bagous spp.	1	-	2	9	1	-	2	-	2	aquatic plants

Species	Minimum numbers of individuals									Habitat or food
	Column 4	Column 39	Column 40	A32 040	A14 W1	Column 55				
Depth (cms): to:	40 60	0 20	5 20			30 40	15 30	0 15	Total	
Sample weight (kg):	2	3	4.5	5	3	2	3	3	8	
Cidnorhinus quadrimaculatus (L.)	-	-	-	1	1	1	1	-	2	*Urtica* spp.
Ceuthorhynchidius troglodytes (F.)	-	-	-	-	-	2	-	-	2	*Plantago lanceolata* L.
Ceutorhynchus atomus Boh.	-	-	-	-	-	-	-	1	1	Cruciferae
C. erysimi (F.)	-	-	-	-	-	1	-	-	1	Cruciferae
Ceuthorhynchinae indet.	-	-	2	2	-	1	1	2	4	various herbs
Curculio cf. *venosus* (Grav.)	-	-	-	1	1	-	-	-	-	*Quercus* sp.
Tychius sp.	-	-	-	-	-	-	-	1	1	mostly Leguminosae
Mecinus pyraster (Hbst.)	-	-	-	-	-	-	1	-	1	*Plantago lanceolata* L. and *P. media* L.
Gymnetron pascuorum (Gyl.)	-	-	-	-	-	1	-	-	1	*Plantago lanceolata* L.
G. veronicae (Germ.)	-	-	-	2	-	-	-	-	-	marsh and aquatic *Veronica* spp.
Rhynchaenus cf. *quercus* (L.)	-	-	-	1	-	-	-	-	-	*Quercus* leaves
Rhynchaenus sp. (not *pratensis*)	-	-	-	1	-	-	-	-	-	leaves of various trees
Ramphus pulicarius (Hbst.)	-	-	-	-	-	1	-	-	1	*Salix* spp., also *Populus* and *Betula* spp.
SCOLYTIDAE										
Scolytus cf. *intricatus* (Ratz.)	-	-	-	-	-	-	-	1	1	mainly young Quercus, also other hardwood
S. scolytus (F.)	-	-	1	-	-	-	-	-	-	mostly *Ulmus* spp.
Hylesinus oleiperda (F.)	-	-	1	-	-	-	-	-	-	mainly *Fraxinus* twigs
Leperisinus varius (F.)	-	-	-	2	-	-	-	-	-	mainly *Fraxinus*, also other trees
Acrantus vittatus (F.)	-	-	-	?.	-	-	1	-	1	mainly decaying *Ulmus*, also other trees
Xyleborus saxesini (Ratz.)	-	-	-	2	-	-	-	-	-	dead wood
PLATYPODIDAE										
Platypus cylindrus (F.)	-	-	-	1	-	-	-	-	-	recently dead hardwood esp. *Quercus*
Totals	71	23	192	313	109	146	171	166	483	

Table 8.4   *Other insects (for abbreviations see text).*

Species and anatomy		Minimum numbers of individuals									Habitat or food
		Column 4	Column 39	Column 40	A32 040	A14 W1	Column 55				
							30 40	15 30	0 15	Total	
Depth (cms): to:		40 60	0 20	5 20			30 40	15 30	0 15	Total	
ODONATA											
Agrion splendens (Har.) or *virgo* (L.)		-	-	-	1	-	-	1	1	2	nymphs A
Odonata gen. et sp. indet.		1	-	-	-	-	-	-	-	-	nymphs A
HEMIPTERA											
Pentatoma rufipes (L.)		-	-	1	1	1	-	-	-	-	deciduous trees esp. *Quercus* sp.
Drymus sylvaticus (F.)		-	-	-	1	-	-	-	-	-	T
Scolopostethus sp.		-	-	-	1	-	-	-	-	-	T
Anthocorinae gen. et sp. indet.		-	-	-	1	-	-	-	-	-	T
Gerris sp.		-	-	-	3	-	-	-	-	-	A- on water's surface
Megophthalmus scabripennis Ed. or *scanicus* (F.)		-	-	-	-	-	-	1	-	1	esp. grasses
Aphrodes bicinctus (Schr.)		-	-	-	-	-	1	2	1	4	grasses
A. flavostriatus (Don.)		-	-	-	-	-	1	1	1	3	grasses
Aphidoidea gen. et sp. indet.		-	-	-	1	-	1	-	-	1	
Homoptera gen. et sp. indet.		-	-	-	1	1	-	-	1	1	
TRICHOPTERA											
Ithytrichia lamellaris Eat. or *clavata* Mort.	larval case	18	3	20	164	16	3	6	7	16	A-running water
Orthotrichia sp.	larval case	-	-	-	-	1	-	-	-	-	A-stagnant or slowly moving
Trichoptera gen. et sp. indet.	larva	3	6	27	92	15	20	16	19	55	A
Trichoptera gen. et sp. indet.	larval case	1	-	-	-	-	-	1	-	1	A
HYMENOPTERA											
Lasius flavus gp.	worker	-	-	-	-	-	1	4	8	13	mounds in old pasture and at the edge of woodland
L. niger gp.	worker	-	-	-	-	-	2	-	-	2	T
L. flavus or *niger* gps.	female	-	-	-	-	-	1	2	-	3	T
Apis mellifera L.	worker	-	-	-	-	-	1	-	-	1	nest in hollow tree or hive
Hymenoptera gen. et sp. indet.		-	-	2	6	3	4	5	4	13	T
DIPTERA											
Chironomidae gen. et sp. indet.		+	-	-	+	+	-	+	+	+	A
Bibionidae gen. et sp. indet.	adult	-	-	-	-	-	-	2	-	2	FVC
Diptera gen. et sp. indet.	puparium	1	1	1	2	1	-	1	1	2	
Diptera gen. et sp. indet.	adult	2	1	-	3	2	7	4	3	14	

CHAPTER 9

Palynology and palaeoenvironment

ROB SCAIFE

Introduction and aims

Pollen investigation by Greig (Greig 1991, 1992) of the waterlogged deposits found in the rescue excavation provided detailed information relating to the Middle to Late Neolithic and Late Bronze Age periods. The further excavations reported in this volume have demonstrated that archaeology and sediments present in other areas of the floodplain span a larger part of the Holocene period. These sedimentary units afforded the possibility of further pollen analysis, and thus the potential for providing a vegetational and environmental history of the Thames floodplain around Runnymede. No single sedimentary sequence spans the whole time period represented and, therefore, a total of eight individual pollen profiles have been examined and the data integrated. The oldest of these dates back to the early Boreal period at *c.* 8000–7500 BC (*c.* 9000–8500 BP) and the most recent to the Middle to Late Bronze Age, late second millennium BC (*c.* 2900 BP). From these profiles, a picture of vegetation change has been constructed which establishes the Boreal pine-hazel vegetation of early Holocene Flandrian chronozone Ib/c; the mid-Holocene climax woodland (Atlantic; Flandrian chronozone II) and the disturbances and modification of this woodland by first Neolithic and then Bronze Age communities in the later Holocene (Flandrian chronozone III). Site-specific data are presented in a broad chronological order with a brief discussion of the inherent characteristics of each column and, where appropriate, pollen assemblage zones are defined. An attempt has then been made to provide a vegetation and environmental history spanning the periods delimited above.

Methodology

Sub-samples for pollen analysis were extracted from monolith columns which were taken during excavation. All of the samples contained a substantial inorganic/minerogenic component and were highly alkaline. Given these factors, rigorous pollen extraction procedures were in most cases required. Extraction procedures followed those outlined by Moore and Webb (1978) and Moore *et al.* (1991) but with the addition of micromesh sieving (10 µm) where necessary. Samples were decalcified with 10% HCl and deflocculated with 8% KOH. Coarse debris was removed through sieving at 150 µm and clay by micromesh (10 µm). Remaining silica (silt fraction) was digested with 40% hydrofluoric acid. Erdtman's acetolysis was carried out for removal of cellulose. In some cases, absolute pollen frequencies were calculated using the addition of a known number of exotic markers (Stockmarr 1971). The concentrated pollen and spores were stained with safranin and mounted in glycerol jelly. Pollen was identified and counted using an Olympus biological research microscope with phase contrast facility at magnifications of × 400 and × 1000.

These extraction procedures were not always successful, especially in the upper sections of the alluvial units. In general, where sediments were taken from above the present water table, pollen was absent. This is attributed to fluctuating water levels and oxidation of the microfossils in these predominantly calcareous sediments. This preserving environment differs from Greig's earlier columns in that the circumstances then gave access to much deeper profiles below the water table (also enabling preservation of seeds). However, the eight profiles (Table 9.1) yielded pollen and enabled the plotting of pollen diagrams. Because of the variable microfossil preservation in the different columns, different total pollen counts were made. Pollen sums are in some cases based on counts of as low as 100 grains, where pollen was sparse, and a standard of 400 or greater where preserving conditions were more favourable. Calculations are based on percentage of dry land pollen – t.d.l.p., that is, the pollen sum.

Table 9.1 *Columns selected for pollen analysis.*

Column no.	Area	Samples	Chronozone	Parcel
31	19	P3-4	Flandrian I	1C
4	13	P9-13	Flandrian I	1E
39	22	P5-6	Flandrian I/II	1D
40	22	P7-8	Flandrian II	1F-1G
59	32	P2	Flandrian II	1H
60	32	P1	Flandrian II	2D (2F)
5	14	P63-64	Flandrian III	2D-2E
55	31	P9-11	Flandrian III	4B

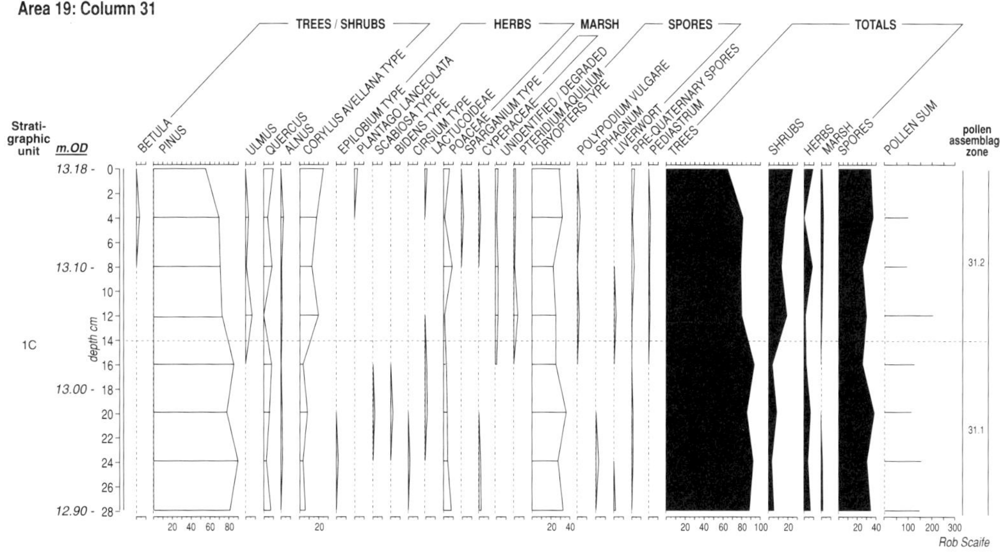

9.1 *Pollen diagram for column 31, Area 19: early Boreal. Details of pollen sum calculations are given in the text.*

Alnus has not, however, been excluded from the pollen sum. Marsh and aquatic totals and spores are calculated as a percentage of this sum. These percentage data have been plotted using *Tilia* and *Tilia* graph. Plant taxonomy/ nomenclature in general follows that of Stace (1991) and for pollen, that of Moore and Webb (1978), Moore *et al.* (1991), but modified according to Bennett *et al.* (1994) to take into account the taxonomy of Stace (1991).

Each column analysed is described below and where pollen assemblage zones have been recognised these are defined. A brief interpretation is also provided for each column followed by a fuller discussion given in relation to the overall vegetational and environmental history.

Results

Column 31: ERB89, Area 19, grid 34.0/15.6

Lithostratigraphy: Two overlapping monoliths were taken from between 12.88 and 13.18 m OD. Lithologically, these are homogeneous, highly calcareous marls containing silt and clay, representing the downward continuation of context 19.809.

Pollen biostratigraphy (Fig. 9.1): Samples were examined at intervals of 4 cm. Because of the highly calcareous nature of this unit and the fluctuating ground water table noted above, pollen preservation was poor. Absolute pollen frequencies were correspondingly small, being in the order of hundreds of grains rather than the normal thousands of grains per ml. A pollen sum of 100 grains at minimum, but averaging 200 grains, was counted for each level. Some evidence of change is seen in the pollen profile and two pollen assemblage zones (p.a.z.) may be tentatively recognised. For the purpose of

description these have been designated from the base upwards as 31:1 and 31:2 and are characterised as follows:

POLLEN ASSEMBLAGE ZONE 31:1
(28–14 cm; 12.90–13.04 m OD)
Pinus–Quercus

This pollen zone comprises four levels, spanning the lower half of the pollen profile. This has been delimited from subsequent levels by the dominance of *Pinus* which attains maximum values of up to 90% of the total pollen. Other arboreal taxa include *Quercus* (to 8%) and sporadic/single occurrences of *Alnus*. *Corylus avellana* type (10%) is the only shrub. Non arboreal/shrub pollen are represented by small percentage values (7%) and include consistent Poaceae (7% at the base of the zone) and the presence of *Epilobium* type, *Scabiosa* type, Asteraceae group Lactucoideae, *Bidens* type, *Cirsium* type. Few marsh taxa are present with only occasional occurrences of Cyperaceae. *Dryopteris* type (monolete) is dominant in the spore category (to 33%).

POLLEN ASSEMBLAGE ZONE 31:2
(14–1 cm; 13.04–13.18 m OD)
Pinus–Quercus–Ulmus–Corylus avellana type

Pinus percentages decrease slightly to 60–70%, but it remains dominant with *Quercus*, the latter continuing at similar percentages to those in the spectra of zone 1. The incoming of *Ulmus* is significant, after which it occurs sporadically. Single grains of *Betula* and *Alnus*, are noted. Shrubs are represented only by *Corylus avellana* type which, along with *Pinus*, is dominant (20%). Herb pollen are higher in number and diversity. Poaceae have the highest percentages (to 9% at 8 cm). Sporadic occurrences of *Plantago lanceolata* and Asteraceae group Lactucoideae, and

169

wetland taxa *Typha angustifolia/Sparganium* type occur. Spores of ferns are represented by *Pteridium aquilinum*, monocolpate/*Dryopteris* type (to 20%) and *Polypodium vulgare*. Derived pre-Quaternary spores and freshwater algal *Pediastrum* are also noted.

INFERRED VEGETATION

This column is thought to have the oldest palynological sequence represented at Runnymede, being of early Holocene date from the eighth millennium BC (*c.* 9000–8000 BP) and representing Flandrian chronozone Ib; that is, the early Boreal period. Pollen assemblage zone 1 shows clearly the dominance of *Pinus* (*P.* cf. *sylvestris*) which is characteristic of the period around 8000 BC (9000 BP) in southern and south-east England (Godwin 1975a, 1975b; Scaife 1980, 1982, 1987; Birks *et al.* 1975; Birks 1989; Bennett 1984) and probably replaced, through its competitive ability, the *Betula*-dominated woodland of the earlier Flandrian chronozone Ia. The relatively low values of *Ulmus* and *Quercus* are similarly characteristic of this period, although it is not clear whether oak and elm were lesser constituents of the local flora or whether we are here seeing a long-distance/extra-regional component coming from early Holocene expansion of these trees into southern England. It seems, however, that in this section there is evidence of the start of the rise to dominance of deciduous woodland dominated by *Quercus* and *Ulmus* with other taxa including *Corylus avellana*. Sporadic, single occurrences of other tree taxa including *Betula* and *Alnus* are not considered important since these are anemophilous and were probably from extra-local sources. Thus, the overall vegetation of the better-drained soils adjacent to the river floodplain comprised predominantly pine woodland, possibly with hazel as an understorey (discussed further below) but with some evidence for oak and elm woodland expanding into this area during the late ninth millennium BC (*c.* 9500–9000 BP).

The environment of the floodplain itself is indicated by the presence of marginal aquatic herbs including *Typha angustifolia* type (which includes *Sparganium*) and algal *Pediastrum* which indicates a freshwater environment bounded by reed swamp.

Column 4: ERB85, Area 13, grid 34.0/44.5

This sequence has been dated to the early Holocene, Boreal (Flandrian chronozone Ib) at 7790 ± 80 BP (BM-2550) from stratified wood remains. This makes this profile later than the interpreted date of column 31 above. Samples for pollen analysis were obtained between 12.71 and 13.51 m OD.

Lithostratigraphy: This sequence comprised a waterlogged homogeneous, grey-brown silt/clay with lenses of orange mottling and occasional shell layers. The unit was also somewhat organic but with no apparent macro-vegetative structure. The lower levels grade into black sandy material.

These sediments were apparently laid down in a palaeochannel under fairly low-energy conditions. Contexts 13.555, 570, 556 and perhaps the base of 557 are represented.

Pollen biostratigraphy (Fig. 9.2): Pollen preservation was generally good, contrasting, therefore, with column 31. This is largely due to the more organic character of the clayey silt and possibly more continuous waterlogging at greater depth. Absolute pollen frequencies were highest in the lower levels of the profile (attaining values of 400,000 grains per ml at 215 cm). These values decline upwards in the profile with absolute frequencies diminishing to 2–3,000 grains per ml. A pollen sum comprising a minimum of 300 grains was counted for each level, except for the upper two levels (185 cm and 190 cm) for which counts of 100 and 150 grains only were obtained. Changes in the overall pollen stratigraphy are not definitive enough to warrant any pollen assemblage zonation and it is therefore described as one unit/pollen assemblage zone: *Pinus–Quercus–Corylus avellana* type. It is noted, however, that *Pinus* and *Corylus avellana* type may show a small increase in the upper half of the stratigraphy (185–205 cm) while *Quercus* may show a relative decline. The negligible changes may imply that the sedimentary unit accumulated over a relatively short time-span.

Pinus (to 38%), *Quercus* (40%) and *Corylus avellana* type (to 51%) are dominant. *Ulmus* is consistent throughout with values of 5–6% TP (of total tree pollen). *Alnus* is present in the uppermost level (185 cm) to 5% and single occurrences of *Tilia*, cf. *Populus* and *Prunus* type are noted. In addition to *Corylus avellana* type, *Salix* (to 7%) is important, especially in view of its entomophily and usual underrepresentation in pollen spectra. Herb taxa are more diverse than seen in column 31 with Poaceae (to 10% at 205 cm) and the presence of Ranunculaceae, *Chenopodium* type, Fabaceae (*Lotus* type, *Vicia sylvatica* type), Apiaceae, *Rumex*, *Mentha* type, *Galium* and Asteraceae types. Wetland and aquatic herbs are present and include *Filipendula ulmaria*, *Succisa* type, *Nuphar lutea*, *Potamogeton* type, *Typha angustifolia/Sparganium* type, *Typha latifolia* and Cyperaceae. Spores comprise *Pteridium aquilinum*, monolete *Dryopteris* type (dominant) and sporadic *Polypodium vulgare* and an interesting botanical record of *Ophioglossum vulgatum*.

INFERRED VEGETATION

The radiocarbon dating of this sequence is of particular importance for the early Holocene vegetation of Runnymede. The substantially higher values of *Quercus*, *Ulmus* and *Corylus avellana* type suggest that this profile post-dates by some 1,000 years that of column 31, oak and elm having become fully established and, in the process, having ousted pine through superior competitive ability. High values of *Corylus avellana* are typical of the Boreal period with some researchers regarding its early expansion as a possible function of Early Mesolithic activities (see

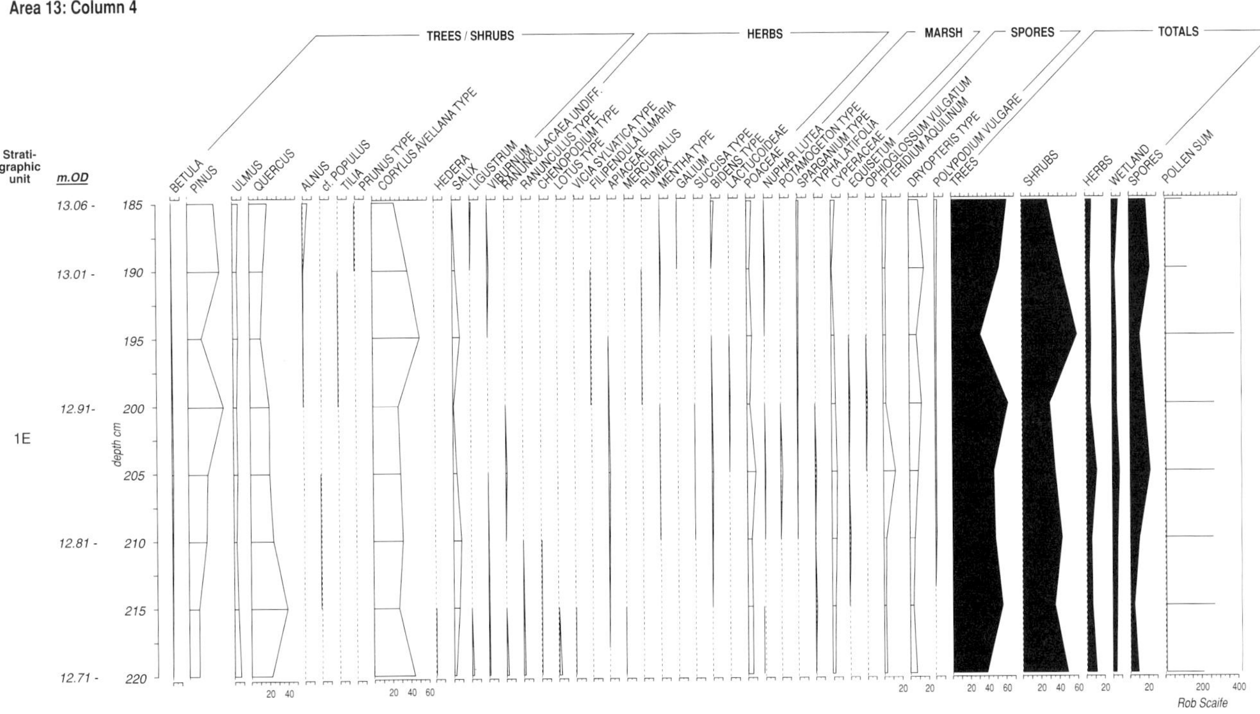

9.2 *Pollen diagram for column 4, Area 13: late Boreal.*

general discussion). The dominant herb taxon is Poaceae. However, herb pollen is sparse with the only real representation coming from taxa growing on the floodplain. The low-growing fern *Botrychium lunaria* (moonwort) is unusual as this taxon is somewhat rare outside Late Devensian floras and grows today in dry grassland usually in upland areas (Stace 1991). It is possible that its occurrence here is a relic from the Late Devensian/early Holocene. This lack of diversity reflects the heavily wooded character of the interfluves and the consequent small pollen catchment of the site. The date of the uppermost sample, in which *Alnus* percentages rise sharply, is not clear. Such a rise in the late Boreal at *c.* 6000 BC (*c.* 7000 BP) has in the past been used as one of the factors in ascertaining the Boreal–Atlantic transition, better attested at Runnymede in column 39 (below). A single occurrence of *Tilia* was also found at 195 cm and it is possible that this shows the arrival of the species during the late Boreal prior to separation of England from the continent as the Straits of Dover were inundated at the culmination of the Flandrian transgression (Godwin 1975a, 1975b).

During this period the wetland/floodplain was dominated by *Salix*. This taxon is frequently underrepresented in pollen spectra and percentages recorded here attest to the local importance of willow. Areas of reed swamp with *Sparganium* spp. or *Typha angustifolia*, *Typha latifolia* and Cyperaceae are also in evidence. It is probable that many of the herb pollen taxa present come from this community but, unfortunately,

the pollen morphology does not allow separation to species or even generic level which might place the types more clearly within an ecological group(s). However, the macrophytes, *Nuphar lutea* and *Potamogeton* type are evidence for a true aquatic habitat.

Column 39: ERB89, Area 22, grid 37.5/40.0

Lithostratigraphy: This profile of 31 cm spans 12.70 to 13.01 m OD. Highly humified basal peats are overlain by sand and silt and calcareous marls/silts. The stratigraphical units were identified as follows:

0–9 cm	White to pale grey calcareous silt/marl with fine sand (context 22.216).
9–12 cm	Greyer silt containing more organic material (context 22.215).
12–21 cm	Buff-coloured sand/silt, occasional specks of black organic material (context 22.214).
21–26 cm	Pale brown to grey sand and silt containing black organic fragments (context 22.213).
26–31 cm	Highly humified black peat (context 22.212).
31 cm >	Underlying sand?

Pollen biostratigraphy (Fig. 9.3): Pollen was surprisingly well preserved in this column and has provided evidence of the important transition from Flandrian chronozone I to chronozone II. Two pollen assemblage zones are recognised, divided at 18 cm depth (12.83 m OD).

171

Area 22: Column 39

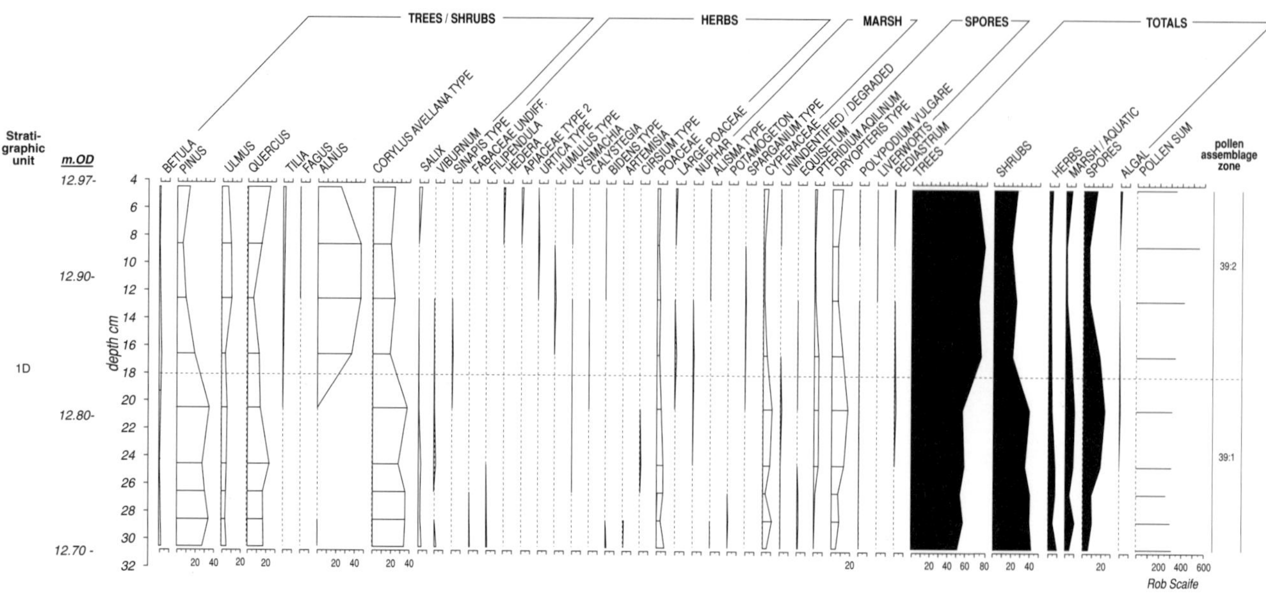

9.3   *Pollen diagram for column 39, Area 22: Boreal/Atlantic transition.*

POLLEN ASSEMBLAGE ZONE 39:1
(30–18 cm, 12.70–12.83 m OD, five levels)

Arboreal and shrub pollen are dominant (50% and 40% respectively). Tree pollen percentages are dominated by *Pinus* (to 35%) with *Quercus* (to 25%) and *Ulmus* (5%). Shrubs comprise *Corylus avellana* type (35%), *Salix* (5%) and sporadic *Viburnum*. Herbs are few but Poaceae are present continuously (to 8%) and aquatic/wetland taxa, with Cyperaceae (10%) and sporadic occurrences of *Nuphar*, *Alisma* type and *Potamogeton* type. Spores of ferns comprise monolete *Dryopteris* type (to 20%) with *Pteridium* and *Polypodium vulgare*.

POLLEN ASSEMBLAGE ZONE 39:2
(18–4 cm, 12.83–12.97 m OD, four levels)

This zone is defined by a very marked increase in *Alnus* (to 50%), an expansion of *Ulmus* (11%) and the first presence of *Tilia* and *Fagus*. *Pinus* percentages are reduced (statistical). *Quercus* appears to be increasing towards the top of the profile. The shrubs, *Corylus avellana* type (20%), *Salix* and herbs Poaceae (5%) have reduced values. Some large grains of the latter (> 45 µm) were recorded but were not of cereal type (i.e., thick exines, enlarged pores/annulus). Small numbers of algal *Pediastrum* are noted.

INFERRED VEGETATION

This pollen profile shows clear environmental change in the woodland communities on both the drier soils of the interfluves and on the river floodplain. The clearest aspects are a marked expansion of *Alnus* in p.a.z. 39:2 which appears to replace *Salix* in p.a.z. 39:1. *Salix* is poorly represented in

pollen spectra and even the small numbers found here suggest on-site growth of a willow community associated with a typical ground flora of Cyperaceae. This community was replaced by alder (?carr) which rapidly became dominant. The ground flora is poorly represented but *Humulus* type (including hop and hemp but with the former most probable here), *Lysimachia* and Cyperaceae are found. The sporadic presence of full aquatic taxa (*Nuphar*, *Potamogeton* type and *Pediastrum*), combined with the lithostratigraphy of zone 39:2 and the top of zone 39:1, gives direct evidence for fluvial conditions, with the alder community perhaps fringing the river/floodplain.

Terrestrial/dry land vegetation appears dominated by *Pinus*, *Quercus*, *Ulmus* and *Corylus avellana* type. In zone 2, *Pinus* values are reduced, but *Ulmus* expands. The sudden expansion of *Alnus* affects these relative frequencies, but even if *Alnus* is excluded from the pollen sum, *Pinus* values are still reduced. The *Ulmus* percentage is, on the other hand, understated and it thus indicates a real ecological change. Furthermore, the start of records of *Tilia* and *Fagus*, which are poorly represented in pollen spectra, clearly indicates a changing woodland flora.

The sharp rise in *Alnus* was, in the past (e.g., Godwin 1940, 1956, 1975), used as the indication of the Flandrian I/II transition (Boreal/Atlantic) at *c.* 7000 BP. It is now recognised that this change was asynchronous (Brown 1988) but nevertheless broadly around this date for southern England. That column 39 represents the transition is supported by the increasing importance of *Tilia* and *Fagus* which became significant elements of the Atlantic woodland. Here, the stratigraphical unit examined is not deep and being of largely minerogenic character, the period represented by the pollen spectra is perhaps not great.

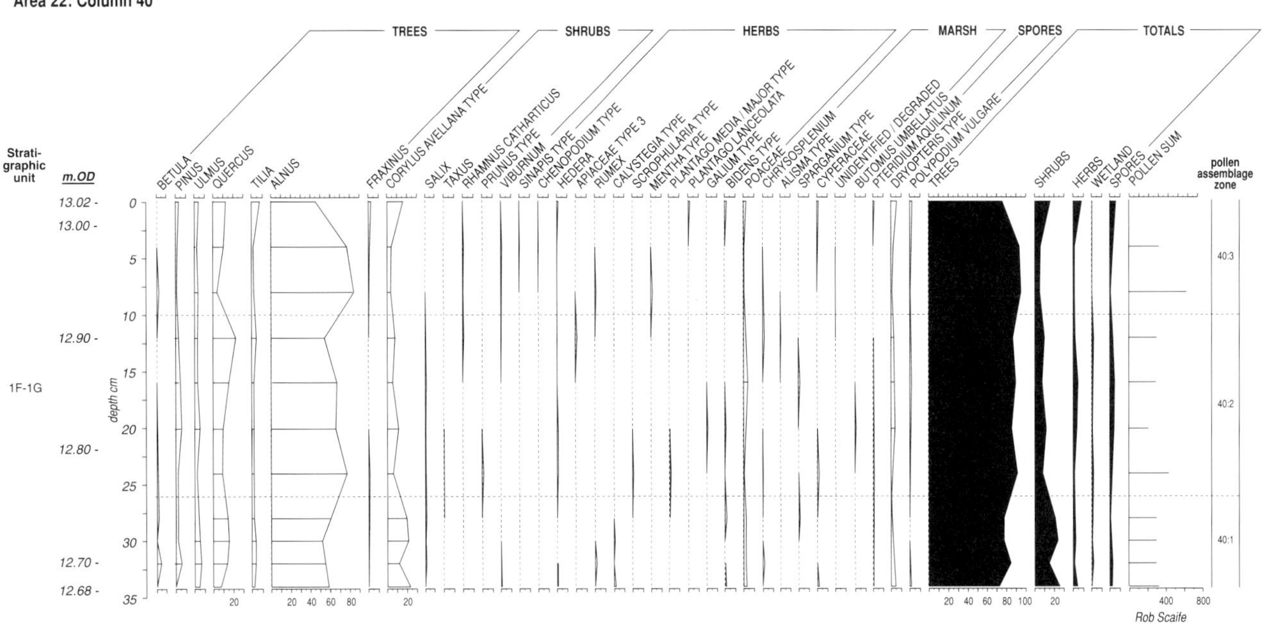

9.4 *Pollen diagram for column 40, Area 22: Atlantic.*

The possibility of hiatuses between the stratigraphical units in this profile has been considered but the overall consistency of the pollen of all taxa apart from *Alnus* suggests that if present, any temporal gaps are small. In conclusion, therefore, column 39 illustrates this change from the late Boreal woodland of pine, oak, elm and hazel dominance to one in which lime/lindens and beech were expanding; the latter ultimately becoming dominant in some areas. The wetter valley bottom also shows a typical change from a willow-dominated community to one of alder growing as patches of carr or as trees along the river bank. This environmental change, although asynchronous, reflects a series of environmental changes, including climatic change in response to increased humidity and dynamic vegetation changes resulting from this and successional factors (see below).

Column 40: ERB89, Area 22, grid 41.0/40.0

Column 40 from Area 22 is younger than those described above, but may be compared with columns 39 and 59. Stratigraphic bracketing argues for a *c.* sixth-millennium BC date, falling within Flandrian chronozone II (the Atlantic period). The pollen sample sequence lies below the drawn section at 12.68–13.02 m OD.

Lithostratigraphy: A total of 34 cm of horizontally bedded sands and organic silts was described and sampled. The stratigraphy was more complex than in the previous units described above and is summarised as follows:

0–4 cm Buff-coloured marl containing freshwater molluscs (context 22.186).

4–12 cm Dark grey organic silt. No plant macro-structure (context 22.222).

12–14 cm Shelly band: numerous freshwater molluscs (context 22.221).

14–19 cm Transitional silt horizon (context 22.220).

19–26 cm Grey silt/clay. No visible plant structure (context 22.219).

26–26.5 cm Sand layer (context 22.218).

26.5–34 cm Grey, organic, fine silt to base of profile (context 22.217).

Pollen biostratigraphy (Fig. 9.4): Pollen was in general well preserved but sparse in terms of absolute pollen numbers present. Pollen sums of 300 grains or more were obtained, except at 20 cm (12.82 m OD). Three pollen assemblage zones have been delimited and are characterised as follows:

POLLEN ASSEMBLAGE ZONE 40:1
(34–26 cm, 12.68–12.76 m OD, four levels)

This basal zone is delimited by higher values of *Corylus avellana* type (to 24% at 34 cm) and *Tilia* (to 5%). Arboreal pollen is the dominant group and *Alnus* (to 60%) is important with *Quercus* (15%), *Ulmus* (5%) and relatively low percentages of *Betula*, *Pinus* and *Fraxinus*. Shrubs are dominated by *Corylus avellana* type with sporadic occurrences of *Salix* and *Viburnum* cf. *lantana*. Herb pollen taxa are few with only Poaceae having a continuous representation. Percentages/numbers are, however, small (<5%). Spores of ferns are relatively few with minor representations of *Pteridium*, *Dryopteris* type and *Polypodium vulgare*.

173

POLLEN ASSEMBLAGE ZONE 40:2
(26–10 cm, 12.76–12.92 m OD, four levels)

Alnus percentages show some increase (75% at 24 cm). *Quercus* also appears to show increase to the top of this zone (22% at 12 cm). *Corylus* values show a corresponding decrease. These fluctuations are, however, minor. There remains a consistent representation of *Pinus*, *Ulmus* and *Tilia*; the former is possibly of little significance because of its copious pollen production and anemophily. As for p.a.z.40:1, *Corylus avellana* type is the dominant shrub, but there are also records of *Salix*, and *Rhamnus catharticus*, *Prunus* type, *Viburnum* and *Hedera*. Herb assemblages remain of similar character to the preceding zone but *Plantago media / major* type is also present. A minor peak in Poaceae at 16 cm is noted. Mire herbs present include *Chrysosplenium oppositifolium*, *Alisma* type, *Typha angustifolia* type, Cyperaceae and cf. *Butomus umbellatus*. Spores remain as in zone 40:1.

POLLEN ASSEMBLAGE ZONE 40:3
(10–0 cm, 12.92–13.02 m OD, three levels)

Alnus values are highest in this assemblage zone (to 85% at 8 cm). *Fraxinus* is relatively more important than in preceding zones. *Quercus*, *Ulmus* and *Tilia* remain at previous levels, with the latter having its highest value at the top of the zone (7% TP). In the herb group, a single occurrence of *Plantago lanceolata* is noted at 0–1 cm and is part of a slight increase in diversity at the top of the assemblage zone with *Sinapis* type, *Chenopodium* type, *Bidens* type and Poaceae.

INFERRED VEGETATION

The pollen assemblages and vegetational characteristics are typical of the mid-Holocene (Flandrian chronozone II) Atlantic period. Woodland is shown to be dominant throughout the pollen sequence and would have been growing on the interfluve. Its composition was mixed deciduous with the principal constituents comprising the thermophiles, *Tilia* and *Fraxinus*, which are substantially underrepresented in the pollen spectra. *Quercus*, *Ulmus* and *Corylus* were, however, also important elements and it is possible that a polyclimax existed with dominant woodland types growing on the different soils of the valley-side soil catena, i.e., *Quercus* and *Ulmus* may have been more suited to the more poorly drained soils (gleyed) of the lower valley slopes, while *Tilia* would undoubtedly have favoured better-drained terrace soils. It is suggested that the uppermost levels may show evidence of increased human disturbance because of the presence of dry land ruderal taxa (e.g., *Plantago lanceolata*, *P. media / major* type, Chenopodiaceae, etc.)

Extremely strong *Alnus* representation, both as pollen percentage and absolute frequency, shows the clear dominance of alder carr woodland on the wetter floodplain.

Other shrub and herb taxa also grew in this community, including the shrubs *Salix*, *Rhamnus catharticus*, and possibly *Viburnum*. Some herbs are diagnostic and include *Chrysosplenium oppositifolium*, *Alisma plantago-aquatica*, *Typha angustifolia / Sparganium* type, *Butomus umbellatus* and Cyperaceae. Many other herbs were not definable to lower taxonomic level but may also be autochthonous components (i.e., to the wetter floodplain).

Column 59: ERB89, Area 32, grid 55.0/39.0

Lithostratigraphy: The unit examined lay between 12.90 and 13.10 m OD and comprised medium and coarse sand and silt with numerous freshwater molluscs, overlain by detrital peat and alluvial silt.

0–3 cm	Clay/silt; buff-coloured (context 32.056, upper).
3–15 cm	Dark brown/black highly humified detrital peat (context 32.040 and 056, lower).
15–20 cm	Medium/coarse orange sand. Contained freshwater molluscs (context 32.062).

Pollen biostratigraphy (Fig. 9.5): Pollen was only found in the organic detrital peat and the upper silts but was, however, relatively well preserved compared with other columns examined. Radiocarbon measurement at 5130 ± 50 BP (BM-2657, *c.* 4040–3790 BC) confirms that this organic unit is of late, mid-Holocene date (Flandrian chronozone II).

The pollen assemblages are diagnostic of this period and geographical region. This profile, therefore, relates closely to column 40 (above) which is also of Atlantic age, but may be slightly earlier. Two pollen assemblage zones have been tentatively recognised and are characterised from the base upwards as follows:

POLLEN ASSEMBLAGE ZONE 59:1
(15–9 cm, 12.95–13.01 m OD, three levels)

This is characterised by high values of *Alnus* (67%) which is the dominant taxon. Along with *Quercus* and *Corylus avellana* type these are the principal trees and shrubs present. However, *Ulmus*, *Tilia*, and *Fraxinus* are consistently present at low levels. There are few herbs present but these include a single record of *Plantago lanceolata*. Spores of ferns are similarly few in number but include *Dryopteris* (monolete) type and sporadic *Pteridium aquilinum*.

POLLEN ASSEMBLAGE ZONE 59:2
(9–0 cm, 13.01–13.10 m OD, five levels)

This assemblage zone differs from zone 1 with reduced levels of *Alnus* (30 to 40%) and expansions of *Tilia* (to 8%), *Ulmus* (to 3%) and *Pinus* (to 6%). Herbs are more diverse than in the preceding zone. Poaceae and Cyperaceae are dominant with a small increase in their percentages throughout the zone (to 8% and 14% respectively). *Plantago*

174

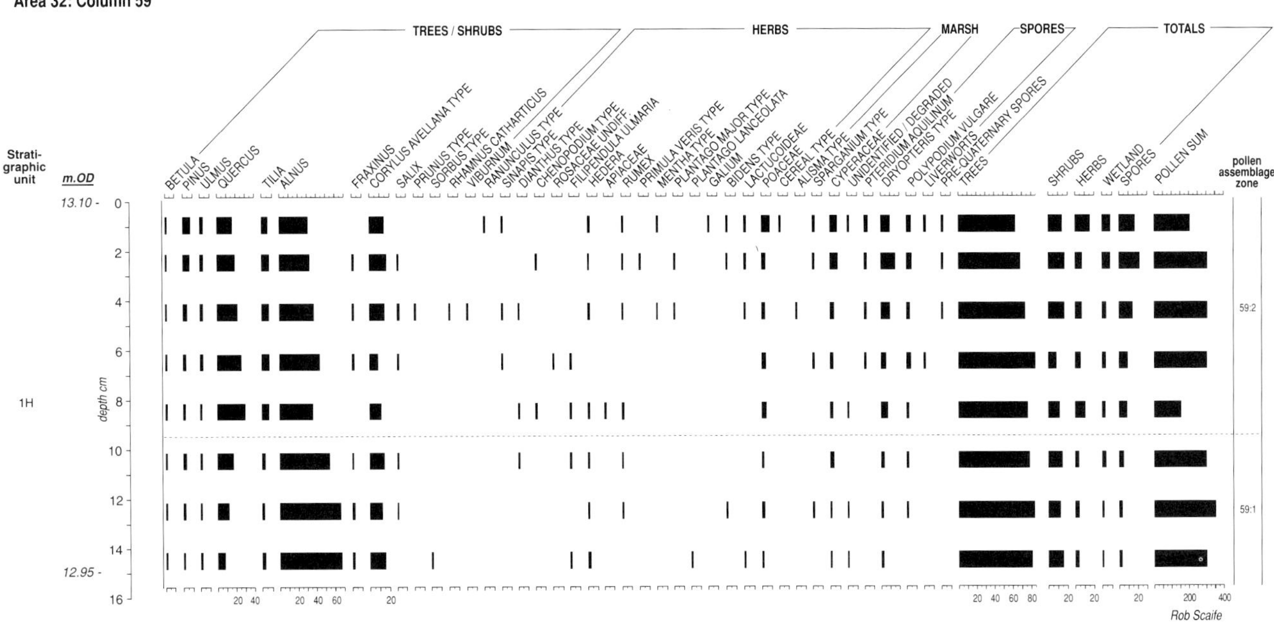

9.5 *Pollen diagram for column 59, Area 32: Early Neolithic.*

major type and, importantly, cereal type are noted at the top of the sequence. Spores of *Pteridium aquilinum* are recorded for the first time in this sequence and there are increases in *Dryopteris* type and *Polypodium vulgare*. The expansions of these taxa in the upper part of the zone may relate to the stratigraphical change from peat/organic deposits to alluvial clay/silt.

INFERRED VEGETATION

The radiocarbon date suggests that we are here dealing with a pollen sequence which, by comparison with other regional data, might yield evidence for the earliest Neolithic (discussed in greater detail below). Nevertheless p.a.z. 59:1 and the larger part of p.a.z. 59:2 undoubtedly relate to the Atlantic period (Flandrian chronozone II). The arboreal vegetation communities of this period spanning *c.* 6000–4000 BC (7000–5000 BP) represent the middle Holocene, climax woodland on mature and stable soils (likely brown earth soils). The arboreal vegetation evidenced here is typical of this period and comparable with Runnymede columns 39 and 40 attributed to earlier stages of this period. The vegetation similarly shows dominance of the principal woodland elements – *Quercus*, *Ulmus*, *Tilia*, *Fraxinus* and *Corylus avellana*. *Alnus* was also extremely important, either forming carr woodland on the floodplain (Alnetum) or strong local growth along the edges of the river channel(s). The small percentages of *Pinus* and *Betula* are not considered to reflect important elements of the local flora since their high pollen productivity and anemophilous pollination tendency allow their extra-local/regional derivation.

In the upper levels of zone 59:2 there is an increase in herbs associated with disturbance and agriculture (*Plantago lanceolata*, *P. media/major* and especially cereal-type pollen in the highest level). Opening of the forest canopy by human activity may also have extended the pollen catchment and allowed the input of extra-local or extra-regional/long-distance pollen. Increases in *Pinus* percentages may be a function of this canopy opening although it is not clear whether the alder woodland of the sample site was affected as much as the interfluve vegetation. The primary elm decline occurred in Britain at *c.* 5000–5300 BP and is discussed below. This event is frequently associated with Early Neolithic activities, although an insect vector carrying fungal disease is regarded as the essential cause of this roughly synchronous event (Girling 1988). Given the radiocarbon assay obtained, and the evidence of cultivation in the upper levels of the profile, an elm decline might have been expected, but is not present within the time window of column 59.

The floodplain community was typically dominant alder carr woodland containing a rich shrub and herb layer similarly evidenced in column 40. In contrast, however, there is no evidence for aquatic habitats. This may be a taphonomic factor since pollen transport and the size of the pollen catchment in the densely vegetated environment of alder carr may have been extremely small.

Column 60: ERB89, Area 32, grid 59.0/39.0

This pollen profile is stratigraphically later than column 59 as well as the Middle Neolithic settlement. It may be attributed to the immediate post-elm decline showing a

175

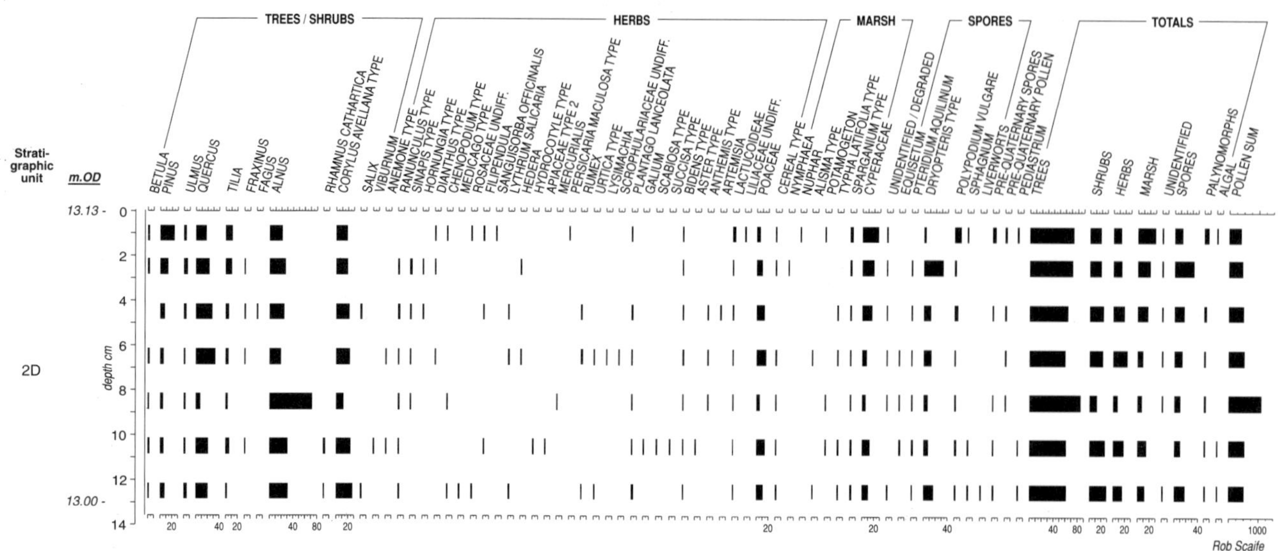

9.6 *Pollen diagram for column 60, Area 32: later Neolithic.*

largely wooded environment but with some herbs and cereal-type pollen attributed to human disturbance.

This pollen sequence probably overlaps that studied by Greig (1991, fig. 112 – column SS1; Fig. 2.37) on the site of the M25 rescue excavation (Area 6). This profile along with column 5, Area 14, is, therefore, of importance in understanding the character of the floodplain, the interfluves and the possible effects of Neolithic occupation on the environment and agricultural background.

Lithostratigraphy: The sequence spans 12.93 to 13.13 m OD.

0–6 cm	buff/grey fine sand and silt layer containing molluscs (context 32.048).
6–12 cm	darker, organic unit comprising finely laminated sandy/silt; the upper level showed some evidence of Fe staining (possibly gleying) (context 32.047).
12–20 cm	medium and coarse sand containing flints to 2 cm and comminuted shells (context 32.038).

Pollen biostratigraphy (Fig. 9.6): Pollen analysis was attempted for all of the stratigraphical units described. However, well-preserved pollen was only recovered from the organic and finer-grained upper units (0–12 cm) and not the coarser basal sands (12–20 cm). Although there are some apparent changes in the pollen stratigraphy, these are not considered significant enough to define pollen assemblage zones; especially as the changes noted are associated with those in the lithostratigraphy.

The overall characteristics of the pollen spectra are as follows: arboreal and shrub pollen are dominant (around 60% and up to 26% respectively). Dry land herbs are in the range 15 to 20% with marsh/wetland taxa expanding to 20% at the top of the sequence. Trees are dominated by *Alnus* which peaks to 70% at 8 cm. Other noteworthy species are

Quercus (to 30%), *Tilia* (to 10%), *Pinus* (to 20%) and *Ulmus* (<5%), with *Fraxinus* and *Fagus* also present. Percentage values of *Tilia* and *Pinus* are highest in the upper levels. Shrubs are characterised by *Corylus avellana* type (to 25%) and sporadic *Rhamnus catharticus*, *Salix* and *Viburnum*. There is a moderately diverse assemblage of herbs, Poaceae having the highest values. Many of these types are characteristic of damp meadow or fen habitats and as such are included in the general herb category. More typical ruderals include *Plantago lanceolata*, *Sinapis* type, *Chenopodium* type and *Artemisia*. Cereal type occurs throughout the profile. Damp meadow/pasture type include *Filipendula ulmaria*, *Sanguisorba officinalis*, *Hydrocotyle*, *Lythrum salicaria*, *Lysimachia*, *Succisa* and possibly pollen types present which are not identifiable to lower taxonomic levels. Definite aquatic and marsh taxa are dominated by Cyperaceae, but also present are *Nymphaea*, *Nuphar*, *Alisma* type, *Potamogeton* type, *Typha latifolia* and *Typha angustifolia*/*Sparganium* type. Some algal *Pediastrum* is present.

INFERRED VEGETATION

This short pollen sequence lies within parcel 2D which has been dated to the late fourth to early third millennium BC (Chapter 10). By comparison with other Runnymede pollen spectra a Neolithic age closely post-dating the 'primary *Ulmus* decline' is indicated. A predominantly wooded environment is shown, but with small percentages of *Ulmus* compared with spectra attributed to Flandrian II. Small but significant numbers of cereal pollen and *Plantago lanceolata* are present to the base of the profile which implies prehistoric (Neolithic) disturbance and agriculture in the environment. However, the overall importance of woodland in the environment is evident with *Quercus*, *Corylus avellana*, *Tilia*, and *Fraxinus* present. Slightly higher percentages of *Ulmus* in the bottom level of the diagram may be tentatively attributed to the latter stages of the *Ulmus* decline. Similarly, higher values at the top of the profile may be a phase of later regeneration. The expansion of *Pinus* at the top of the profile is enigmatic, with percentage values to 21%. This may be the result of the reworking of older sediments containing *Pinus* or the known effects of overrepresentation of saccate pollen grains in fluvial environments.

The lithostratigraphy shows a marked change from sands deposited in a high energy fluvial environment in which pollen was not preserved. The organic unit (6–12 cm) are peaty silts and apparently deposited under a lower-energy environment, on the channel edge. The local and on-site vegetation is not altogether clear, with *Alnus* values generally in the range which might be interpreted as local but not on-site growth of carr woodland, or possibly trees fringing a river channel. Reed swamp vegetation and aquatics are present (*Typha/Sparganium*, *Alisma* type, *Nuphar*, *Nymphaea* and *Potamogeton* type), but similarly in numbers less than might be expected for 'on-site' growth. It is concluded that the depositional environment was an area of grass/sedge fen with the associated taxa noted. This was clearly fringed by alder close by, with aquatic-type pollen coming from the river or areas of standing water (e.g., meander channels).

Column 5: ERB84, Area 14, grid 78.0/30.0

This column is complicated by the presence of an erosion line in the section at *c*. 12.90 m OD, which is the interface between parcels 2D and 2E, both of later Neolithic date. Two radiocarbon dates, 3740 ± 80 BP (BM-2436) and 3770 ± 60 BP (BM-2435), from beaver-gnawed wood in the channel edge deposits above the erosion line (parcel 2E) date the upper part of the pollen profile to 2340–2040 BC.

Lithostratigraphy: Monoliths P63 and P64 were dominated by a brown, humic peat with a high silt content between 12.75 m OD and 13.07 m OD. Underlying these silty peats are coarse sands and gravels of fluviatile origin (P65) containing freshwater molluscs, but not suitable for pollen analysis. This merges into coarser sands and gravels down to a depth of 12.47 m OD (see section – Fig. 2.23). Above the organic horizon, is a buff calcareous silt.

Pollen biostratigraphy (Fig. 9.7): Well-preserved pollen was recovered from the organic horizon (12–38 cm, 13.07–12.81 m OD) and from the upper part of the shell sands (38–48 cm, 12.81–12.71 m OD). Pollen was, however, absent in the upper calcareous silts where high alkalinity and oxidation were the likely causes of its destruction. A total of sixty-four taxa was recorded and thus represents a much greater pollen diversity than encountered in the early and middle Holocene profiles described. Pollen counts of between 300 and 400

grains per level were usually attained, but counts were lower from levels where preservation was poorer. Three tentative pollen assemblage zones have been defined. These are characterised from the base of the section upwards as follows:

POLLEN ASSEMBLAGE ZONE 5:1
(44–33 cm, 12.75–12.86 m OD)

This basal zone comprises predominantly trees and shrubs (to 50% and 40% of total pollen respectively) with the dominant taxa including *Quercus* (to 35%), *Tilia* (to 8%) and *Alnus* (to 30%). *Pinus*, *Ulmus* and *Fraxinus* are also noted. Shrubs are dominated by *Corylus avellana* type (to 38%) with sporadic occurrences of *Salix* and *Cornus*. Herb pollen is less diverse than in subsequent zones but include notably Poaceae (to 15%) and cereal type, which occurs throughout the profile, but increases to a minor peak at the top of the zone. The latter is associated with a peak of *Plantago lanceolata*, *Sinapis* type and sporadic occurrences of other herbs – *Chenopodium* type, *Rumex* and Asteraceae types. Aquatic and marginal aquatic taxa are present with dominant Cyperaceae, *Myriophyllum spicatum*, *Potamogeton* type, *Lythrum salicaria*, *Typha angustifolia* type and algal *Pediastrum*. Spores of *Pteridium aquilinum*, monolete *Dryopteris* type and *Polypodium vulgare* are also present.

POLLEN ASSEMBLAGE ZONE 5:2
(33–27 cm, 12.86–12.92 m OD)

This pollen assemblage zone is characterised by high values of *Corylus avellana* type peaking to 70%. There is also a corresponding increase in *Alnus* (to 32%) and decrease in *Quercus*, *Tilia* and *Pinus*. The latter reductions may be a statistical function of increases in the former. Dry land herbs, marginal aquatic taxa and spores remain essentially similar to the preceding zone but there is some indication of increasing diversity. Cereal and *Plantago lanceolata* are fewer than in p.a.z. 5:1, which again may be attributed to the within-sum dominance of *Corylus avellana* type.

POLLEN ASSEMBLAGE ZONE 5:3
(27–12 cm, 12.92–13.07 m OD)

With the exception of a single level/peak of *Ulmus* at 20 cm, arboreal and shrub pollen values return to those characterising p.a.z. 5:1. There is, however, an increased diversity of shrub taxa with *Taxus*, *Rhamnus catharticus*, *Sorbus/Crataegus* type, *Hedera helix* and *Malus* type present. Herbs are dominated by Poaceae (to 20%). Of note are increases in cereal type and *Plantago lanceolata* between 28 cm and 24 cm. Marginal aquatic plants are more important with *Potamogeton* type, *Typha angustifolia* type (including *Sparganium*) and Cyperaceae. Spores of *Pteridium aquilinum* (10%), *Dryopteris* type (to 19%) and *Polypodium vulgare* (6%). Noteworthy also are increases in pre-Quaternary palynomorphs and freshwater *Pediastrum*.

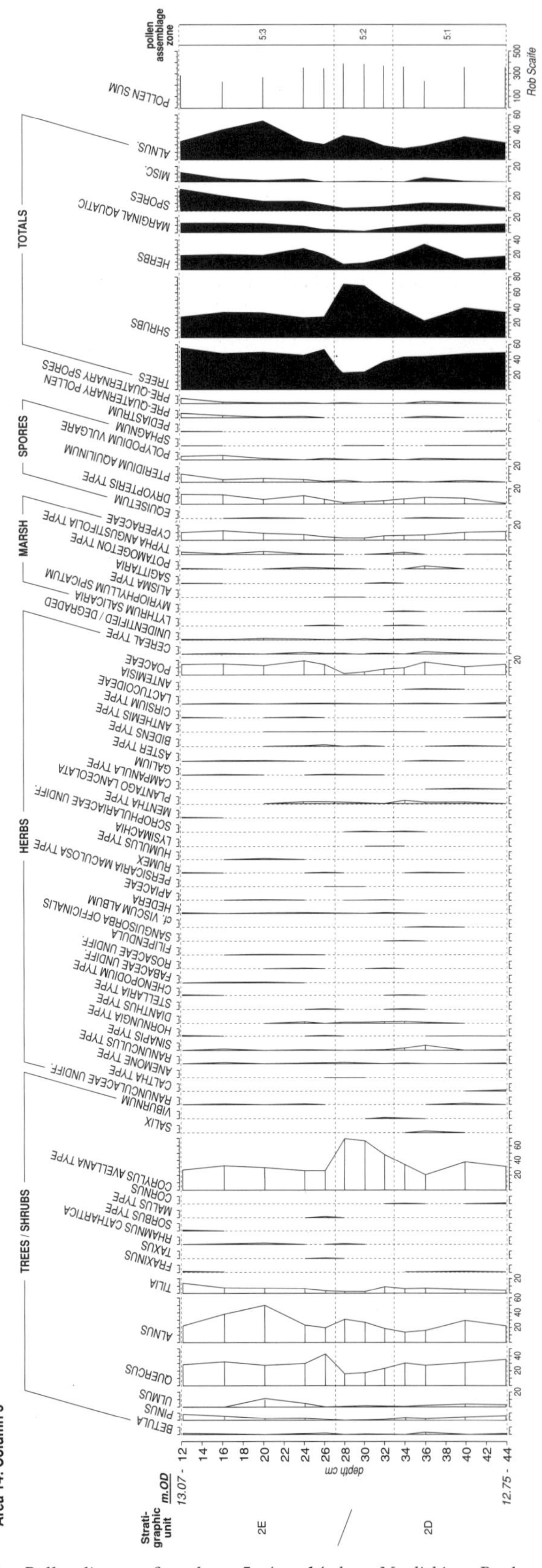

Area 14: Column 5

9.7 *Pollen diagram for column 5, Area 14: later Neolithic to Beaker period.*

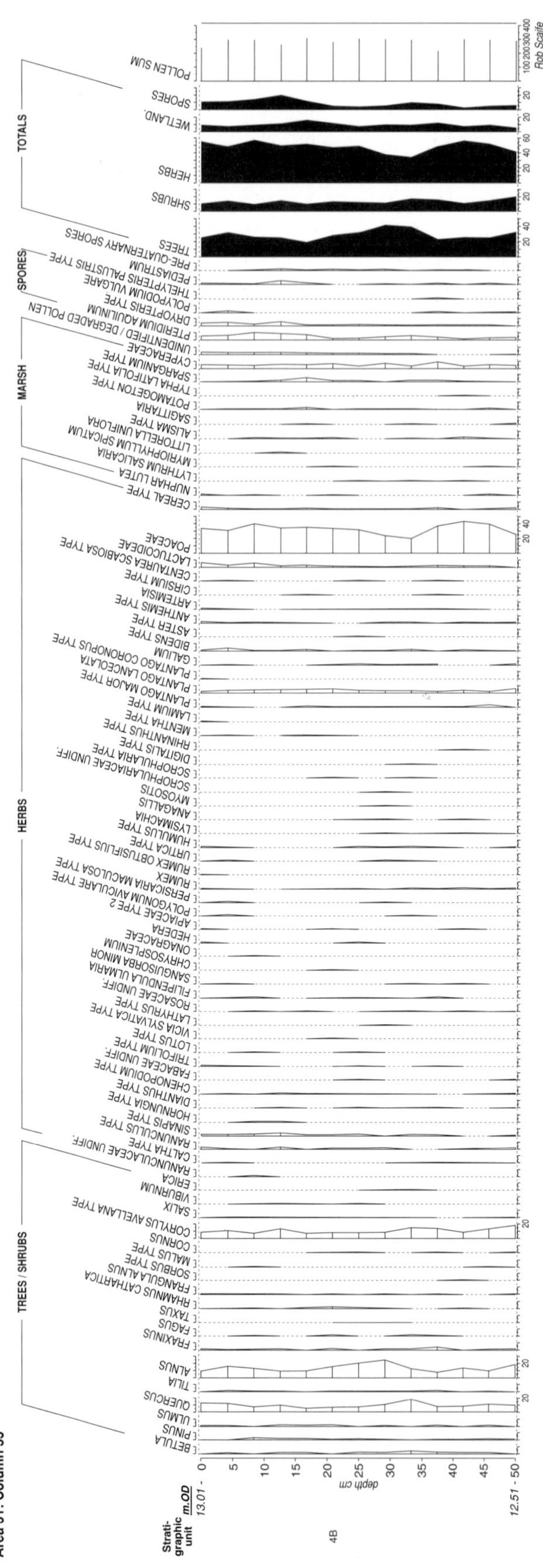

Area 31: Column 55

9.8 *Pollen diagram for column 55, Area 31: Middle / Late Bronze Age transition.*

INFERRED VEGETATION

The pollen spectra obtained from this section provide valuable data on the transition from mid-Holocene (Atlantic; Flandrian II) climax woodland to the more open agricultural environments shown for the Bronze Age. Although this pollen sequence is important in that, with column 60, it demonstrates the first clear evidence of arable/crop cultivation (see minor cereal record in column 59) and the first major human influences on the environment with significant woodland reduction, the overall characteristics of the principal woodland elements remain broadly similar.

The erosional interface between parcels 2D and 2E falls at *c.* 12.90 m, the top of sample monolith P64, and is mirrored in the pollen sequence by the change from p.a.z. 5:2 to 5:3. The temporal extent of the hiatus is not clear but the stratigraphic correlations place the whole profile later than Middle Neolithic occupation and therefore well after the 'primary *Ulmus* decline'. Indeed, there is no apparent reduction of elm pollen percentages in the diagram, and frequencies recorded here are relatively small (1–2%) compared with spectra attributed to pre-*Ulmus* decline periods.

Although a range of sedimentary environments is present in this pollen column, thus perhaps giving rise to variable pollen taphonomy, the vegetation throughout illustrates one still dominated by woodland. As already noted, *Tilia* and *Fraxinus* are frequently poorly represented in pollen spectra due to their entomophily (Andersen 1970, 1973) suggesting, therefore, that their importance here may be underestimated from the pollen values recorded. *Quercus* and *Tilia* were the principal elements on drier, better-drained soils, possibly with an understorey of *Corylus*. Small percentages of *Fraxinus* are, nevertheless, significant and as noted in other post-elm decline sequences, it expanded as secondary woodland, taking the place of *Ulmus*.

Alnus was important on large areas of the floodplain adjacent to the river channels and backwaters. The values of *Alnus* of up to 50% of total pollen are not as high as might be expected if typical and dominant alder carr woodland was growing 'on site'. The presence of Cyperaceae and other marsh/marginal aquatic taxa suggests the presence also of sedge fen. The floodplain habitat is, therefore, interpreted as a mosaic of possible fen carr woodland and wetter sedge fen. On the evidence of the marked changes in lithostratigraphy, the complexity of the fluvial system would have resulted in significant fluctuations in the areas of growth of these wetland plant communities.

Column 55: ERB89, Area 31, grid 86.0/25.0

Three monolith sections were taken which span Middle–Late Bronze Age waterlogged deposits between 12.51 and 13.01 m OD. This was regarded as an important sequence of sediments leading up to the phase of Late Bronze Age occupation. Twigs and beaver-gnawed wood occurring at *c.* 12.80 m OD in the column have been dated to 2900 ± 50 BP (BM-2661) (1260–930 BC) and this essentially precedes the waterlogged sequence from Area 6 (Greig 1991).

Lithostratigraphy: In comparison with most of the lithostratigraphical units examined, this profile comprised numerous minor layers and lenses aggrading against a palaeobank. One layer contained charcoal fragments. Overall, the profile was rich in comminuted shell and was thus highly calcareous. However, its position below the water table was favourable for pollen preservation. The stratigraphy as described from the three monolith sections was as follows:

0–3 cm   Pale, buff silts containing comminuted shell fragments (context 31.648).

3–18 cm   Brown/grey silts containing charcoal. Wood/twig fragments at 12 cm and 16 cm (context 31.644).

18–23 cm   More organic silts containing some monocotyledonous remains. Also containing round pebbles to 2 cm. Wood fragments at 20 cm apparently correspond with beaver-gnawed wood (contexts 31.665 and 669).

23–25 cm   Lens of fine to medium sand (context 31.669).

25–44 cm   Grey silts with comminuted shell fragments. Small stones noted. Wood/twig fragments at 39 cm (context 31.642).

44–45 cm   Horizon containing a greater number of comminuted shell fragments.

45–50 cm   Shell in silts (context 31.666).

Pollen biostratigraphy (Fig. 9.8): A total of thirteen pollen levels have been examined at 4 cm intervals between 0–1 cm and 48 cm. Pollen was well preserved in spite of the alkalinity of the sediments and a pollen sum of 300 was obtained at each level. The taxonomic diversity and floristic make-up differs markedly from the earlier sequences discussed above, with much greater representation of herb pollen dominated by Poaceae (to 42%) and much-reduced levels of tree pollen. Although diverse, the spectra are relatively homogeneous with little variation in the abundance of individual taxa throughout the pollen biostratigraphy. Herb types include many indicators of pastoral/grassland communities and arable habitats. From 0–18 cm, there are some minor expansions of *Sinapis* type, Asteraceae (*Bidens* type and Lactucoideae), *Potamogeton* type and spores of *Pteridium aquilinum*, *Dryopteris* type and algal spores of *Pediastrum*. These are, however, minor and have not been used to define a pollen assemblage zone. Arboreal pollen comprises taxa present in earlier periods but at lower percentage levels, including *Quercus* (15%) and *Corylus avellana* type (to 20%) as the principal species, with lesser records of *Tilia* (1 to 2%) and *Fraxinus* (1 to 2%). Of importance is the occurrence of *Fagus*, albeit sporadically. Shrub pollen types are more diverse, with *Rhamnus catharticus, Frangula alnus, Salix, Viburnum, Sorbus* type, *Malus* type, *Taxus* and *Cornus*. Marsh and aquatic taxa are well represented with rooting macrophytes and fen/reed swamp type. The former include *Nuphar lutea, Potamogeton* and *Myriophyllum spicatum*. Fen swamp taxa include *Lythrum salicaria, Alisma* type, *Typha angustifolia/ Sparganium, T. latifolia* and Cyperaceae.

INFERRED VEGETATION

Compared with the profiles of greater age, this sequence shows a further reduction in woodland and the corresponding expansion of open areas. However, elements from the earlier periods remain *Quercus, Corylus, Fraxinus* and *Tilia*. It is postulated that small areas or localised stands of woodland remained. Of note, is the occurrence of *Fagus* since increased woodland clearance would expand the pollen catchment and allow the greater representation of taxa from further distances. This is especially plausible with *Fagus* pollen since it is a taxon which does not disperse widely because of its large and relatively heavy pollen grains. The fact that some woodland survived into the Late Bronze Age was found by both Greig (1991) and Robinson (1991) and is discussed further below. The expansion of herb types and percentages attests to a mixed pastoral and arable environment, although it is not possible to state the extent of each activity. However, it can be noted that while cereal-type pollen and associated arable weeds (e.g., *Polygonum* sp.) are continuously present, it is those taxa typical of pastoral or grassland habitats which predominate – for example, Poaceae, *Plantago lanceolata* and *Rumex*.

The floodplain habitat during this phase shows some changes over the preceding Neolithic and earlier sequence. Although still an important element, alder carr (with willow), or trees fringing river channels, appears to have been less dominant in the local region. Aquatic and fen taxa are present indicating expansion of the reed swamp elements with Cyperaceae, *Typha angustifolia/Sparganium, Typha latifolia* and associated fen types.

The Holocene vegetational history

One of the primary aims has been to establish a vegetation and environmental framework for the Runnymede site which spans as much of the Holocene period as possible. Aggradation phases have been identified which appear to represent, at intervals, the period from the middle of Flandrian chronozone I through to the Late Bronze Age stage of chronozone III. Thus the time-span covered by the eight profiles is *c*. 8000–1000 BC (*c*. 9000–2800 BP).

Flandrian chronozone I: the Boreal Mesolithic period

The earliest units obtained come from Area 19 (column 31), Area 13 (column 4) and the basal pollen zone of Area 22

(column 39). From the pollen sequences obtained it is clear that there is no evidence for the Late Devensian or early Holocene (Flandrian chronozone Ia pre-Boreal) periods. However, it is by no means certain that sediments of this age are not present beneath those reached in excavation. For the Devensian/Holocene transitional phase, recourse must be made to other research carried out in the region of West London where organic sediments have been located in the Devensian floodplain gravels of the Thames. These analyses provide the environmental picture for the broader region under study.

GENERAL VEGETATION

Gibbard *et al.* (1982), Gibbard and Hall (1982) and Kerney *et al.* (1982) provide data relevant to the Late Devensian from sites at Kempton Park, Sunbury, the Lower Colne, West Drayton marshes and Isleworth. Organic channel fills of the River Colne and at West Drayton have been dated at 11,230 ± 120 BP and 13,405 ± 170 BP. Pollen data from these sites typify what is known about the Late Devensian environment, which was essentially an open, treeless, herbaceous environment of the Loch Lomond stadial (zone III), the only possible woody vegetation being *Salix*, such as found at West Drayton. The openness of the vegetation of this period (until about 10,000 BP) indicates that environmental conditions were basically harsh, which resulted in a diverse range of ecological communities. Although problems exist in defining the plant communities present, by using the ecotypes as proxy for temperature/ climate it is clear that the pollen taxa of the closing stages of the Devensian cold stage show the presence of shrub, dwarf shrub, tall herb, short turf, fen mire, aquatic and disturbed soil plant communities. Thus, there was a diverse range of plant communities with varied phytosociological characteristics. This diverse Late Devensian flora, with its wide range of phytogeographical elements reflects the variety of the local microclimatic, edaphic and geological conditions present.

The change from open herbaceous elements of tundra environments of the Late Devensian (zone III) occurred with rapid temperature amelioration at *c*. 10,000 BP (Coope 1977). Once the growth threshold for most plants was exceeded, successional changes in vegetation were initiated. Expansion of *Juniperus*, followed rapidly by *Betula*, marks the start of vegetation migration from glacial refugia, with seral successions and the ultimate dominance of various vegetation communities during the early Holocene (Flandrian chronozone I).

Available data from the region, which provide some idea of the vegetational characteristics of the earliest Holocene, come from Enfield Lock (Chambers and Mighall 1991; Chambers *et al.* 1996). Although the site is 40 km away in north London, it yielded a date of 9546 ± 56 BP (UB–3350) and substantial *Betula* percentages, typical of the early

Holocene period generally in southern Britain. Thus, although data are not specifically available for Runnymede Bridge, there seems to be little doubt that the dry land areas of the interfluve would have been stabilised by early Holocene pioneer *Juniperus* scrub and *Betula* woodland from 10,000 BP.

On the other hand, sediments at Three Ways Wharf in the Colne valley, to the north of Runnymede (Lewis *et al.* 1992), appear to have started accumulating in the Boreal period, are comparable with columns 4 and 31 at Runnymede, and similarly do not provide evidence for the Devensian/Holocene transition.

The wider data from southern Britain is also relevant. Increased humidity and removal of periglacial conditions would have resulted in a stabilised fluvial regime, with reduction of inorganic inwash from the previously non-stabilised interfluves. *Salix* communities and sedge fen were the principal colonisers of floodplain habitats (Scaife 1980). Although there are some data to suggest early colonisation of these habitats by *Alnus* (Devoy 1979; Scaife 1980, 1982; Waller 1993; Bennett and Birks 1990), this evidence is extremely localised in the lower zones of rivers. At Runnymede it would seem that the development of major stands of *Alnus* occurred in the early to mid-seventh millennium BC (*c.* 8000–7500 BP) (Chapter 10) .

The vegetation changes noted above reflect a complex reaction of communities to variations in external factors and are problems of dynamic phytosociology; that is, the succession of tree taxa and associated ecology in response to rapid temperature rise and migration from glacial refugia. Thus, the *Pinus* domination of column 31 (Area 19), column 4 (Area 13) and the lowest zone of column 39 (Area 22), show the early migration and expansion of *Pinus* in southern England in relation to the north of Britain (Erdtman 1928; Godwin 1975a, 1975b; Scaife 1980; Birks 1989). While *Pinus* appears to have been present from the early Holocene (Flandrian Ia) in the south, it was subordinate to *Betula*. By *c.* 8000 BC (9000 BP; early Boreal, Godwin's pollen zone VI), however, its competitive ability over *Betula* resulted in its widespread dominance in southern England (Godwin 1975a, 1975b; Scaife 1980, 1982) as has been observed at other sites in the region – Three Ways Wharf (Lewis *et al.* 1992) and Cothill, Berkshire (Clapham and Clapham 1939).

Small percentage values of *Quercus* and *Ulmus* are typical of the known vegetation development for the period and region. Godwin (1975a) suggested that *Quercus* had reached the present-day coastline by the pre-Boreal period (Flandrian Ia) and, through its competitive ability over *Betula* and *Pinus*, made strong and early inroads into southern Britain. As with *Pinus*, the migration is elucidated using isopollen maps based on ^{14}C chronology (see Birks 1989). In pollen assemblage zone 31:1, 5% *Quercus* is noted and while *Pinus* was dominant locally, this marks the start of the expansion to ultimate regional dominance of *Quercus*, along with *Ulmus*. In column 4 *Quercus* percentages attain 40% of total pollen with *Pinus* (38%) and *Corylus avellana* type (to 51%); *Ulmus* is also present (to 6%). This pollen sequence, although thought to represent a short phase of accumulation, undoubtedly shows the period of vegetation transition from pine domination (with hazel) to deciduous oak and elm with hazel, during the middle and later Boreal period. A date of 7790 ± 80 BP (BM-2550) provides a valuable chronology for this transitional vegetation phase for this region, and from these data it is suggested that the dry land environment of the interfluves comprised a mixed coniferous and deciduous woodland. The question of whether the woodland stands of pine and oak/hazel were discrete ecological units, with ecotonal gradations between, or whether they were a homogeneous mixture of oak, elm and pine, remains. Given the admixture of *Picea/Abies*, *Fagus* and *Quercus* in the low alpine zone and the Black Forest of Germany today, the latter seems plausible. Certainly the diminishing values of *Pinus* throughout the Boreal period (Flandrian Ib and Ic) attest to its progressive demise due to the competitive ability of *Quercus* and *Ulmus*. It is noted above that there are sporadic occurrences of thermophilous arboreal taxa in column 4 (Area 13), including *Alnus* and *Tilia*. While these are only sporadic occurrences (i.e., between the rational curves for these taxa represented in subsequent periods), they undoubtedly indicate the migration of these taxa into the region, thus poised to become highly significant during the mid-Holocene (see below).

In conclusion, the pollen diagrams constructed for columns 31 (Fig. 9.1) and 4 (Fig. 9.2), and the lower zone of column 39 (Fig. 9.3), clearly represent the Boreal period of dynamic vegetation succession under what has been widely discussed as the Boreal continental period *c.* 8000–6000 BC (9000–7000 BP; Godwin 1975a; Lowe and Walker 1984; Jones and Keen 1993).

POSSIBLE ANTHROPOGENIC INFLUENCE

In terms of the human environment, the discussion above relates to the period of the earlier Mesolithic. Much interest has been expressed in the potential for human activity on the vegetation during this period. These essentially hunting/foraging communities have been seen to have some effect on fragile ecosystems such as the upland montane zone (e.g., North Yorkshire; Simmons 1975a, 1975b) and on the poor soils of the lowlands, e.g., Iping Common (Dimbleby in Keef *et al.* 1965). However, the possibility of distinguishing human influences on the heavily wooded environment of the lowland zone is highly problematic and there is unlikely to be evidence in the relatively poor pollen-preserving environment of Runnymede.

Two factors are, though, worth mentioning. Firstly, the often-discussed importance of *Corylus* at an early stage of the interglacial has been thought to be due to Mesolithic activity, directly or indirectly. This has been much debated (see Smith 1970; Jacobi 1978; Simmons and Tooley 1981) and is not further discussed here, except to note that hazelnuts may have been used as a food resource (Scaife 1992) and some *Corylus* peaks in pollen diagrams are associated with charcoal (see Smith 1970). It seems more plausible that the dominance of *Corylus* at an early stage of the Holocene is a function of migration rates from refugia (Deacon 1974).

Secondly, the presence of ruderals (albeit sporadically) may be cited as evidence for disturbances, possibly regularly trodden areas. In the Boreal pollen sequences discussed, there were sporadic occurrences of ruderals. These include *Plantago lanceolata*, *Scabiosa*, Chenopodiaceae, etc., and consistent records of grasses (Poaceae). Since these are all light-demanding taxa (heliophytes), such herb pollen percentages and taxa are unusual after the onset of Holocene closed forest; a marked diminution of these heliophilous elements (from the Devensian) occurred at many sites. The range of herb taxa present (especially in column 4) is moderately diverse. It is unfortunate that pollen morphology does not allow separation to lower taxonomic levels and thus the definition of more specific ecological groups. Consequently, it can only be postulated generally that the herb taxa present formed part of the floodplain habitat (see below) or that there were open 'glades' in the woodland, perhaps resulting from annual disturbances of human activity.

THE BOREAL FLOODPLAIN ENVIRONMENT

The autochthonous pollen components provide evidence for a diverse range of aquatic, fluvial and floodplain micro-habitats in the early Holocene. Although sporadic records of *Alnus* are noted especially in column 31, the earliest section, this is seen as being of extra-local origin. Regardless, it is clear, given the high pollen productivity of *Alnus* and its anemophily, that it was not a locally important element of the floodplain taxa at this time. Its presence during the early part of the Flandrian in the lowland coastlands of southern England has been addressed elsewhere (Devoy 1979; Scaife 1980; Bush and Hall 1987). Conversely, however, *Salix* is largely underrepresented in pollen spectra, being entomophilous. Although not present in column 31, which may be due to poor pollen preservation and relatively small pollen sum, there is a constant record in columns 4 and 39. This undoubtedly shows the importance of *Salix* growing on or along the fringes of the river floodplain. This is likely to have been nearby, although the possibility of fluvial transport from upstream also needs to be considered.

Aquatic macrophytes and reed swamp taxa are also present. Column 31 has a more restricted range of such taxa than column 4, but small numbers of freshwater algal cysts of *Pediastrum* there do attest to a freshwater habitat. In columns 4 and 39, *Nuphar lutea* and *Potamogeton* type also attest to freshwater habitats, still or slow-flowing water.

Pollen of Cyperaceae, *Sparganium/Typha angustifolia* type, *Typha latifolia* and probably some Poaceae (e.g., *Phragmites*) are indicative of reed swamp possibly fringing areas of open water. Wet pasture may also be indicated by *Filipendula ulmaria*, *Succisa* and also taxa which are not identifiable to species or even generic level, e.g., *Ranunculus* type, Rubiaceae, Apiaceae, Poaceae, etc.

Flandrian chronozone II: the Atlantic Mesolithic period

The often quoted 'climatic optimum' (Atlantic period) is usually regarded as a time of vegetation stability with higher temperatures and humidity, a period in which the maximum extension of Holocene woodland occurred. It falls between *c.* 6000 and 4000 BC (7000–5000 BP). Two points must be noted. Firstly, it is now known that early Holocene temperatures were higher during the pre-Boreal and early Boreal periods but were associated with a drier climate; an earlier 'climatic optimum' thus seems plausible. Secondly, the original delimitation of the period was based on supposed climatically determined changes, the rapid expansion of *Alnus* at *c.* 6000 BC (7000 BP) at the beginning and the decline in *Ulmus* at *c.* 4000 BC (5000 BP) at the end; the cause of these changes is now thought to be less straightforward. The asynchroneity of the former (Smith and Pilcher 1973; Birks 1989) and the broadly synchronous but human/insect causation of the latter, mean that the boundaries of the Atlantic would only correspond approximately with these frequently quoted dates. Traditionally also, the Atlantic period has been viewed as one of mixed deciduous forest or Quercetum mixtum with a strong thermophilous woodland component comprising *Tilia*, *Ilex*, *Fraxinus*, *Lonicera*, *Hedera* and *Viscum* in a woodland largely dominated by *Quercus*, *Ulmus* and *Alnus*. This broad assumption has also been questioned by many researchers in recent years. It is now realised that a polyclimax existed with different communities dominant, according to variations in lithology, pedology and geographical position. Thus, in the south and south-east of England *Tilia* was apparently the dominant woodland element on better-drained soils, while *Alnus* undoubtedly was a dominant element (alder carr) in the valley bottom habitats (Baker *et al.* 1978; Scaife 1980, 1987; Greig 1982; Waller 1994).

At Runnymede, three pollen sequences date to the Atlantic period. Column 39 produced a very marked *Alnus* expansion in p.a.z. 39:2 which defines locally the Boreal/Atlantic transition. Column 40 (Area 22) falls within the period with neither upper or lower markers. Column 59 (Area 32), however, has an expansion of herbs, including *Plantago* species and cereals in the upper levels (p.a.z. 59:2, subzone b) and may be associated with the first appearance of agriculture. The pollen spectra in pollen assemblage zones 59:1–2 and 40:1–3 are broadly similar; this is not unexpected as the two samples come from some 15 m apart and precede the Middle Neolithic horizon.

The vegetation characteristics typical of the Atlantic period for this region are seen in all three profiles. Contrasting with the Boreal sites discussed above *Alnus* is the dominant overall taxon and undoubtedly formed dense alder carr woodland on the wetter floodplain. Values of 60–80% are strongly indicative of such local, on–site dominance (Janssen 1959; Tauber 1965, 1967a; Andersen 1970, 1973). The major expansion of *Alnus* is argued to have taken place here in the early to mid-seventh millennium BC (Chapter 10). It is clearly seen in column 39, and a minor expansion of *Alnus* in the uppermost pollen-yielding level of column 4 may represent the start.

The expansion of Alnetum has been viewed as a consequence of climatic change to one of more oceanicity associated with increased temperatures (1.5–2 degrees), humidity and the attainment of sea levels broadly similar to those today. Bennett and Birks (1990) summarises views expressed by a number of earlier researchers, that the expansion of *Alnus* across the region is largely habitat related. Thus, with sea level change, hydroseral succession and floodplain development, *Alnus* was able to expand its dominance in areas such as Runnymede and many other sites where suitable conditions pertained. This expansion apparently spread out of the localised communities already established in Flandrian chronozone I. *Salix* is consistently present throughout both profiles (Figs 9.3 and 9.4), indicating that willow was also an element (possibly important) of the Alnetum community (typically with *Rhamnus catharticus* and *Viburnum*). Although Godwin (1940, 1975a) has suggested that climatic wetness may have allowed *Alnus* to have been a component on interfluve soils, this seems unlikely; relevant pollen sequences generally come from valley/wetland profiles where it was the dominant element (thus masking any possibility of identifying *Alnus* from the interfluves).

Other wetland elements include Cyperaceae, *Sparganium* type, *Alisma plantago-aquatica*, *Filipendula ulmaria*, Apiaceae types and *Chrysosplenium oppositifolium*. These are all taxa which may be attributed to the carr woodland environment. *Alisma* may, however, indicate shallow, standing water. *Butomus umbellatus* is an important pollen record and is today a rare plant in Britain, growing in ponds, canals, ditches and river edges (Stace 1991).

CLIMAX WOODLAND

A typical range of tree taxa is recorded in the profiles of Atlantic age. Interpretation of the pollen percentages/assemblages of the dominant tree taxa (which reach 90% of the total pollen) must take into account differential pollen production and dispersal characteristics of the taxa present. The dominance of *Alnus* and its disproportionately high

percentage values are explained in terms of its high pollen production and anemophily. In contrast, there are records of *Tilia* and *Fraxinus* that produce relatively small numbers of grains which are insect pollinated (entomophilous). Thus, these taxa are typically markedly underrepresented in pollen spectra (Tauber 1965, 1967a; Andersen 1970, 1973). Numerical conversions have been calculated using modern pollen dispersion studies (Andersen 1970, 1973) but are rarely used to calibrate the raw pollen data. If such 'r' values (Andersen 1967), or transforms, are considered in relation to the pollen data presented here, it is clear that *Tilia* and *Fraxinus* were important or even dominant elements of the mid-Holocene climax forest. The dominance of *Tilia* in this period has been argued from many analyses of southern and eastern English sites (Moore 1977; Scaife 1980, 1987, 1988; Waton 1982a, 1982b; Waller 1994) and more locally at Hampstead Heath (Greig 1989, 1992) and in east London at the site of Rotherhithe similarly dated to the mid-Holocene period (Scaife unpublished data and Sidell *et al.* 1995). It is now considered that *Tilia* (likely *Tilia cordata*) was the dominant or at least co-dominant tree taxon over large areas of south-east and eastern England from the beginning of Flandrian chronozone II until its demise as a result of forest clearance during the later prehistoric period.

Fraxinus is likewise a poor pollen producer and thus its recurrence, albeit at low levels, during the mid-Holocene at Runnymede is significant. Expansion is often taken to be associated with Neolithic forest clearance during and after the elm decline. The picture provided by the Runnymede results is of *Fraxinus* as an existing component of the forest which subsequently expanded rapidly in range and importance following human interference. *Ulmus*, *Quercus* and *Corylus avellana* were undoubtedly also prominent constituents of the woodland. However, while past discussion may have regarded these latter as important elements of mixed deciduous woodland, it is also plausible, and perhaps more likely, that these taxa were significant on the less-well-drained soils fringing the lower interfluve slopes and the floodplain (i.e., the colluvial/alluvial interface). *Betula* and *Pinus* have almost constant occurrences but their high pollen production and wind dispersal allow that these were extra-local and possibly even extra-regional vegetation elements.

HERB SPECIES

As with the Boreal period, a range of herb taxa is present including ruderals such as *Plantago media/major*, *P. lanceolata*, *Rumex*, *Chenopodium* type, etc. Many taxa not identifiable at more specific taxonomic levels may be referable either to the mire/aquatic floodplain vegetation, or to openings and glades within the woodland. It has long been thought that the climax forest of the mid-Holocene was dense, closed-canopy deciduous forest (Quercetum mixtum noted above). However, there is evidence that this prevailing idea may be erroneous. Rose (1974) postulated that Atlantic forest may have contained high lichen species diversity due to glades caused by grazing herbivores. Other possibilities might include natural tree throw, man-made pathways and glades caused by the death of trees (Scaife 1980). These may have been locally ephemeral but continuously changing, thus acting as refuges for many heliophilous herb taxa.

The period under discussion corresponds with the later Mesolithic. As for the Early Mesolithic, definite evidence of anthropogenic modification to the environment is elusive and only the presence of disturbed ground species might be presented as a possible indication (Smith 1970; Turner 1970). Artefacts on the site do suggest some local activity, a point of interest in the context of an environment dominated by dense and inhospitable alder carr swamp. Winter periods are likely to have seen overbank flooding and standing areas of water across the floodplain zone.

Flandrian chronozone III: the later prehistoric period

The period from *c.* 4000 BC (5000 BP) can be regarded as the most interesting from the archaeological viewpoint since it is during this phase that intensive occupation took place. Chronologically this period covers the post-*Ulmus* decline, i.e., the sub-Boreal and sub-Atlantic periods. Five pollen diagrams are available for this chronozone. Two were produced previously by Greig (Greig 1991, 1992), one dating to the Middle and Late Neolithic (SS1), the other spanning part of the Late Bronze Age (WF1). In the current investigation, two further profiles span the Late Neolithic (column 5 in Area 14, column 60 in Area 32) and another falls around the Middle–Late Bronze Age transition (column 55 in Area 31).

The above sequences show a marked contrast in the vegetation compared with earlier records on the site. In the profiles from chronozone III, we are seeing the direct, profound and increasing impact of humans on the environment. As might be expected, the Neolithic pollen spectra SS1 (Greig 1991) and columns 5 and 60 are transitional between the dominant woodland vegetation of the Atlantic and the largely anthropogenic communities of the second and first millennia BC. The Neolithic and Bronze Age data are discussed separately.

THE NEOLITHIC

The elm decline, taken to define the beginning of this chronozone, was unfortunately not found within any of the Runnymede pollen columns, either in the research campaign or in the earlier work (Greig 1991, 1992). Within the lower Thames basin, this important event is identified in pollen sequences from Hampstead Heath (Girling and Greig 1977; Greig 1982), Bryan Road, Rotherhithe (Sidell *et al.* 1995), where it has been dated to 5040 ± 70 BP, Beta-58577),

Epping Forest (Baker *et al.* 1978), Stone Marsh in the Thames estuary (4930 ± 110 BP, Q-1336; Devoy 1977, 1979) and Mar Dyke, Essex (Scaife in Wilkinson 1988) with a date slightly above declining elm pollen percentages of 4650 ± 90 BP (HAR-4523). These dates demonstrate the very broad synchroneity of the *Ulmus* decline and its value as a palynological marker. It is unfortunate that the excavated sediments at Runnymede do not span this event although the column 59 date (sequence already discussed) appears to be only a little earlier.

Very broad dating of the Runnymede parcel 2D and 2E sequences comes from the cultural horizons that bracket it stratigraphically, while radiocarbon dates on two isolated contexts, one each in Areas 6 and 14, suggest that the sediments span similar date ranges. The Area 6 sequence may, however, be a little earlier, starting perhaps just before Middle Neolithic activity came to a close on the site; the overall correlations are shown in Fig. 2.37.

In both Greig's pollen sequence and those established here, the tree taxa are dominant, as in the preceding Atlantic period, and will be discussed first. Overall, these pollen spectra show continued dominance of tree pollen with *Quercus*, *Tilia* and *Corylus avellana* being dominant on drier soils, along with remaining *Ulmus* and *Fraxinus*. In the wetter parts of the floodplain, alder carr (Alnetum) maintained the dominance established in the preceding Atlantic period. *Rhamnus catharticus*, *Viburnum*, *Humulus* type (this includes *Cannabis sativa* but here the likelihood is for *Humulus*) are also associated with this fen carr woodland (see below). Greig (1991, 1992) recovered waterlogged seeds and fruits, as well as sporadic pollen, of some of these taxa including *Taxus baccata*, *Crataegus monogyna* and *Prunus spinosa*. Having seeds to compare with the pollen record highlights the problems of interpreting pollen spectra when some taxa are very poorly represented due to pollen production (especially entomophily), differential dispersion and taphonomic factors. Robinson (1991) has also emphasised the continuity of woodland from the presence of phytophagous Coleoptera which show the existence of *Alnus*, *Quercus*, *Corylus*, Prunoideae, Pomoideae, *Salix/Populus* and *Tilia cordata*. These same species are well represented in the pollen spectra. The coleopteran assemblages also suggested significant open grassland and possible pastoral use, but not an overgrazed habitat. This is commensurate with the pollen data obtained.

From Figs 9.6 and 9.7, it is immediately apparent that while tree pollen is important, there is a marked expansion in herbaceous diversity, probably attributable to the activities of Neolithic people. However, the extent of alder carr in the immediate environs of the site may have had a substantial filtering effect on herb pollen from local sources. Broad proportions of trees, shrubs and herb pollen (excluding *Alnus*) suggest that there was a mixed environment of woodland, possibly scrub and cleared land. The continuous presence of cereal pollen implies that arable cultivation was taking place somewhere in the catchment. However, there is also a strong representation of pasture species such as *Ranunculus* type, Fabaceae, *Sanguisorba officinalis*, Poaceae, *Plantago lanceolata* and others. Pastoralism is always more difficult than arable activity to establish in the pollen record since many of the taxa are constituents of 'natural' grassland such as floodplain communities.

It is interesting to note that in many of the profiles now examined in southern Britain, there is evidence for ephemeral 'Landnam' clearances such as classically described by Iversen (1941, 1949) but of longer duration than the 30–40 years originally postulated (but later revised by Iversen). Thus, while it was once considered that the arrival of Neolithic subsistence agriculture resulted in widespread woodland clearance especially on shallow soils (such as chalklands), it has become increasingly apparent that in most areas only partial clearances took place. In many cases the clearings were later abandoned as, for example, at Gatcombe Withy Bed, Isle of Wight (Scaife 1980, 1987) and Hockham Mere, Norfolk (Sims 1973, 1978; Bennett 1983) with subsequent forest regeneration taking place. Typical secondary woodland elements include *Fraxinus*, *Ilex* and secondary *Ulmus*. This is particularly important for the Late Neolithic for which there is strong regional evidence for widespread woodland in the pollen record (Barker and Webley 1978; Scaife 1987). It has been suggested that this indicates a changing agricultural economy to one based on woodland pastoralism. At Runnymede, however, there is no evidence of such 'Landnam' type clearance or phases of woodland regeneration. In part this may be due to the short duration of the organic profiles it was possible to sample.

The Neolithic floodplain environment

It has already been argued that alder carr was widespread along the valley floor. The ecological requirements for such a community to flourish are quite narrow (McVean 1953, 1956; Brown 1988). The resulting community may contain a diverse range of lower shrub taxa and trees depending on the relative height of the water table and length of time that the community has a flooded surface. Typically, *Alnus* will tolerate 2–3 months maximum per year (winter) with its roots submerged. A rich fen shrub flora can include *Taxus baccata*, *Viburnum lantana/opulus*, *Humulus lupulus*, *Frangula alnus*, *Hedera helix* and *Lonicera periclymenum* (Tansley 1949). The majority of these taxa has been recorded in Runnymede pollen spectra and it is concluded that such a rich fen carr with drier and wetter habitats existed locally during the Neolithic. Pollen of aquatic taxa has also been recorded, including *Myriophyllum spicatum*, *Alisma* type (cf. *A. plantago-lanceolata*), *Sagittaria sagittifolia* and *Potamogeton*. Reed swamp/poor swamp taxa include *Lythrum salicaria* and *Typha angustifolia/Sparganium* type.

The Bronze Age

The later Neolithic profiles have been considered as transitional between the dominant woodland of the mid-Holocene (Atlantic) and the later prehistoric period of the Bronze Age and Iron Age with its more open, cleared landscape. Bronze Age pollen assemblages from columns 55 (Area 31) and WF1 (Area 6; Greig 1991, 1992) clearly show marked reduction in tree pollen and correspondingly higher non-arboreal pollen percentages with agriculturally associated weeds and the more consistent occurrence of cereal pollen.

In contrast to the 50% average tree pollen and 30% shrub pollen of the Neolithic, these groups decline by the end of the Middle Bronze Age to 20–30% and 15–20% respectively. The expansion of herbs to 50% or greater implies much larger expanses of open land which included both pastoral and arable. Although cereal pollen and segetals (*Sinapis* type, *Persicaria maculosa* type, *Polygonum* spp.) are present, more dramatic is the expansion of Poaceae and other species best attributed to pasture/grasslands (Fabaceae spp. – *Lotus* type, *Trifolium* type, *Vicia* type). From this, it may be concluded that human population expansion and/or demographic changes within the region necessitated more cleared land. Clearance for agriculture was possibly largely for pastoral usage in the Runnymede area. Cereal cultivation was, however, practised as evidenced by its pollen and segetal spectra as well as charred plant macrofossils (Greig 1991).

The remaining woodland

All of the principal tree pollen taxa described for earlier Holocene periods are recorded in the latter phases, albeit with very reduced percentages. They include *Quercus*, *Alnus* and *Corylus avellana* type, in each case barely attaining at maximum 20% of pollen. It is significant that there are increases in both the diversity of shrub taxa and the percentages of lesser shrub types as compared to the Neolithic records. Greig (1991) and Robinson (1991) have previously shown the survival of at least some woodland in the Runnymede region as late as the Late Bronze Age. Robinson's coleopteran analysis has yielded phytophages which provide evidence for woodland/trees (*Quercus*). These elements are consistent with the pollen data obtained from the slightly earlier column 55.

The woodland communities on the floodplain were dominated by *Alnus* and *Salix*, with other typical carr woodland elements, much as described for the Neolithic (*Taxus baccata*, *Rhamnus catharticus*, *Frangula alnus* and *Viburnum lantana*). The ground marsh herb flora is evidenced by Cyperaceae (*Carex paniculata?*), *Chrysosplenium oppositifolium*, and *Caltha palustris*. While it could be suggested that alder fringed the river channel, the presence of these typical carr taxa suggests that discrete areas of *Alnetae glutinosae* remained in localised stands on the floodplain or its margins.

Quercus, *Corylus* and *Tilia* woodland remained on the drier interfluves although their pollen percentages are markedly diminished. A noteworthy response, though, is the significant expansion of secondary woodland elements, especially *Fraxinus* and *Fagus*, after allowance is made for their underrepresentation. The Bronze Age spectra of columns 55 and WF1 are the first consistent records of *Fagus* at Runnymede and, taken at face value, the species appears to have been a relatively late coloniser in this region. This is, however, a local phenomenon since it has been recorded from the middle Holocene (column 39) and in Sussex by Thorley (1981) and more frequently from the Neolithic onwards (Scaife 1980). Taphonomic factors may also have played an important role and absence of pollen does not imply the absence of the plant in question in the pollen catchment. It is possible that *Fagus* was present, but that its pollen was only transported onto the floodplain once the existing woodland canopy was opened up, effectively increasing the area of the pollen catchment.

Betula, *Pinus* and *Ulmus* are also recorded consistently but in small percentages and, therefore, probably small numbers. In the case of *Betula* and *Pinus*, these may be of extra-local origin (Tauber 1967a) and need not be relevant to the interpretation of local valley-bottom vegetation communities. *Ulmus*, however, is more problematic in that the small, consistent occurrence may represent occasional local or extra-local growth.

Almost as significant as the *Ulmus* decline in the pollen records is that of the 'lime decline' in southern England. There is now a substantial corpus of data dealing with this topic. The significance of the sharp reduction in *Tilia* pollen frequencies often associated with the Late Bronze Age period was in the past regarded as a climatic response to climatic deterioration at *c.* 500 BC; that is, the change from the sub-Boreal to the sub-Atlantic. While there is much evidence for climatic deterioration during the first millennium BC, it has long been held that the lime decline is a response to woodland clearance/deforestation (Turner 1962). The anthropogenic clearance of lime would presumably be asynchronous across a region and associated with an expansion of weeds of cultivation. This has in fact been shown from a number of pollen sites/studies with radiocarbon measurements. Dates range from as early as the Neolithic in the Isle of Wight (Scaife 1980, 1987) and Sussex (Waller 1993) to as late as the Saxon period in Epping Forest (Baker *et al.* 1978). Nevertheless, the majority of dates currently fall in the Middle to Late Bronze Age period.

In addition to these anthropogenic arguments, Waller (1994) has shown that in coastal areas, progressive inundation causing habitat change resulted in marked changes in pollen taphonomy; that is, certain species only survived further away from the sample site, thus resulting in a decline in pollen. At Runnymede, there is clearly a reduction in the amount of *Tilia* pollen in the Bronze Age spectra (column 55) compared with the Neolithic (column 5,

Area 14). The two sequences in Area 6 (Greig 1991) also show this difference. However, unlike the pollen sequences from Hampstead Heath (Greig 1992; Girling and Greig 1977), at sites in east London – at Ferndale Street, North Woolwich, Rotherhithe (Scaife, unpublished report), Beckton (Scaife, unpublished report) and Barking (Scaife, unpublished report) – there is no clear decline of *Tilia* in the pollen diagrams attributed to the Middle Bronze Age and it is assumed that the 'typical' decline occurred earlier. There is, though, little doubt that the decline of *Tilia* woodland at Runnymede was as a result of increased human impact, especially during the Middle Bronze Age. Undoubtedly, increased land pressure and changes in agriculture and land use management (Ellison and Harriss 1972) resulted in the widespread demise of lime woodland in southern and eastern England. One crucial question is why *Tilia* woodland, which would have flourished on better-drained and perhaps more base-rich soils, should have remained with little change in many regions during the Neolithic and earlier Bronze Age. Such land would have been ideal for agricultural purposes and thus woodland clearance. It seems likely also that *Tilia* was specifically exploited for bast fibres or its nutritious leaves which may have been used as animal fodder (Troels-Smith 1956).

Pollen taphonomy and alluviation

Until recent years, few attempts have been made to produce meaningful results and vegetation histories from the pollen analyses of river alluvial sediments. In spite of a small number of papers suggesting the value of such studies (Grichuk 1967; Smith 1970; Burrin and Scaife 1984; Scaife and Burrin 1992), there has been much scepticism of the technique, assuming that secondary/reworked pollen and long-distance fluvial transportation negate meaningful results. Intercalated organic/peat deposits have thus usually been the foci for analyses rather than any underlying and overlying lithofacies. While there are certainly problems associated with analysis, it has now been demonstrated that useful pollen data can be obtained from alluvium deposited during overbank flooding or during phases of channel aggradation. Such sedimentary/alluvial units (parcels of Needham 1989, 1992) frequently appear to have accumulated in response to allogenic factors such as climatic change, vegetation change or human activities in the drainage basin. This makes the deposits potentially more useful indicators of environmental changes at a local level than might be found in the stable organic/waterlogged organic peat formation. The problems of taphonomy and interpretation of the pollen and spore assemblages are complex and are not dealt with in detail here (see Burrin and Scaife 1984; Moore *et al.* 1991; Scaife and Burrin 1992).

However, the factors which are pertinent to the interpretation of floodplain development should include inter/multi-disciplinary examination of the biological, lithological and archaeological evidence. If these data can be integrated, a coherent picture of floodplain archaeology and environment might be obtained. One problem in this composite pollen study arises from the temporal dimensions of the pollen sequences presented. While there has been a fortuitous representation of pollen sporadically through Flandrian chronozones I–III, it has been found that certain diagnostic features/phenomena such as the *Ulmus* decline at *c.* 5000 BP and the later prehistoric lime decline are missing in the profiles. From the discussion of woodland change during the different chronozones, it is clear that these events occurred 'between' the periods represented by the pollen columns. The nature of sedimentation involved allows the possibility that these columns, being the profiles beneath water tables, accumulated rapidly and, therefore, have only a short temporal span. They may thus be short time-windows in potentially a much longer sequence of deposition. If this is the case, questions might be posed regarding the nature of the organic accumulation and its causes, for example whether later prehistoric activity on the interfluves may have been responsible for influencing the depositional environment and character of sedimentation.

Conclusion

One of the primary aims of this research project was to establish as full a chronology of vegetation and environmental change as possible from this floodplain site. Since no single excavated profile has a long temporal sequence of organic deposits, the study presented here has relied upon pollen obtained from sediments and from the thinner organic horizons also encountered. Given that the alluvial sediments on this site are frequently highly calcareous silts and marls containing numerous freshwater and terrestrial Mollusca, the conditions were overall not conducive to pollen preservation; this is especially so where the water table has fluctuated, such as in the upper contexts of each profile. Pollen, albeit in less than satisfactory numbers, has generally been obtained from those units lying below the water table.

A total of eight pollen profiles has been examined, dated to different Holocene stages using a combination of evidence (Chapter 10). From the above discussions, it will be apparent that some of the crucial and diagnostic phenomena (e.g., the elm and lime declines) are not present in the current pollen profiles. In spite of this, however, careful selection of sites has allowed the vegetation and environmental reconstruction of the early Holocene (Boreal; Flandrian Ib–c), the mid-Holocene supposed climatic optimum (Atlantic; Flandrian II) and the later prehistoric period (Neolithic and Bronze Age; sub-Boreal; Flandrian III). The latter is of particular importance for its relevance to reconstruction of the environment and economy of the later prehistoric occupations at Runnymede. With the earlier work of Greig (1991, 1992) there are now four pollen sequences reflecting the Neolithic environment and two for the Bronze Age. The attempts at obtaining pollen from these marginal situations seem to have been rewarded by the environmental record obtained. Furthermore, the value of these data lie in the possibility of correlation with other environmental aspects examined (i.e., sediments, molluscs, insects and plant macrofossils). The use of transform functions and multi-proxy data to provide environmental reconstruction is becoming important in palaeoecological investigation.

CHAPTER 10

An alluvial site history: environment, topography and human interaction

A relatively comprehensive and coherent chronology has been possible for the alluvial sequence at Runnymede Bridge. This is partly due to the presence of two excellent marker horizons, given by the in situ archaeological deposits, in the fourth and first millennia BC respectively. However, the alluvial history pieced together by persistent investigation has a long span and, although most parcels of sediments can be directly related to others stratigraphically, cultural material is generally rather sparse. In a few cases it has been necessary to confirm relative dating, but the more pressing matter has been to obtain as good an absolute chronology as possible. The independent dating of most phases of sedimentation has depended on the radiocarbon and archaeomagnetic methods (Chapter 4). Further assistance for sequencing the early Holocene deposits came from the pollen assemblages, which, at the site-specific level, can indicate relative chronology through the vegetational succession revealed. The dating evidence available for each alluvial parcel, and indeed the major cultural horizons, is discussed below in the individual sections; overall conclusions on chronology are summarised in Fig. 2.1.

The main objective of this chapter is to weave together the important implications of the various cultural and natural inputs to the sequence. It concentrates on understanding the site topography, site and regional environment and human activity in the catchment. In the process it sets out the evidence for the location and orientation of river channels as they are encountered. However, linking the latter observations together to reconstruct the long-term change of the river system is reserved for the next chapter, since the arguments for any particular sequence of change are highly dependent on seeing continuity through time. Similarly, a description of the present-day topography of the local region around Runnymede Bridge is also left to Chapter 11, for which it is an essential backdrop.

The pattern of change in the water table also proves to be important for the hydrological and topographical history of the site. Although water tables have a gradient, increasing in height away from contemporary river channels, this is not likely to be very significant within the small plot of the site and no attempt has been made to compensate for such differences. The existence from an early stage of two channel courses, flanking either side of the site and often both active, will have further diminished within-site gradients.

Alluvial parcel 1

Parcels 1C/1J: eighth millennium BC; Boreal

The stratigraphically earliest deposits from the research excavations – the marls comprising parcel 1C in Area 19 – while not dated directly by dating methods, have yielded an early Boreal pollen suite appropriate to approximately the eighth millennium BC from the basal deposits (in radiocarbon years, *c.* 9000–8000 BP). Most of the marl sequence recorded was above the water table and represents later deposition than the pollen record. However, a limit is given by the overlying deposits (1K), dated by both archaeomagnetic and ^{14}C dating to the later seventh millennium BC.

Other deposits found on the site that seem to be as early as these were encountered in rescue excavations: probably some of those low in Area 7, parcel 1A (see below), but also now the lowest in Area 5 (5.140–5; Needham 1991, fig. 7). These latter layers are now regrouped as parcel 1J, having been radiocarbon dated to around the first quarter of the eighth millennium BC using an accumulation of branches trapped in the sediments. The wood included willow/poplar, hazel, oak and *Prunus* sp. (Gale 1991). Although an early record for *Prunus* sp., the species is in fact documented throughout the Holocene flora; indeed, pollen data suggests initial colonisation in the Late Devensian (Godwin 1975a, 196). The deposits here were a sequence of very organic silts, some peaty, and clayey horizons; they lay between 11.60 and 12.00 m OD, over a metre lower than the marls of parcel 1C. They were presumably in a channel, but the nature of deposits suggests a fairly slack-water environment.

If the deposits of 1C and 1J, which are 100 m apart, were strictly contemporary they would give important information on the Boreal (Flandrian Ib) topography and the conditions under which the respective deposits were formed, the marls being much more elevated than the channel deposits. However, the snail evidence for 1C generally supports a varied wet, tussocky and reedy environment and does not indicate much drying out; indeed, Evans and Evans suggest open gently flowing water early on, rather than a quiet-water lake, becoming a bit more swampy towards the top, while Limbrey is also in favour of marl formation having taken place in an entirely aquatic environment (Chapters 5 and 7). Since the marl attained an altitude of 13.80 m OD in the recorded sections, and up to 14.00 m OD in the walls of Neolithic cut features in Area 20, it indicates a high, or steadily growing water table. This makes it perhaps more likely that marl deposition followed on from parcel 1J during the course of the eighth millennium, involving a steady vertical build-up of sediment. (Unfortunately 1J was directly overlain in Area 5 by the base of a Late Bronze Age river channel.) The best interpretation, therefore, is that during at least some of the Boreal period a very high water table was present locally. One can envisage an environment of impeded drainage with very wide water bodies through which water flowed at differential rates, but generally all rather slow (the likely cause of this lake is discussed in the next chapter). Lewis *et al.* (1992), in reviewing other early Holocene wet sites in the lower Thames basin, have shown that impeded drainage was a regular feature of the valley bottoms.

The lack of surviving organic detritus in the marl above the present-day water table is to be expected, since this water level appears to have been pretty stable for millennia and all organics above would have decayed. More surprising is the on-site observation, backed by Scaife's lithological description (Chapter 9), that there were no organics in the marl below 13.00 m OD. This suggests that at sometime after deposition there would have been a fall in the water table to a lower level. Since we can establish from other profiles that the level was restored to about 13.00 m OD by the Atlantic period and subsequently never fell appreciably below that, it is evident that the low-point must have occurred later in the Boreal. This can be tied in with the observed low organic preservation level of 12.50–12.75 m OD in Area 7, thus having implications for the chronology of the earliest parcel 1A deposits (below).

Higher up the river system, along the Kennet tributary at Thatcham, marl deposition was discovered in excavations on a Mesolithic site; its start is dated to the pre-Boreal (Wymer 1962; see also Healy *et al.* 1992). An early suggestion was that it formed in a floodplain lake caused by beaver damming, but Brown argues that many shallow water-filled depressions would have dotted valley floors, the result of Late Devensian fluvial deposition (1987; 1997, 206).

During this period the lower Colne valley was experiencing widespread peat growth wherever the Late Devensian gravel surface was low-lying (Gibbard 1985, 91). At the site of Sandstone Pit, Iver, 12 km north of Runnymede Bridge, pollen from the peat layer indicated a late Boreal vegetation (Lacaille 1963). Early peat growth also featured in the upper reaches (Gibbard 1974), and has recently been demonstrated in a new profile only 1 km north-east of the Runnymede Bridge site (unpublished pollen analysis by Dr Keith-Lucas, inf. Steve Ford). An important alluvial sequence has been excavated at Three Ways Wharf, Uxbridge, yielding Late Devensian and Early Mesolithic occupation surfaces on an island at 31 m OD,

between braided streams of the Colne (Lewis *et al.* 1992). Here the water table had fallen after the deposition of Late Devensian gravels and argillaceous sediments thus allowing human settlement, but subsequently, during the Boreal, it became high enough again to promote peat growth on top of the occupied surface. Indeed, the earlier part of the peat sequence is interpreted as forming under shallow water with an environment very similar to that at Runnymede: sedges, grasses and other herbs locally dominant; pine and early colonising deciduous trees in the neighbourhood.

The pine/hazel dominant woodland composition seen in columns 31 and 4 is to be expected of a mature stage in the early post-glacial vegetational succession; Scaife suggests a reed swamp habitat fringing the open water, or wet pasture nearby. This fits with the vegetation structure implied by the molluscs (Evans and Evans, Chapter 7). At this stage in the succession these low vegetation covers would be competing on the wetter ground with *Salix* (willow etc.) which, allowing for its severe underrepresentation in the pollen record, was very significant on the water margins by the late Boreal (column 4; note also that the trunk and branch wood recovered in Area 13 North were *Salix*). But it may be that this was the only significant tree/shrub presence on the valley floor prior to the rise of alder. On drier soils the pine and hazel were accompanied by birch, oak, elm and already scattered lime trees.

Parcels 1D/1E: earlier seventh millennium BC; Boreal/Atlantic transition

Some 1,000 years later, on the evidence of both ^{14}C and the pollen spectrum, as well as tentatively on the archaeomagnetic dating, variable deposits were accumulating in Area 13 (parcel 1E), 25 m to the north-west of the Area 19 marl, but only a few metres to the north-west of marl seen in Areas 13 and 20. The recorded sequence is at a lower altitude than the top of parcel 1C, 12.57–13.41 m OD, and the sediments have the character of channel deposits, evidently close to a bank which would have seen active flow prior to this aggrading phase (Fig. 10.1).

Parcel 1E has organic preservation to at least 12.86 m OD. The deduced high-water level of the early to mid-Boreal may have fallen after down-cutting to below 12.75, as noted above, but was evidently rising again towards the end of the period. By the transition to the Atlantic (parcel 1D), organic material is preserved up to 12.92 m OD, and a little later (parcels 1F–G) up to 12.98 m OD.

Just 5 m to the east of the parcel 1E exposure, another sequence of channel deposits was exposed in Area 22. Although there are some points of comparison in sedimentary terms, along with a similar altitudinal range (12.70–13.62 m OD), the sequences are not identical and correlation thus not certain, hence the separate labelling as parcel 1D. However, the pollen from the base of 1D clearly shows a change from late Boreal to earliest Atlantic, with a

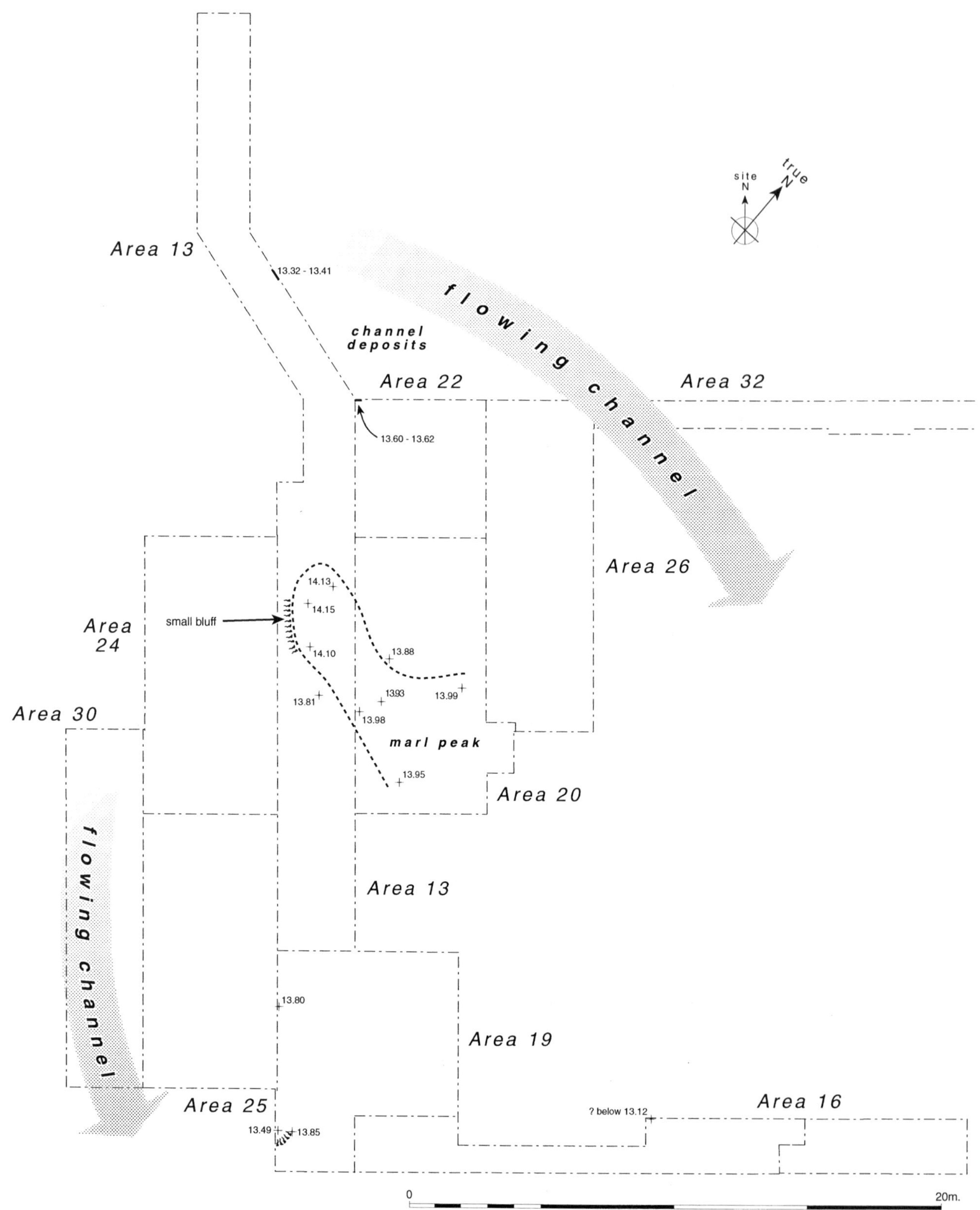

10.1   *The topography of the interior zone in the early seventh millennium BC, before deposition of parcel 1K.*

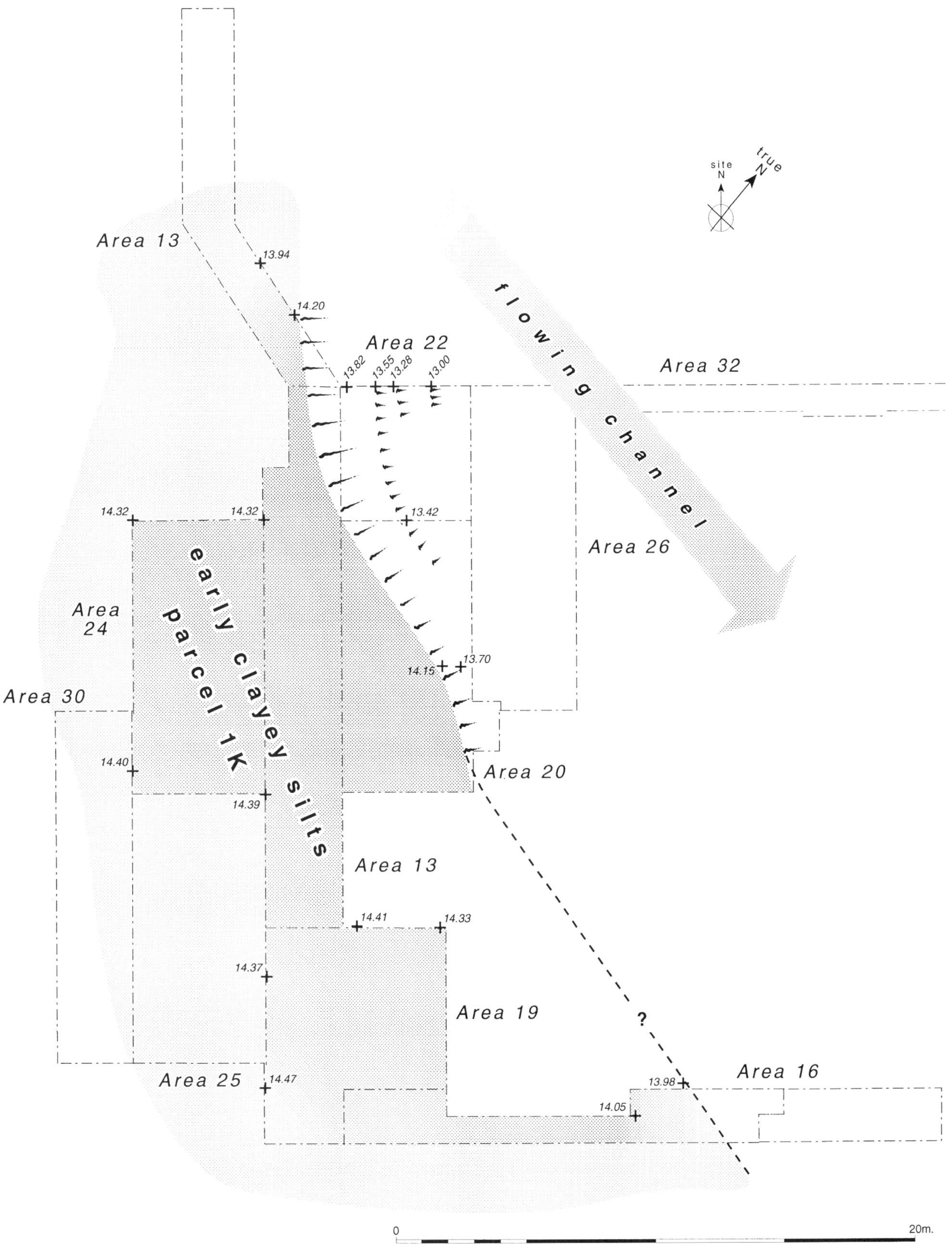

10.2 *The topography of the interior zone in the later seventh millennium BC, after deposition of parcel 1K.*

pronounced rise in alder, a rise only in evidence at the very top of the 1E pollen sequence. This suggests an overlap between the two pollen monoliths, with the former continuing later than the latter. Some corresponding matching can be found between the respective sediment sequences: in particular the grey sand of 13.556 with 22.213, and the pale silts above, 13.558 with 22.216 (Figs 2.2 and 2.33).

Although the two relevant ^{14}C dates give the reverse sequence at their one-sigma calibrations (the sample stratified above the 1D column is earlier than that stratified within the 1E column), at two-sigma levels of confidence the two dates actually overlap. Thus at a general level all the dating evidence is in broad agreement, that is to say the lower parts of parcels 1D and 1E span the transition from Boreal to Atlantic, and fall somewhere within the time-span 7265–6420 BC. Two alternatives may be offered by way of more refined chronology. The first works from the assumption that the whole sequence of silting involved was rapid (which is entirely possible), and that both ^{14}C-dated samples were contemporary with their contexts. On this basis, the best dating for the whole sequence would be in the overlap of the calibrated ranges, i.e., 7000–6615 BC. This would mean that the establishment of alder as a major species locally took place fairly early in the seventh millennium BC (early eighth millennium BP). There is some evidence from lowland sites in the south of England for the local establishment of alder during the course of the Boreal (Scaife, Chapter 9; also Brown 1997, 200).

The alternative is to view the earlier, but later-stratified date to be residual in its context; this is equally possible given that the sample was a small piece of charcoal. In this case independent dating would rely entirely on the Area 13 date, giving a range of 7000–6420 BC for the beginning of the 1E/1D sequence, and the possibility that it continued later, thereby allowing a slightly later date for the rise of alder. Even so, the span is unlikely to extend much beyond the mid-seventh millennium BC (*c.* 7500 BP) because of the dating of the overlying silts of parcel 1K to the later seventh millennium BC. It is hard to escape the conclusion that the rise in alder occurred locally in the middle of the seventh millennium, as in the Kennet valley at Thatcham (Healy *et al.* 1992, 72), and possibly earlier.

The two sequences 1D and 1E may be presumed to have belonged to a single channel, even if there was a recut at some point in between the exposures. Its southern bank would have been very nearby, being defined by the interface with parcel 1C, and must have run broadly west–east. It is doubtless the down-cutting of this channel which accounts for the slope of the 1C/1K interface described below, a slope which can be regarded as a palaeobank. Substantial incision was inevitably associated with the fall in the water table and this could certainly have given rise to the relatively abrupt dry land/channel transition which both Robinson and Evans and Evans have argued, on the basis of the coleopteran and

molluscan assemblages respectively. The latter evidence suggests a greater influence from flowing channels at the expense of the swamps and pools of the early to mid-Boreal, and all this points to the development of freer outflow from the valley bottom in the local reach. The banks would have seen a significant vegetation shift over this period from willow-dominant to alder-dominant.

Parcel 1A: Late Devensian to earliest Atlantic

The main early Holocene silt body exposed during the rescue campaign, parcel 1A (Needham 1991, 26–7), lies 100 m to the south-east of the interior zone. A maximum depth of 1.6 m surviving up to 13.50 m OD includes basal deposits of gravel giving way quickly to organic, then inorganic silts. This sequence probably represents a long time-span, but with the limited evidence available the only further subdivision is to define some of the layers showing in Area 8 as parcel 1L (below and Fig. 2.1).

The gravel may be the top of the Devensian 'floodplain' gravel, which was recorded in a borehole survey ahead of the M25 construction, between 7.00 and 12.75 m OD (Gibbard 1985, fig. 45; Needham 1992). The overlying deposits are not independently dated, other than being pre-Neolithic, and, given the distance to the research excavation profiles, it would be extremely hazardous to attempt any positive correlation. It is noteworthy, however, that no marl as such was recorded, and it is possible that the 'crumbly orange silt' of layer 88, presumably iron-stained, relates to some of the clayey deposits of 1K, e.g., 16.979 and 19.803. An early date for the underlying deposits would in fact be consistent with the boundary for organic preservation at layer interface 8.092/093 (around 12.50 m OD) which is lower than that seen in parcels 1D and 1E and thereafter. It is also worth recalling the small group of flintwork recovered from 8.091/092, post-dating the water table rise, but nevertheless best dated to the earlier Mesolithic (Saville 1991).

Parcel 1K: later seventh millennium BC; earliest Atlantic

The next major unit in the interior zone is a thick and relatively homogeneous clayey silt deposit, parcel 1K. Layers within it can be light grey, at times giving a silvery appearance, or can have distinct orange tones due to mineral staining. This material seems to spread across virtually all of the interior zone and is seen to seal parcels 1C, 1D and 1E in the fully recorded sections. It was also observed across Area 20, particularly in a machine-cut section (SW–NE) across the north-western end (not drawn). Here there was possible evidence that the parts of this parcel interleaved with the uppermost part of parcel 1C, but given the sequence deduced above, it is more likely that this was 'false interleaving' caused by a modicum of redeposition. Other later layers in the channel fills contain some marly elements.

The essential difference between the two thick deposits 1C and 1K, as deduced by Limbrey (Chapter 5), is related to the discussed change in the water table. With its significant lowering, prolific in situ calcite production ceased, but the fine silt that had been washing into the former lake continued to be carried by the river system and was now off-loaded instead on the banks during the ebb of high water. These alluviating banks were seasonally sub-aerial, favouring terrestrial processes (Chapter 5). This is emphasised by the molluscs, which suggest a transition to more terrestrial conditions, although still with regular flooding, especially on the lower-lying areas (column 4; Chapter 7). The molluscan evidence also suggests an essentially treeless environment on the high bank (columns 13 and 31) and this may be specific to these steadily alluviating muddy banks which progressed from a swampy nature, to wetland of some kind (?sedge fen), and then to grassland still subjected to flooding. In contrast, on the lower slopes nearer the river (column 4) there are signs of much more tree cover consistent with the other evidence for developed woodland. The adjacent river was flowing, leaving little scope for standing water.

The deposits underneath the Neolithic in the centre of Area 16 are also of sufficiently similar character to be correlated. The parcel was seen down to the base of that section, 13.12 m OD, but elsewhere started higher according to the pre-existing topography; indeed, it lapped up over the highest extant ground, reaching an altitude of 14.47 m OD in Area 19 where a deep soil profile subsequently developed on top (Chapter 5, especially column 13). From the crest, the 1C/1K interface falls away to the south-west; by Areas 10 and 27 deposits probably belonging to 1K go down to at least 13.30 m OD. By the end of the phase of 1K alluviation, progressive deposition had built up the flanks of the surviving pedestal of marl, thus creating a more substantial island (Figs 2.2 and 10.2.) In the process the northern channel was pushed a little north-eastwards to the position shown in Fig. 10.2. The inferred southern channel was also migrating away, i.e., southward, or at least was being constricted by the southerly movement of its northern bank. Thus was created the long enduring pattern of a central spine aligned WNW–ESE, acting as a watershed between two active channels. Even after the anastomosed river system was transformed into a very different configuration much later, this spine was to form the core of, and indeed the highest land within, an island.

There is no pollen or insect record from these deposits, but dating is provided by both the archaeomagnetic and radiocarbon methods. The former gives a span of 6750–5850 BC in both columns 4 and 31, although the earliest part refers to pre-1K deposits in both sequences. Corroboration comes from a ^{14}C date on charcoal from a feature (19.800) cutting into an intermediate horizon of the parcel, which calibrates to 6370–6000 BC. In broad terms, therefore, deposition can be attributed to the later half of the seventh millennium BC, and perhaps into the sixth millennium.

Later Mesolithic occupation evidence in the top of parcel 1K

The developing silt island attracted some human activity. There is evidence for later Mesolithic activity both on the surface of parcel 1K, that is strictly speaking within the soil profile, and at lower stratigraphic horizons. The lowest such evidence recovered in the research campaign took the form of a deep feature (19.800), discovered by accident when the deep cutting was made in Area 19. Its stratigraphic position and dating evidence have been discussed above in relation to parcel 1K alluviation. It was 0.85 m deep at the section and had a relatively pointed bottom (Fig. 2.6). Although at first sight suggestive of a driven feature, the broader upper profile is probably more consistent with a dug hole with packing around a post-pipe from a decayed post about 0.25 m across. No artefacts came out of the limited amount of fill it was possible to excavate.

The best parallel for such a feature at this date lies in a set of five post-pits excavated beside Stonehenge (Allen 1995a). These are rather larger and evidently held posts around 0.75 cm in diameter on the evidence of three yielding post-pipes. Associated radiocarbon dates and environmental evidence place them earlier than the Runnymede feature, primarily in the ninth millennium BC of the Boreal period.

It may be significant that at the very horizon from which the Runnymede feature was dug, magnetic intensity and susceptibility measurements (around sample 10) reached a peak; Clark wondered (in correspondence) whether this might have indicated upstream clearance, but perhaps instead it can be seen as enhancement due to on-site activity. Also well within the 1K alluvium, rather than in the overlying soil, was a group of burnt and struck flints in Area 20 (20.019); these were located after a shallow machine trench had been dug after removal of the late Atlantic soil.

Later activity is attested to by the occurrence of microliths within the soil profile, albeit generally below the main concentrations of Neolithic debris. The microliths are consistently of later Mesolithic types. On stratigraphic grounds it can be argued that these pre-date the Middle Neolithic horizon, but the question of the possible continuity of microlith use into the earliest Neolithic is raised by at least one of the microlith-containing contexts (19.677; Chapters 3 and 4). Another tentative suggestion of human disturbance to the ecosystem was adduced by Scaife from the pollen records for dry land ruderals (Chapter 9). The evidence came from column 40 p.a.z. 3, which would be coeval with the early existence of the land surface on top of parcel 1K. While there was undoubtedly dense alder carr all around (see 1F–1H below), it does seem that there was still a degree of openness on the central spine of the island; Evans and Evans describe it as 'damp open woodland', stability evidently having allowed some tree growth. Although grazing by animals might have played a part in preventing full regeneration, maintenance of such a patch of semi-open land amid forest over the longer term may imply the added factor of human intervention. In the context of regular use by Mesolithic communities, pit 19.800 could well be interpreted as supporting a marking totem post. The possible social context of such totem posts is discussed by Allen (1995b).

A final feature warranting mention here is a linear ditch running obliquely through Area 20 (20.957 *et al.*) with a terminal in Area 13 (Pl. 3). This is again stratigraphically earlier than the main Neolithic horizon; refuse-rich deposits run uninterrupted across a totally filled ditch. The natural inclination was to regard it as an Early Neolithic feature and the archaeomagnetic dating of an in situ burnt horizon very high in its fill is not inconsistent with such a date (Chapter 4). However, a radiocarbon date on *Fraxinus* sp. charcoal from the basal fill calibrates to the early fifth millennium BC (OxA-3582; Chapter 4), which urges reconsideration. Very few finds came from its lower fill, while the sediment analysis showed a fill which was entirely in character with the surrounding subsoil (i.e., 1K silt), with no indication of significant anthropic contributions during the filling. This contrasted with another pre-Middle Neolithic feature, a pit in Area 24 (Chapter 5). Moreover, the ditch contained four microliths (Chapter 3), one from the burnt horizon.

Parcels 1F/1G: c. sixth millennium BC; mid-Atlantic

These two sedimentary bodies were each deposited against a palaeobank, indicating short-lived southward reversals of a generally northward migrating channel edge. Locally the channel edge filled by these parcels would seem to have been aligned WNW–ESE and any eastward extension would need to be sought to the east under, for example, the eastern end of Area 16 (Fig. 10.2). The pollen monolith (column 40) comes from immediately below the recorded section and may include layers in both 1F and 1G, given the shallow inclination of the erosion interface and the sedimentary comparisons noted in the layer catalogue.

Although Scaife has divided the sequence into three pollen zones, indicating some local ecological variation, there are no radical vegetational differences and an Atlantic vegetation is characteristic throughout. The associated insect assemblage and plant macrofossils are in agreement and argue for very dense alder woodland flanking the river system with other woodland compositions beyond. Such limited occurrences as there are for more open-loving species of plant and Coleoptera could in fact be explained in terms of more airy bankside habitats. This heavily wooded picture seems at variance with the snail evidence from the contemporary land surface alongside (see above), but the molluscs from the 1F–1G channel do also indicate wet woodland, and the more limited 'open ground' indicated in this channel edge location may in fact be mudflats or similar habitats.

The apparent discrepancy with the higher ground can in fact be explained in terms of sampling sites, sediment origins and different catchment scales for the different organisms (as summarised by Evans and Evans in Chapter 7). The channel fills covered here (and indeed also the later fill, 1H) are marginally upstream from the high bank, hence all ecofacts which had fallen into the river channels and then settled out with the sediments, which might well account for the majority in these samples, will have come from upstream of the island. The direct airborne pollen contribution will have reflected a larger area, and this would seem in general to have been largely wooded, masking any tiny proportion of open land. The snail communities are known to reflect habitat on a very small scale, as emphasised by Evans and Evans, and indeed this point is also made well by the difference showing between the contemporary sample sequences for the 1K silt (above) and, again, those for 2B alluvium (below).

The only independent dating came from an archaeomagnetic column through the silts of parcel 1G above the monoliths. The conclusion that the span of samples 12–1 was 6500–5950 BC is, however, at variance with the stratigraphic and vegetational evidence, unless the whole of the sequence 1D/E–1K–1F–1G is to be compressed within the seventh millennium BC. In this event, it would still be difficult to reconcile the apparent concordance between the archaeomagnetic profiles of columns 4 and 40. For the present, the archaeomagnetic dating here is treated with reserve.

Parcel 1H: early fourth millennium BC; late Atlantic; inception of agriculture

This parcel obviously represents the continuation of the trend towards northward migration of the channel. The consequence was to broaden the island in this direction.

Stratigraphic interrelationship with the earlier deposits to the south-east was not established (no deep cutting was made in Areas 26 and 32 West; Fig. 2.2). Nevertheless, a radiocarbon determination on branch wood from context 32.040 gives a calibrated range of 4040–3785 BC. Strictly this is a *terminus post quem*, but evidence that these sediments cannot be much later than the radiocarbon date comes from the overlying Middle Neolithic occupation, broadly dated to the middle centuries of the fourth millennium BC.

The dating can thus be applied to the first signs (in column 59, p.a.z. 2) of a slight opening up of the local canopy, accompanied by indications of agriculture. Such initial signs are generally found around the beginning of the fourth millennium BC in Britain. In detail, the changes from p.a.z. 1 to p.a.z. 2 in column 59 could all be consequences of the loss of some of the overhanging alder cover. The pollen evidence for essential continuity of the dense alder carr in the earlier unit (p.a.z. 1) is backed up by the insect evidence from sample A32.040, although plant macros of *Rhamnus*

194

cartharticus are viewed as evidence for some areas of more open woodland (Robinson, Chapter 8). The substantial reduction in alder seen in p.a.z. 2 would directly cause the percentage increase in all pollen coming from beyond the river banks, providing that they were not also subject to significant reduction at the same time. Increases are seen in oak, lime and pine as well as in the herbs and grasses. This allows the possibility that the catchment already featured more open patches of land before the deposition of the layer which first shows it – 32.056. It may also be relevant that the pollen assemblage change coincides with a break in the lithology, the later deposit incorporating a greater component of material from a wider area. This would have similar implications for the possible pre-existence of the slightly opened woodscape.

The on-site clearance inferred from this basal part of parcel 1H (p.a.z. 2) is also clearly reflected in the accompanying mollusc column a little higher. In the channel deposits themselves, aquatic species and groups 6–8 predominate, as would be expected, but as the continuing aggradation lifted this channel edge above the water table, these progressively declined in phase with a sustained rise in open-country species of snail (Evans and Evans, Figs 7.7 and 7.8). A relatively sharp bank/channel interface is inferred. Taken on its own, the molluscan evidence might simply reflect the slow colonisation of these newly created mudbanks by tall vegetation, a feature also of the 1K alluviation. However, this profile dates to the early centuries of the fourth millennium BC, before the main occupation, but at a time when there was clearly human activity from time to time. It has already been seen that even the long-established surface of the high bank had not reverted entirely to wood or scrub and the overall impression is that the island was being maintained as open land deliberately, or incidentally through grazing. In this latter context, it may be highly significant that right at the beginning of the sequence, just before any convincing evidence for on-site clearance, a high representation of dung beetles was found, whereas grassland-specific species were not well attested. Robinson (Chapter 8) suggests that woodland grazing by large herbivores was responsible for the coleopteran assemblage and, although wild species of cattle and pig are plausible culprits, the introduction to the local ecosystem of their domestic counterparts is suggested by the livestock-rich Middle Neolithic culture present just three centuries later.

Parcel 1L: ?Neolithic

It can be argued that certain layers at the southern end of the parcel 1A exposure in Area 8 (8.085, 086, 087 and probably the underlying part of 8.093) were much later than the remainder. Organic preservation continues up to around 13.00 m OD, indicating that they post-date the water table rise at the beginning of the Atlantic period. These sediments

lie to the south-west of a junction between two sedimentary sequences that was not adequately explored but in retrospect suggests an erosional interface. Some indication of broad dating might come from two clusters of stakes, one stake from which is radiocarbon dated to the earlier Neolithic (HAR-6133), although it is possible that they were driven through pre-existing layers. The layers are overlain by deposits attributed to parcels 2F and 2B and should therefore pre-date *c.* 3000 BC.

The implications of this interpretation are significant for the channel system in documenting a channel flowing to the south of, and cutting into, the early silt spine (parcels 1C/1K) (Figs 11.3 and 11.4).

Alluvial parcel 2

It became apparent from as early as 1984 that beneath the Late Bronze Age occupied surface in the riverside zone there was a sequence of channel fills overlain by overbank clay-loam alluvium very similar to that encountered in the rescue campaign. As stratigraphic relationships were found and scientific dating obtained, the broad contemporaneity of these two sequences was confirmed. However, they were separated by a distance of 70 m at their nearest exposures and, given also the recurrence over time of similar sediment types and the presence of minor recuts, one felt cautious about the prospects for a total correlation.

It was decided to apply a distinct set of labels to the channel sediments exposed during the research excavations. In the rescue report the equivalent channel fills were collectively labelled parcel 2A (Needham 1991, 27), but a consistently present erosion scarp was later used to divide them into two major blocks, 2A and 2C (Needham 1992). In the riverside zone the channel deposits again proved to be split into two by a major erosion phase, seen running through Area 14 (Fig. 2.35), and this was used to define parcels 2D and 2E. The correlation ventured below in fact places this division later than the 2A/2C boundary. The overlying overbank alluvium in the rescue zone remains as originally labelled, parcel 2B, and this is now also applied to the matching deposits across the research area (Fig. 2.1).

In 1989 the deep cutting in Area 32 revealed these later Neolithic channel deposits running up to a palaeobank (Fig. 2.13). The latter was overlain by a continuation of a highly distinctive gravel bed, termed parcel 2F, which covered most of the dry bank. This profile finally offered a better interrelationship between the deeper channel fills and the gravel, to be described in more detail below. In broad outline, the chronological relationships of these various parcels is as shown in Fig. 2.1.

Where the channel fill 2E in Area 14 tapers out it shows that there is a hiatus at the layer 14.114/115 interface; this serves to define the boundary between parcels 2D (below) and 2B (above). However, it also demonstrates that a very similar beige/pale grey clay, being a gleyed horizon

(Limbrey, Chapter 5), occupies the top of both parcels 2D and 2E, thus separated in time. This makes it impossible to be sure about the correlation of similar layers detached from the Area 14 profile, namely 21.202 and 32.048, but the projected alignment of the 2D/2E erosional interface (Fig. 11.6) would suggest that they belong to the earlier parcel, 2D. Some support might come from the irregular surface of 32.048, which suggests erosion and perhaps, therefore, a hiatus before 32.052.

Such correlations cannot be precise in temporal terms, even where continuity of the layer is proven. This is because the particular depositional horizon (e.g., the 14.114/115 interface) may have undulated, whereas the gleyed horizon would be closer to a horizontal plane, determined by fluctuations in the water table. In general, however, good sequence correlations were found between Areas 14 and 32, allowing definition of parcel 2D (Fig. 2.35). Similarly, running in the opposite direction (south-east), good correlations for the 2E sediments were established throughout the close-set exposures of Areas 14, 18 and 31.

Overall the channel components of parcel 2, extending over 150 m from Area 32 to Area 4, indicate a broad channel system of the later Neolithic, running roughly parallel with the modern Thames. It can be referred to as the Neolithic northern channel and is flanked by dry banks to the SSW.

Neolithic occupation surface and parcel 2A; early to mid-fourth millennium BC

Evidence for Middle Neolithic occupation, in the form of many cut features and extensive spreads of in situ refuse, all associated with a soil profile, provides an invaluable marker-horizon across the whole of the interior zone. To the south-west it was truncated in Area 28, to the north-west in Area 13 North, and to the north-east in the middle of Area 32 and between Areas 16 and 14. In fact on the longer-established dry banks, there is a longer sequence of activity, starting in the later Mesolithic (discussed above). The immediate environment continues to feature grassland, more or less dry, though still with some indication of woody vegetation. The lower slopes were inevitably wetter; indeed, the relevant sample from column 10 (Sn 19) in Area 16 had such a high proportion of aquatic shells that it has been classified as 'river channel', although its altitude (*c.* 13.70 m OD) is no lower than the top of column 59. This is a little curious since a remnant 'B' horizon of soil was present here (Chapter 5), associated with a spread of cultural debris which, it has been concluded, were effectively undisturbed by fluvial action. The phosphate columns nearby (24 and 25) also showed a peak characteristic of occupation activity, although this in itself does not rule out the deposition of refuse in a wet spot.

Independent dating evidence for Neolithic activity within the interior zone comes from thermoluminescence and archaeomagnetic dating, but in addition there are the broader correlations of the ceramic assemblage. Of less

direct application are a few radiocarbon dates from lower-lying ground to the east (Areas 4, 6 and 8), from sediments and structures yielding broadly comparable pottery.

One of the earliest 'structures' stratigraphically is a burnt spread in situ in the upper fill of the early ditch (20.908.3). Archaeomagnetic measurement could not provide an actual date, but conformed to the proposition that the spread pre-dated the hearths in Area 19 and, furthermore, suggested a *terminus post quem* of 3950 BC. Similarly, measurements on the two hearths (19.208, 19.258) could only be used to show conformity with the presumed dating of the surrounding refuse deposits (3700–3350 BC), rather than provide independent dating. Hence no information is offered regarding the more detailed relationship between different zones of Neolithic activity on the site. On the other hand, thermoluminescence dating of two burnt flint clusters (19.677, 20.993), although very imprecise, give ranges centred on the expected range. These, again, stratigraphically precede the hearths and the main horizon of refuse. The ceramics associated with the latter (not yet fully studied), although mainly in an 'early decorated' style, show some indications of development towards earliest Peterborough style (Longworth and Varndell 1996). Peterborough pottery is now thought to span *c.* 3400–2500 BC (Gibson and Kinnes 1997).

The most useful dating from the earlier campaign now comes from the brushwood structure in Area 4, having the benefit of a new date on the brushwood itself, in layers associated with an early decorated assemblage (BM-2773, Chapter 4). The calibrated range, 3680–3375 BC, is the best dating of these layers and does in fact overlap with both of the earlier determinations on associated piles, even though these are somewhat divergent from one another. The contextual integrity of the samples dated in Area 6 is less good, since both timbers and associated inorganic artefacts were refuse in accumulating channel deposits, and the possibility of redeposition from a somewhat earlier context could not be wholly ruled out. Nevertheless, the broad ranges for two prostrate timbers, 3970–3520 and 3790–3380 BC, are not inconsistent with the Area 4 dating, especially when allowance is made for some age-offset (Needham 1991). The pottery is again early decorated, with some tendencies towards Peterborough (Kinnes 1991).

The associated archaeomagnetic dating gave an earlier range (4200–3650 BC), but refers largely to underlying sediments which would seem, therefore, to be broadly contemporary with parcel 1H, 100 m to the north-west. Indeed, the correlations suggested (Fig. 2.37) show that all the early 2A aggradation, including the basal tufa bed, must be early fourth millennium BC and conceivably a little earlier. It is of note that stratigraphically early contexts in parcel 2A in the Area 7 section, and preceding the main cultural horizon, yielded possible burnt material, recalling the earliest Neolithic evidence on the higher ground of the interior zone.

Parcels 2C, 2D and 2F: late fourth to early third millennium BC; later Neolithic channel and flood deposits

Overlying the Neolithic occupation horizon on the higher ground was an extremely unusual deposit, a more or less continuous layer of gravel apparently deposited during overbank flooding. This is in such contrast to the overall alluvial sequence, with the obvious exception of riverbed gravels, that it was felt important to designate it separately, as parcel 2F. The roughly contemporary channel deposits are separately designated as 2C and 2D, as outlined above. Tracing the gravel layer into flanking channels is only unequivocal in the centre of Area 32 where a critical land/water interface was encountered. It is probable that the thick, gravelly deposit recorded in the rescue campaign in Area 8 (Needham 1991, 37, fig. 12, layer 84), also represents this layer, dipping into the edge of a contemporary southern channel.

Throughout the interior zone, however, the gravel layer occupies a terrestrial environment that seems not to have been otherwise subjected to any alluvial deposition for a long time before or after. The consequence would have been the resumption of vigorous faunal activity. This is believed to account for the persistent infiltration of gravel into the underlying cultural soil containing in situ refuse deposits, a process supported by Limbrey's analysis of the soil matrix surrounding the gravel (Chapter 5). The result is a blurred interface between upper soil with gravel, containing only reworked artefacts, and a lower soil profile, essentially clean of gravel and rich in undisturbed occupation material. For a long time during the excavation campaign it was believed that, despite the direct layer contact, there was a long interlude of over a millennium between occupation and the gravel incursion (see, for example, Needham 1992). The stratigraphic bracketing allowed a long time-window from late-fourth to early-second millennia BC. A date late within this bracket was argued on the basis of the small assemblage of Beaker and Early Bronze Age pottery and two barbed-and-tanged flint arrowheads (Chapter 3), since these came from a variety of stratigraphic contexts, including two just beneath the base of the gravel (Tables 3.3–3.5). However, this now needs careful reconsideration in the light of new dating evidence, particularly from the flanking channel deposits.

Area 32 provides the key sequence: in section RH17 (Fig. 2.13) the gravel capping the Neolithic ground surface is seen to spill, in more diffuse form (32.050), down a palaeobank which marks the limit of a phase of southward cutting. Extending away from the foot of the bank, and just below the water table, the top of a dense but thin gravel deposit was observed (32.038). The contiguity of these elements across the land–bank–channel edge topography forms an effective horizon, even allowing for the possibility of some time depth and reworking of deposits. It post-dates

parcel 1H and the Neolithic surface, and pre-dates part at least of parcel 2D and all of 2B.

It is theoretically possible that there was an interlude between this lateral incision and the gravel, but it is rather more likely that they were intimately associated: scour followed by massive deposition. Such a large-magnitude flood event would inevitably have had some effect on the ground surface of the floodplain. It is almost certain that the upper part of the soil profile was widely disturbed, given the quantities of more weathered cultural debris entrapped in the gravel bed. More specifically, the lower slopes were vulnerable and truncation is clear in Area 16 East (Limbrey's column 22) and the centre of Area 32 (archaeological observation). It is possible, furthermore, that surface occupation debris had been entirely stripped from the centre of Area 16, given that no corresponding peak in phosphate values was found in column 9. The Neolithic levels were not reached by excavation in this part of the trench, but a shallow cut feature was visible in the wall of a modern pit (Fig. 2.7; context 16.877) and yielded much of a Neolithic pot (Fig. 3.3).

Dating evidence for 2D comes from two sources. In Area 32 itself, the datable part of the archaeomagnetic column begins just within the top of this alluvial unit (high in 32.048); the date suggested at this point was *c.* 3350 BC (Chapter 4). This runs surprisingly close to the current date bracket for the main occupation and would imply that not only the gravel incursion, but also further channel sedimentation (much of parcel 2D) had taken place quickly during the third quarter of the fourth millennium BC.

Notwithstanding the problems of detailed correlation outlined, it may be noted that a radiocarbon date on an in situ group of butchered bone in context 6.040 of parcel 2C, correlated to high in parcel 2D, gave 3350–2500 BC. This supports dating the prior gravel deposition to the later fourth millennium BC, or the early third millennium at the latest. Another attempt at dating these channel sediments was made on two stray bones in layer 14.116, essentially the same layer as the butchered bone group. This came out somewhat earlier than expected, 4250–3650 BC, and is perhaps best regarded as a *terminus post quem*, deriving from reworked material. However, again there is nothing to push the dating of the sedimentation of parcel 2D to later in the third millennium BC.

Although gravel occurring within channel beds is obviously of considerably less significance than that thrown up over normally dry banks, it is perhaps noteworthy that on the correlation offered (Fig. 2.37) a persistent presence of gravel can be identified in positions correlatable with 32.038. This even extends to Area 4, 140 m to the south-east, where a deposit containing large cobbles of gravel and wood pieces (4.118), formerly interpreted as a deliberate dump (Needham 1991, 28), might now better be interpreted as a relict of the main gravel event. It seems to follow an erosional interface, obliquely truncating 4.119 and 120a south-westwards, and could well represent heavy particles left stranded on a hummock as the flood subsided. That it has apparently been eroded on both sides by later fluvial action is hardly surprising given the low-lying context. Whatever its origin, the deposition of the cobble-rich layer seems to have come soon after the cultural activity dated 3680–3375 BC and discussed above. Area 7 nearby also shows a gravel layer (7.072) at essentially the same stratigraphic horizon.

On balance, the major gravel event seems best dated roughly to the last third of the fourth millennium BC. Even so, the starting date for the archaeomagnetic bracket (which begins somewhat above 2F-equivalent deposits) seems a little too early, and this might be due to inherent imprecision in the technique at present.

This leaves to be resolved the question of the Beaker/ Early Bronze Age pottery and flintwork on the high bank. It is difficult to countenance the possibility that there were two well-separated phases of gravel deposition, given the almost continuous nature and uniform stratigraphic position of the horizon. However, there may be some significance to the distinction between the uniform blanket of gravel over the highest ground (Areas 13, 19, 20, 22 and 24) and its distribution on the flanking slopes, as seen in Areas 10, 16 and 32, with patches free of significant gravel; this patchiness would presumably be due to the currents running either during original deposition, or soon afterwards.

One feature of the Beaker/EBA pottery is its small size, and most is also fairly abraded. It is spread vertically from the base of the Late Bronze Age deposits through to the Middle Neolithic deposits. However, the great majority catalogued so far – eighteen sherds (Tables 3.4 and 3.5) – comes from various levels within the gravel bed. A further nine sherds come either from the transition to the Middle Neolithic layers, or from 'cut' features that contain gravel in their fills. Of three lower still, two are from the top of a refuse-rich deposit, otherwise entirely of earlier material. Above the gravel are eight sherds from the 2B silts, but mainly from low down, and just one from the base of the Late Bronze Age soil. Whatever process or processes were responsible, it is clear that the predominant resulting association is with the gravel 2F.

Given the condition of the sherds, one possibility is that they had migrated downwards as a result of normal soil processes, but it is hard to see why, if much later than the gravel deposition, they should have so thoroughly permeated it. In the circumstances a better explanation is that the gravel and the putatively later pottery was mixed by tilling and then, after the arable phase, moved together through the soil profile, tending to concentrate in a bed. On this interpretation, the episode of cultivation would have occurred in the Early Bronze Age or later, since some urn sherds were in and immediately below the gravel. Although sought, no ard or spade marks were ever observed in the underlying soil, but they could have vanished with subsequent microturbation.

Cultivation, as an alternative to later flood reworking, could also be the mechanism by which stringers of gravel were dragged from the main body across the edge of the lapping 2B silt (e.g., Figs 2.2 and 2.11).

The unusual nature of the gravel incursion demands enquiry of its cause. The first issue is the question of appropriate sediment supply. Gravel would always be a feature of riverbeds, albeit probably intermittently distributed (very little is evident in the early to mid-Holocene channels on the site itself). It would be subject to transportation along the bed when spate flow reached sufficient energy, but the quantity evidently deposited over the mid-Holocene island and its flanks would suggest the gathering up of bed gravel from some considerable reach. The alternative is that the source was predominantly gravel terrace material, which was being stripped by the flood in question. A second consideration concerns the factors that might have triggered this flood deposit at a time when overbank sedimentation was otherwise imperceptible, and a third is why the gravel was offloaded at this particular location, for it may be assumed that this scale of deposition would not have been continuous or even dominant along the valley.

The first two questions could actually have a common underlying cause. Precipitation in the Atlantic period was obviously high, but most heavy rainfall was absorbed by the dominant forest cover which prevented rapid run-off. The earlier Neolithic then saw the destabilisation of this hydro-ecosystem – not only the clearance of some woodland cover, but also the conversion of that land in part to ploughed soil. Although the earlier Neolithic clearances seem only to have affected a minor part of the local landscape, it might be conjectured that they were situated preferentially in the lower valley, both on the first terraces and on lower-lying ground. This proposition stems partly from the distribution of causewayed enclosures along the middle reaches of the Thames, for their siting indicates a preoccupation with the valley floor and its resources (Needham and Trott 1987; see also Case and Whittle 1982, 2–3 for higher up the Thames). A more specific point comes from the pollen record. Scaife suggests that the reduction in alder percentages between the Atlantic profiles (columns 39 p.a.z. 2; 40; 59 p.a.z. 1) and the later Neolithic (columns 60, 5 and Area 6 SS1 – where alder fluctuates between 19% and 38% of total arboreal pollen, adjusted from Greig 1991), while not indicating decimation of the pre-existing alder carr regionally, should imply its disappearance from the immediate proximity of the site. This evidently occurred during the earlier Neolithic and may be attributed to the activities of the occupants within and around the settled area. If we envisage a repeat of this pattern up and down the river system, it is easy to see how substantial areas of the banks themselves would have become much more vulnerable to erosion even by normal high river flow, quite aside from floods. Destabilisation of tracts of land close to the river system could have allowed for the first time the rapid collection of flood waters in the valley floor, but more critically, the substantial erosion of soil and latent gravel banks. This situation recalls, for example, that in the Maxey area, Cambridgeshire, where soil micromorphological studies have indicated that there was sufficient disturbance in the Neolithic period to cause significant soil erosion (French 1988).

Nevertheless, this clearly was a massive flood (or rapid succession of floods), and it seems plausible that the precipitation responsible affected more than just the immediate catchment. If it had fallen simultaneously on the upper Thames, the Wessex uplands and the Chiltern Hills to the north, the meeting of floodwaters could have created havoc in the confluence zone between Ankerwyke bend, just upstream of the site, and Staines. With a much larger quantity of water to be absorbed by the floodplain below, the backing effect would have been to slow velocity and thus carrying capacity, resulting in the deposition of the heavier elements of the load.

The only definitive evidence for human visitation to the site during the later (pre-Beaker) Neolithic is the in situ cluster of butchered bone recovered in Area 6 (6.F125) in a layer correlated with upper parcel 2D. There is very little artefactual evidence, the most diagnostic pieces being six transverse arrowheads. These come from a range of stratigraphic contexts (Table 3.3), two of them above Late Bronze Age in situ levels, strongly suggesting redeposition. The earlier stratified examples lie between the main gravel bed of 2F and the Late Bronze Age dark earth. Taken at face value these would help to confirm the early dating of the gravel, but that conclusion would take no account of the contradictory evidence of the Beaker sherds and barbed-and-tanged arrowheads. Moreover, the intervening deposit is overbank flood silt 2B, shown to post-date the late third millennium BC. Thus either they are evidence for the rather late use of tranchet arrowheads, or they represent items washed in with the flooding. Although none are intact, they are generally quite fresh, with just localised edge damage rather than any more extensive rolling. They cannot, therefore, have been subjected to turbulation with gravel, but they might well have fluvio-dynamic properties which allowed them to be transported by silt-carrying currents. It seems likely that they had not come from far away.

The environmental evidence gives a broader backdrop as to human activity nearby. The earlier work on pollen (Greig 1991) indicated that there was some contraction of both open land and cereal cultivation locally after the Middle Neolithic settlement phase. Nevertheless, the new pollen evidence from parcel 2D presents some arable indicators in the form of ruderal- and cereal-type pollen (Chapter 9), and the reconstruction of a small-scale shifting mosaic of cleared and uncleared land still seems valid. During this period there was at least one minor peak in cereals and associated herbs, towards the top of column 5, p.a.z. 1. Fluctuations in the tree pollen percentages are also apparent, but not in any

systematic fashion. Again, this might reflect small-scale erratic disturbance of the cover as land-use patterns shifted. As already noted, alder values are lower than expected for an undisturbed alder carr along the valley bottom and the site itself seems to have been still free of significant scrub or woodland. This finds confirmation from the molluscs, which indicate that land in the immediate vicinity of the sampling sites (all channel fills) was relatively open, with sparse woody vegetation, though perhaps less open than earlier in the Neolithic period (Evans and Evans, Chapter 7). The low-lying ground in particular may have represented the damp meadow/fen-like habitat drawn by Scaife from among the moderately diverse herb evidence.

Parcel 2E: later third millennium BC; Beaker period channel deposits

As already set out above, channel deposits post-dating an erosional interface showing in Area 14 are defined as parcel 2E. It has in fact been possible to suggest a correlation with features of the rescue sequence in Areas 4, 6 and 7 where evidence for truncation is followed by comparable sedimentary sequences (Fig. 2.37). This would link parcel 2E to the uppermost part of parcel 2C. The erosion in Area 4 includes a scarp (4.116/117 interface), as in Area 14.

Dating evidence for this channel fill comprises three radiocarbon dates from two contexts and is coherent. Two measurements were run on different fractions of a sample from a dense 'mat' of branch wood and twigs, much of which bore the distinctive marks of cutting and bark stripping by beavers (context 14.121; Chapter 3; Pl. 7). Lenses of wood frequently build up in channel edges or lags and most often they are doubtless natural accumulations of driftwood, potentially comprising detritus of mixed age. In the case of context 14.121, however, the high proportion of beaver-gnawed material and the consistently fresh appearance of wood and gnawing scars all suggest that, if not in situ, this wood saw little reworking by the river. It may best be explained as a store of food (Coles 1992).

The two measurements were in good agreement and combined give a calibrated date of 2355–2035 BC for the context, and, presumably, therefore, the enveloping sediments. The third date was on a prostrate trunk of alder in Area 31 (31.676) and, while its context could not be recorded in detail (beneath the water table), it appeared to sit within equivalent sediments. The date of 2480–2200 BC overlaps the previous one to a good degree. The overlying layer in Area 14 (14.120) yielded two red deer antlers, one showing the shed burr, but neither with any signs of working.

The beaver activity represented by the gnawed wood low in parcel 2E coincides approximately with Scaife's column 5, p.a.z. 2, which shows a pronounced, but temporary peak in hazel pollen, perhaps due to more prolific flowering of an understorey relieved of some of its covering canopy. It is perhaps possible that this was an effect of beaver deforestation.

A small rise in the water table is associated with this stratigraphic unit relative to parcel 2D, on the evidence of sections RH29 and RH30 (Figs 2.23–24). It is tempting to suggest that this might relate to the formation of a lake locally due to beaver damming downstream. Although it is not yet certain what the behavioural pattern of beavers would have been in prehistoric Britain, a suggestion has been made that the Mar Dyke in Essex was an ancient beaver dam (Coles 1992, 97). Whether or not there was ponding at Runnymede, it is clear from the insects that bulrush reed swamp was present along the water margins, although Evans and Evans felt there was little molluscan evidence for swamp.

After the vegetational disturbance represented in column 5, p.a.z. 2, the composition returns to something broadly similar to that typifying parcel 2D and there is another minor peak in pollen types indicative of agriculture. However, there is no sign of any sustained onslaught on the woodland in the catchment before the end of the third millennium BC; if anything the immediate environs of the site were tending towards woodland regeneration. This trend was observed from the Coleoptera in the later Neolithic sequence in Area 6 (Robinson 1991) and appears, on the current results, to continue through to the Beaker period. Associated with this was a steady decline in dung-feeding beetles, so that woodland regeneration can be attributed to reducing grazing pressure. All in all the picture is of a long-lived (over 1,000 years) economic exploitation pattern which preserved a balance maintaining substantial woodland cover; this is the woodland pasture economy inferred by Scaife and others.

In the rescue excavations evidence was found for some truncation of the later Neolithic parcel 2C channel fill, prior to its covering by 2B. It is possible that a further indication of extensive planar erosion lies in the increase in pre-Quaternary spores and pollen towards the top of the column 5 pollen profile, thus high in parcel 2E.

Parcel 2E or 2B

One set of sediments in Area 21 falls between parcels 2D and 4B in the stratigraphic sequence (21.203–205), but cannot be correlated with certainty with either 2E or 2B. In general, their character and altitude would fit in with the 2B phase alluviation, but 21.204 had a bluish tinge not noted elsewhere. There is also an optically stimulated luminescence date of 2600 ± 1000 BC for the overlying layer 21.205, but this is extremely imprecise and was measured as part of a pilot project on the technique (by Eddie Rhodes). It should probably not be used to favour correlation with 2E rather than 2B.

Parcel 2B: earlier second millennium BC; Early to Middle Bronze Age alluviation

The gradual filling of the southern part of the northern channel system over the course of the previous 2,000 years led to a progressive transition from water to mudflats, to damp ground, then to drier ground, as alluviation steadily raised the surface (Chapter 7). The alluvium in question is consistently fine-grained, generally a clay-loam, and gives the initial impression of homogeneity. The coarser fractions present, fine sand to medium silt, tend to be calcite dominated (Chapter 5). It does, however, feature some zonation through the deeper sequences which cover the low zones in the former topography. The maximum depth is 1.2 m in Area 32, or 1.35 m in Area 6, and this thins as the base rises over the banks and onto the higher ground (Fig. 2.2). The silt body also thickens south-westwards, beyond the high spine, as seen in Areas 10, 27 and 28, but later truncation in this part of the site (by the channel filled with parcel 7) has precluded unequivocal evidence that it was dipping into another former channel in this direction. It has already been argued that parcel 1L to the south-east is likely to be the fill of a channel active in or around the Neolithic period and the overall channel pattern reconstruction below implies the long-term continuity of a southern channel.

These 2B silts are bereft of organic components, having formed above the water table, and moreover very few artefacts have been recovered. The noteworthy occurrence of four transverse arrowheads in parcel 2B where it covers the higher ground (Table 3.3) has been discussed above in the context of parcel 2D, but these do not necessarily closely date the deposit. The same goes for a sherd of incised pottery from Area 16 East likened to Grooved Ware (Longworth and Varndell 1996). Dating, therefore, has to rely on the stratigraphic bracketing and an archaeomagnetic column in Area 32. The latter extended down into the top of parcel 2D (discussed above) but most of it refers to 2B with the possibility that the upper sediments are 4B. The overall span suggested was 3350–1450 BC. It has already been argued that the start date is too early, and this is reinforced if the correlation of layer 32.052 with 14.114 is correct, since the latter must post-date parcel 2E of the late third millennium BC. The end date, on the other hand, seems more consistent with other evidence; this body of alluvium had largely accumulated before parcel 4B filled the adjacent channel towards the end of the second millennium BC.

Looking in more detail at the stratigraphic succession, the distinction between parcels 2B and 4B close to the palaeobank running through the riverside zone (described below) was clear. However, the sandy upper sediments of 4B become progressively finer as they lap over the 'levee' surface of 2B (Fig. 2.2). It is possible that this later silting regime would have contributed a little silt to the top of 2B more widely across the island without being detectable. Faunal turbation on the terrestrial surface of 2B would have helped obscure a minor component. Soil formation on top of 2B would have followed the slowing of accretion late in the second millennium BC; its uppermost profile must thus be accepted as being exposed to activities and processes during the parcel 4B accumulation.

The part of the parcel 2B surface still extant between later channels had a domed profile along a line from south-west (Area 27) to north-east (riverside zone). In the latter zone, alongside the steep bank of the 4B channel, the ground surface seems to have been relatively level, at around 14.00 m OD (Fig. 10.3). It then rose very gradually inland, for example reaching 14.20 by the north-east end of Area 16 and 14.30 at the north-east end of Area 32, eventually attaining a high point of about 14.60 m OD in Area 19. Across the spine, in a south-west direction, it fell again to 14.35 by the south-west end of Area 10 and to 14.23 in Area 28, before being truncated by the later river channel (Fig. 2.2).

The broad dating of parcel 2B to the second millennium BC has always suggested the possibility of a connection with the intensification of arable agriculture, supposedly indicated by the tangible bounded-field landscapes of the Deverel-Rimbury culture. This flourished between *c*. 1500 and 1200 BC, but probably emerged earlier, perhaps from the seventeenth century BC and it is possible that the agricultural practices concerned emerged earlier still. It is also clear from the bracketing pollen assemblages of parcels 2E (column 5, p.a.z. 3) and 4B (column 31), that the early to mid-second millennium BC saw a massive change in the balance between woodland and more open ground. All major tree and shrub species declined substantially over this eight-century-long period, while herb taxa increased dramatically to around 40–50% of total pollen (Chapter 9). The insect evidence is in full accord with this general catchment picture (Chapter 8). This correlation provides a good basis for regarding the alluviation pattern of parcel 2B as a consequence, direct or indirect, of clearance and presumed agricultural activities. However, further consideration needs to be given to the part played by the field systems of the period (for the region, see Yates 1999).

Ditch-and-bank or hedge boundaries will largely arrest soil movement leading to lynchet formation on slopes. So the critical factors for enhancement of sediment supply would be either arable agriculture in an open or partially unbounded system of land use, or its practice in areas subjected to flooding. The archaeological evidence for Bronze Age field systems in the Thames valley is so far largely from the gravel terraces back from the river, and these themselves may not have contributed much new sediment to the river system. This could easily be a biased distribution because of the difficulty of detecting field systems on alluvium, especially when buried. One important observation in this respect was made by the author on the opposite bank of the Thames, just east of the Colne Brook. Some cut features, perhaps ditches, were seen in the section of a gravel pit at NGR 020723. Although no close dating evidence was recovered, some antiquity was suggested by the fact that they were overlain by alluvium 0.6 m thick comprising two yellow-brown layers, each covered by a grey-brown layer, the top one of which must have been a palaeosol. The features were cut into a yellowy silt not unlike

parcel 2B, while the overlying material could equate with parcel 6. A few kilometres upstream from Runnymede, extensive archaeological investigations on the valley floor at Eton College Rowing Lake have revealed a Bronze Age ditched enclosure system, albeit on a 'gravel terrace' (Allen *et al.* 1997, 120–1, fig. 3, 125).

Another key site at Phoenix Wharf, in the centre of London, is also well dated (Merriman 1992). Here a low-lying hump of sand was found to have been ploughed during the earlier Bronze Age, before being covered in peat and then alluvium. More recent excavations by Mark Birley (Museum of London Archaeological Service) at Cranford Lane, west London, have revealed a Bronze Age divided landscape which borders the small tributary of the Crane; one alignment of ditch boundaries run down the shallow slope towards the river (Mark Birley and Robert Whytehead, pers. comm.). The cultivation hypothesised above, to have taken place on the Runnymede site in the early to mid-second millennium BC (discussion of parcel 2F), cannot be taken as definite, but there are indications that field systems and/or cultivation did extend onto the valley floor.

There is also growing evidence for the occupation of fairly low-lying areas nearby. There is Deverel-Rimbury pottery among the Bronze Age assemblage excavated alongside the church in Wraysbury village (TQ 00137394); the site is a small knoll beside the Horton Brook (Astill and Lobb 1989). Even closer is an enclosure excavated in 1991 at Church Lammas, Staines, on the tip of the gravel spur between the county ditch and the adjacent stream (TQ 028722; Bird *et al.* 1994, 207–8, fig. 2). It is dated by a good group of Deverel-Rimbury pottery in the ditch fill. The site directly overlooks the Egham–Staines basin described in Chapter 11 and is only marginally elevated above the valley floor.

Where phosphate columns were taken down through thick deposits of 2B alluvium at Runnymede Bridge (columns 9, 24/25), they gave measurements at surprisingly high levels, given no obvious on-site activity. Cowell suggests (Chapter 6) that this might largely be a combined function of the reworking of earlier phosphate-rich soil and the downward migration of later such material, both processes which have been demonstrated. Another factor, however, might be the introduction of inorganic phosphates with the alluvial sediment; such phosphates could have originated via the manuring of fields, the original organic form having been converted to an inorganic one prior to erosion.

To suggest that changing landscapes and agricultural practice were causing changes in fine sediment supply and flood risk does not necessarily mean that the alluvial fall-out will be consistent throughout the region. Accumulation may be favoured in certain areas, for example where confluences give rise to the possibility of conjoining flood waters, causing the sudden slowing of currents and hence the preferential deposition of suspended silt. As yet it has proved difficult to identify bodies of overbank alluvium of comparable date elsewhere in the middle Thames (Needham 1992, 258–9), although very little work has been done in sufficient detail.

Gibbard reports repeated occurrences of brownish silty clay deposits high in alluvial sequences in the middle Thames and lower Colne (1985, 86 ff), though none is closely dated. Close synchroneity should perhaps not be assumed, but such deposits may all reflect overbank deposition during more intensive arable land use from the Bronze Age onwards. In terms of the Runnymede Bridge sequence the brownish silty clays might match better the parcel 6 overbank alluviation.

A dated sequence involving alluvium comes from excavations in the centre of Staines 1.5 km downstream. At the Friends' Burial Ground site a bed of 'yellow sandy clay', layer 285, is dated to about the right period (Crouch and Shanks 1984, 5). It was cut by a pit (264) containing the articulated skeleton of a cow, radiocarbon dated to 2820 ± 100 BP (HAR-number not cited). This gives a *terminus ante quem* of *c.* 1130–840 BC (1 sigma) for the cessation of deposition. Its top surface was seen to undulate between about 13.4 and 14.2 m OD, reflecting the underlying topography of a largely silted channel (channel 1) and an island (island 2), from about 12.6 to 14.1 m OD, but also a second channel to the south where it descended to a greater depth, suggesting that this was initial aggradation against a previously flowing channel edge.

In terms of the Runnymede alluvial sequence this deposit could equate with either 2B, or the channel fills of parcels 3 and 4B, which would be more consistent with the 'sandy' character of Staines, layer 285. The upper surface of the latter is a little lower than that attained by parcels 2B and 4B at Runnymede Bridge, but this would be expected from the downstream location. At Staines the upper surface was also very interrupted by dug features of the early Roman occupation. Roman layers in fact directly overlay this silt body, indicating a prolonged hiatus in deposition. The pre-layer 285 surface at the Friends' Burial Ground site includes channel 1, which was probably by now carrying little flow, but how much earlier the lower, main fill is cannot be determined. Up on the adjacent island a few cut features were found and, although devoid of datable finds, these may relate to the assemblage of Neolithic and Early–Middle Bronze Age pottery recovered from the site (Crouch and Shanks 1984, 125).

The overbank nature of the 2B alluvium and subsequent wetting and drying cycles precluded the survival of pollen in this calcareous environment. Retrospectively it has been possible to identify some contemporary deposits dipping below the water table in Area 31, contexts 31.677, 678, which were unfortunately not sampled. The site-specific environment is, however, documented by the molluscs. These give a much more local picture, and one that does not entirely conform to the aggregate picture for the catchment painted above. Although damp grasslands, mudflats and perhaps sedge fens are implied at different times and places,

a significant component of damp woodland is also indicated throughout the deep column 5 (Area 14), yet not really in columns 10 (Area 16) or 60 (Area 32). These results indicate a patchy distribution of vegetation types, with shrubby or tree types at first concentrated along some banks. However, late in the sequence on the slightly higher ground – column 10, Area 16 – there is actually a reversion from grassland to woodland, bucking the general trend in the catchment during this period. It might be ventured that the cause was a decline in grazing on the site itself. This reversion should date close to the 2B/4B transition, and it is appealing to connect it with renewed isolation of the site as an island. The topographic reconstructions (below) suggest that a breakthrough across a narrow neck of land re-established a vigorous northern channel, and this obviously happened before it began to be infilled with parcel 4B in the later second millennium BC; the uppermost part of parcel 2B where not covered by 4B could have been infiltrated by the subsequent pollen rain reflecting the vegetational response to the topographic change.

Alluvial parcels 3 and 4

The research excavation targeted the riverside zone partly in the hope of identifying the periphery of the Late Bronze Age site as a continuation of that found in Area 6 (Needham 1991). That area had yielded a contemporary edge to the dry land settlement which had escaped any later channel incision. A channel flowing NW–SE had begun to silt up against its south-west edge before occupation of the adjacent bank and continued aggrading during and after occupation, in the process trapping quantities of refuse and a series of waterfront structures. It was already suspected that the reversal from erosion to aggradation in this short stretch of the bank might have been caused by a breakthrough channel across a narrow isthmus between a loop of the main river and a tributary (now the Colne Brook) entering from the north (Fig. 11.7). It was vital to see if any evidence could be found to the west to support this hypothesis and, more specifically, whether similar contemporary channel-edge deposits had survived there.

The main objective was successively achieved in that a channel edge was located whose alignment and dating gave little doubt that it was a continuation of that recorded in Areas 4, 6 and 7. This has certainly helped to clarify aspects of the changing channel configuration. However, the stratigraphic relationship of the Late Bronze Age occupation to the silting of this channel edge was different: whereas in Area 6 occupation featured in the middle part of the sequence, in the riverside zone it essentially formed the top of the sequence (Fig. 2.2; cf. Needham 1992). In other words, the silting in the riverside zone had started earlier, or had accumulated quicker than downstream. This is an entirely expected pattern in the development of point-bars along convex channel edges, but could also arise from the sudden choking of the channel upstream, with slower aggradation in the slack water created downstream alongside the continuing flow of the tributary. The detailed implications for channel changes are reserved for Chapter 11.

The newly reclaimed bank in the riverside zone was, by the time of occupation, already essentially dry, or at worst damp ground. If a channel still flowed here, it must have migrated to the north. Unfortunately, in this direction the Late Bronze Age deposits and some of the underlying sediments were truncated by a Post-Medieval cut, a little inset from the modern river bank (Fig. 10.6). The prospects for recovering further waterfront evidence were thus precluded in this zone. However, there were compensations, in particular the early silting of the channel edge meant that a good sequence of sediments clearly pre-dated the settlement (4B), and some fairly fine sediments were located below the water table, thus preserving excellent environmental remains. This was the first opportunity to document the environmental prelude to the Late Bronze Age occupation.

Since there is little doubt that the two stretches of palaeochannel recovered respectively in the rescue and research excavations joined as a single functioning channel, it seemed sensible to attribute them to the same parcel, parcel 4. However, the chronological asymmetry and the emphasis on different sediment types necessitates their treatment as two distinct sequences. Hence, the previous set of deposits defined in Needham 1991 becomes parcel 4A (the lower part) and 4C (upper), while those documented here for the riverside zone are parcel 4B. Correlations now suggest that the upper deposits in the east (4C) can be related to a sequence in the conjoining channel, flanking the south-east side of the settlement. The latter is a lateral continuation of the fill originally identified in Area 8 and termed parcel 5, and 4C is therefore discussed together with it.

Parcel 3: later second millennium BC

This parcel, only seen clearly in section in Area 8, defines a more southerly channel, apparently flowing and filling after the deposition of both 2F and 2B (layers 8.084 and 083; Needham 1991, 36, fig. 12). It is thus broadly contemporary with the early activity in the parcel 4 channel but became defunct sooner, before Late Bronze Age occupation. Unfortunately, its projected alignment westwards seems to preclude opportunities for easy further exploration, given the modern road network. Certainly no deposits relatable to parcel 3 were encountered under the Late Bronze Age in Area 10 or in the research trenches.

A general WSW–ENE alignment was deduced from the observation of a similar sand deposit beneath the 1976 trench, Area 2 (Needham 1991, 29), where a fall in the topography from 14.10 m OD at the south-west end to

13.46 m OD at the north-east mirrors the gentle dished profile of the top of 8.078 (13.42–14.13 m OD). The reverse slopes might suggest that opposite banks are represented in the exposures; this reconstruction would give a channel no more than 25 m wide, and possibly narrower, depending on the precise alignment. The alternative would be to regard the rise at the south-west end of Area 2 as due to the later, sandy fill riding over a mid-channel bar; this cannot be proven on the extant evidence, but makes most sense when considering the longer trajectory of channel change (Chapter 11).

A further correlation that may be made in Area 8 was previously overlooked. The furthest north-west limit of the later parcel 5 channel is marked by a steep palaeobank which appears in section SA8 – as well as in section SA7 – published here for the first time (Fig. 2.29). However, a remnant of an earlier block of sediments survived beneath the palaeobank, 8.154–8.157. This section does in fact lie on the projected alignment of the parcel 3 channel, while the interleaved shell-sands and organic rich layers of 8.154–8.156, lying between 12.65 and 12.83 m OD, compare well with 8.079, described as 'lenses of dirty sand and organic silt' at 12.55–12.93 m OD. The underlying gravel layers (8.082, 8.157) in the two sections also correspond. Immediately to the east of this exposure, parcel 3 would have been totally removed by the parcel 5 channel.

Parcel 4B: late second millennium BC; *Middle / Late Bronze Age transition*

It was possible to examine these deposits and the underlying palaeobank in a number of exposures distributed through the riverside zone (Fig. 1.4); they were explored most intensively early in the campaign in Area 14. This has allowed a detailed and confident reconstruction of the topography. Interpolated contours for the palaeobank itself (Fig. 10.3) show a fairly steeply cut bank, mostly following a straight line aligned west–east, but with a sudden southward deviation at the eastern end, in Area 31. Having been cut into parcel 2B, the bank was clayey and had some interesting micro-topographic features associated: some mini-stacks occupied the lower slopes, while in Area 14 a 'pot-hole' of 10 cm diameter was found. A pot-hole is a small roughly hemispherical hollow formed by the rotational action of a pebble, which was indeed found in situ (I am grateful to Professor John Allen for explaining this feature). The depth of the channel is not known; where seen at its deepest in Area 31, the bank was still dipping away steeply at 12.50 m OD. However, the lowest gravels encountered to the west (14.402) are horizontally bedded and lie between 12.95 and 13.15 m OD and could represent bed gravels of the active river (the 'basal platform' – Brown 1997, 70), rather than a higher point in the bar. If this is so, then the Area 31 evidence could represent localised deeper scouring along the channel edge. From the layer formation in the recorded

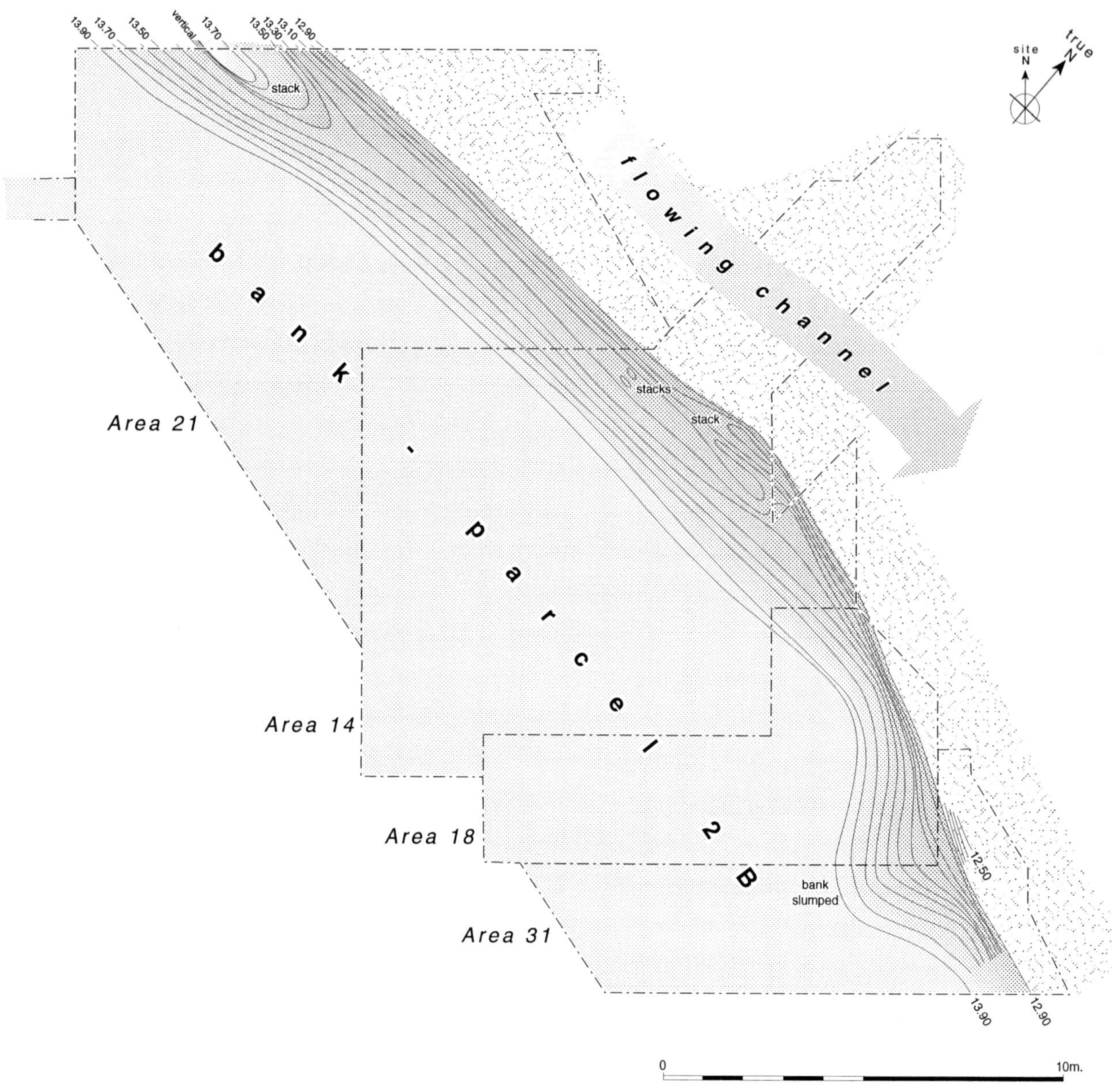

10.3   *The topography of the riverside zone late in the second millennium BC, after channel 4B had cut into parcel 2B.*

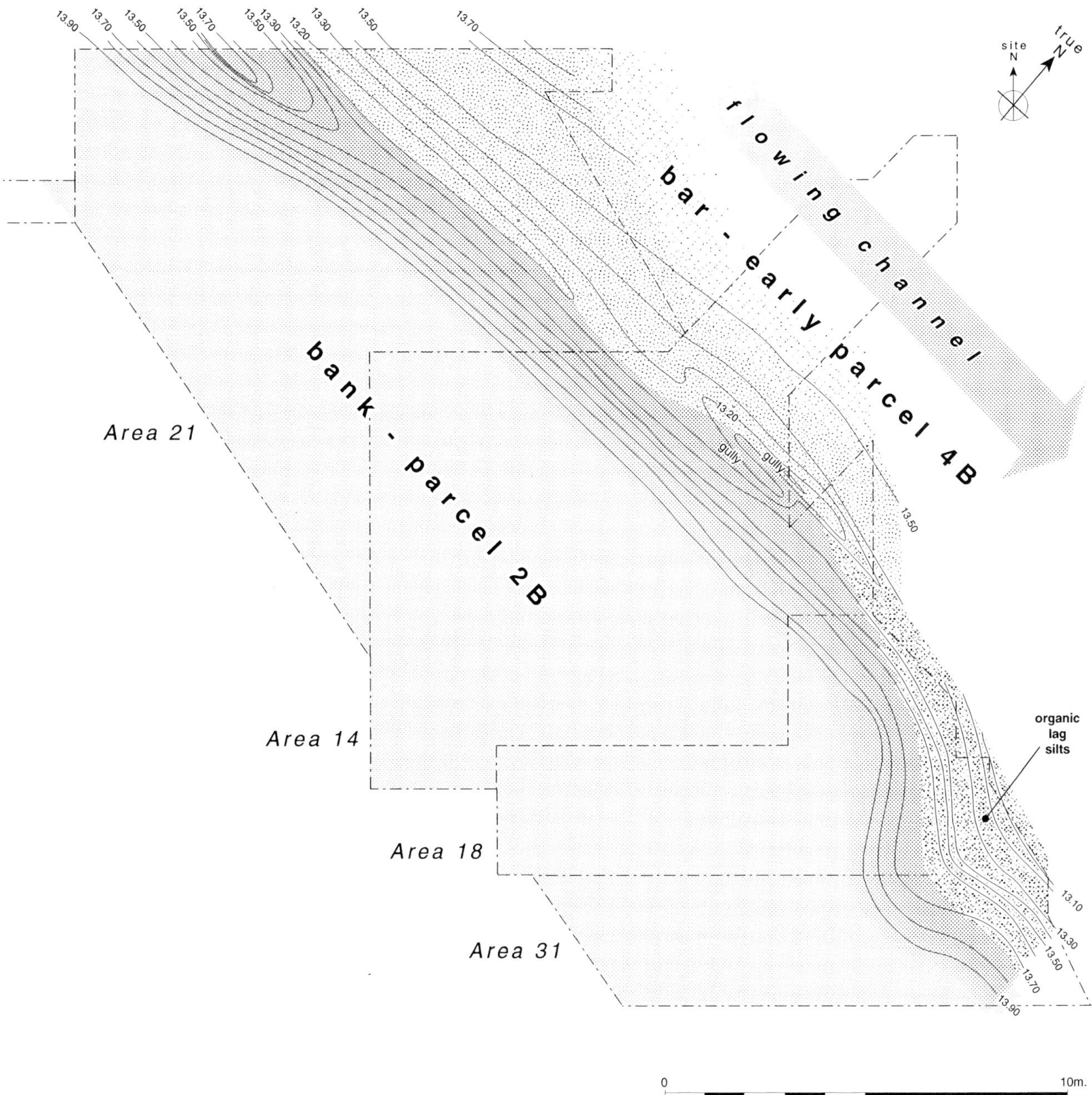

10.4 *The topography of the riverside zone after the formation of the lower parcel 4B fill.*

sections there was no obvious sign of the approach of the opposite, northern bank of this channel. All that can be said, therefore, is that it was at least 11 m wide. Given the relatively recent deposits in Area 15 and the modern course of the Thames, it is rather unlikely that the other bank will have survived.

The abrupt turn in Areas 18 and 31 is probably part of a minor embayment such as can frequently be seen on present-day slow-flowing rivers with muddy banks. However, the sections through it do suggest that it arose partly at least due to slumpage, rather than simply to eddying currents. While slumpage can of course result from the undercutting of banks, it is perhaps worth noting that the overlying sediments contained another concentration of beaver-worked wood (31.665; additional to the Beaker period cluster) and there must be the possibility that this indentation was a chute by which beavers regularly entered the water. A suggestion has been made that a gully feature at the Thatcham Mesolithic site might be a beaver 'canal', but it was rather longer than the Runnymede feature (Coles 1992; Wymer 1962, 335, fig. 5). At the turn in Area 18, the bank is at its steepest and it also shows evidence for either the burrows of small animals, or false interleaving of bank and channel sediments, or both.

The lower part of the 4B fill splits into two zones with very different sedimentary characteristics (Fig. 10.4). Parallel to the straighter stretch of bank a linear bar of gravel and coarse-sand lenses developed. This runs through Areas 21 and 14, the highest excavated point of the bar lying at 13.80 m OD in the northern corner of Area 21, a height only a little short of the bank crest 6 m away. The lenses dip towards the bank, thus creating a trough, or swale, in between which attained an altitude of 13.20 m OD before the coarse sedimentation ceased. The bar clearly extends eastwards beyond the last section in which it featured in Area 14 (section RH28) and it presumably passed to the north of Areas 18 and 31. This whole formation is reminiscent of that seen in the rescue areas, but is more exaggerated in morphology (Fig. 10.4), and while the two zones need not necessarily have formed a continuous ribbon, they are clearly pene-contemporaneous, probably with a temporal drift downstream as argued above.

The deposits found against the palaeobank in Areas 18 and 31 actually occupy the mini-embayment already described and would presumably have been trapped between the bank and the bar. This seems to have allowed the accumulation of mainly fine sediments, often with a strong organic component (surviving only beneath the water table). Sand and gravel does appear at certain horizons, as would be expected, given the interleaving of the two sequences (matrix, Fig. 2.35) and, overall, the marked lensing of the body suggests flow with varying depositional conditions; the environmental remains indicate flowing water. This would argue against accumulation in the relative quiet of a beaver-dammed pond. Any nearby dam would thus have had to be upstream.

The finer-sediment sequence yielded a concentration of woody material suitable for radiocarbon dating, context 31.665. A calibrated range of 1265–935 BC was obtained. As with the earlier concentration in parcel 2E, this material was fresh, with some bearing clear beaver tooth marks; it is extremely unlikely to have been exposed for any length of time. The wood is also of very young growth, thus ruling out any age-offset. A body of sediment 1 m thick accumulated above this horizon before the first Late Bronze Age rubbish was deposited. A date on an unweathered cattle bone gave a date range of 990–800 BC (Table 4.1), suggesting rapid accumulation of the intervening sediment.

The gravel/sand bar contained a modest assemblage of artefact finds, which might at first seem to be a further aid to dating (Table 3.2). However, it transpires that all of the diagnostic pottery involved is Neolithic, while the body sherds are all in fabrics consistent with such a date. The dominant fabric is a more or less crumbly structure (usually exacerbated by the erosive agents), gritted with ill-sorted crushed burnt flint in medium density; large grits are usually present. Exceptions are noted in Table 3.2. This assemblage, dating to the earlier rather than later Neolithic period, must be considerably older than the formation of parcel 4B. By the time of formation, the intact Neolithic land surface nearby had been well sealed by gravel, then fine alluvium, and there is no evidence in the excavated areas of either its re-exposure this late on, or its lateral erosion by the 4B channel. This suggests that the material was eroding from a bank or river deposit upstream. This might hint at more extensive Neolithic occupation in the vicinity. Whatever the history of reworking, it is extremely unlikely that this material had suffered much fluvial action, for otherwise these fabrics would have totally disintegrated.

The deposits with a distinctive bluey tinge low in the Area 31 sequence (31.642, 666, 682) recall the finding of similarly coloured deposits in similar channel-edge situations in Area 6. This colouration apparently derives from the formation of vivianite (ferrous phosphate) by reduction in waterlogged conditions (Brown 1997, 99).

The uppermost layers of parcel 4B are more or less consistent across the whole fill, characterised by shell-sandy silts, but still with some lensing (e.g., 14.167, 168). They are deepest in the trough, thinning as they lap over the bank and the bar. Even so, they did not wholly level the ground prior to occupation, a trough up to 0.45 m deep was still present (Fig. 10.5). The dating of these deposits is constrained between the radiocarbon date cited above and the date of occupation, starting broadly around 900 BC (see below for detail). The question of whether the flood regime had changed by the time occupation began is also considered below.

The upwards fining of the main sequence from gravel bar to sandy silts is presumably in large measure due to the vertical accretion involved, but also perhaps implied in this abatement of energy levels is the shifting of the main

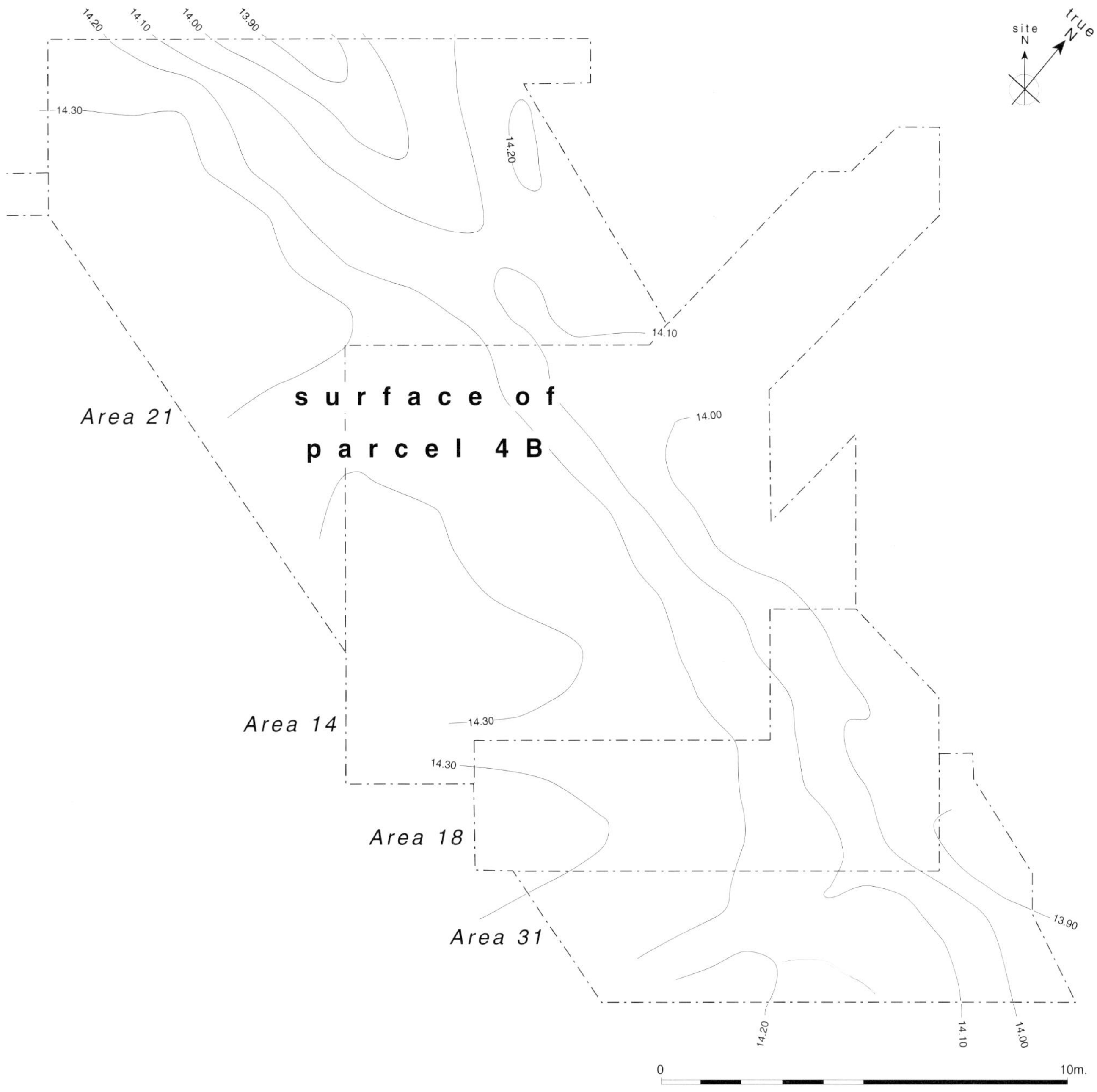

10.5 *The topography of the riverside zone on top of parcel 4B, prior to Late Bronze Age deposits.*

channel still further away. What yet cannot be determined conclusively is whether this newly reactivated channel was simply migrating northwards, or whether in fact the gravel bar had quickly choked its opening off the presumed loop and had already re-established a land bridge to the north (Fig. 11.9). This latter scenario is favoured in the discussion in Chapter 11, but the channel would still have been visible, perhaps as a marshy belt, and it would doubtless have acted as an overflow channel, which could account for the steady later silting.

The finer sediments against the palaeobank and below the water table proved suitable for plant and insect remains, and snails were also analysed. The marked opening up of the landscape when compared to the later Neolithic environment has already been raised in connection with the phase of 2B alluviation. Of great importance at the transition to the Late Bronze Age, when many aspects of society were changing, is to assess the character of the open land and its specific use. Scaife acknowledges that the pastoral/arable balance is difficult to gauge, but he ventures

207

that it is 'those taxa typical of pastoral or grassland habitats which predominate'. Likewise, while not ruling out the possibility of arable in the catchment, Robinson considers few of the beetles present in column 55 to be diagnostic of arable land (Chapter 8). This would certainly tie in with the impression gained from the cultural debris accumulated on the site a little later, that livestock rearing for high meat consumption (especially cattle and pigs) were central to the food economy and, perhaps, the social order (e.g., Serjeantson 1996). Whether this was an emphasis reserved for certain sites or certain habitats requires a broader consideration than can be tackled here.

On the basis of the insect remains from the rescue campaign it was argued that the extensive grasslands that had developed by the Late Bronze Age were not heavily grazed. The extension of the time-frame backwards into the second millennium BC does not alter this conclusion, but one significant difference is that the proportion of dung beetles increases later, that is during the occupation (almost doubles in column WF1b), and this may therefore represent the difference between, initially, the background levels of dung spread throughout the local pastures and, subsequently, relative concentrations due to the on-site coralling of livestock from time to time.

The wood- and tree-dependent Coleoptera are at very low levels in column 55, as expected of this greatly cleared landscape. However, the molluscs do suggest the presence of more woody vegetation than in the succeeding millennium, as well as the mudflat/swamp conditions immediately around the 4B channel edge. The overbank deposits only a little distance up the slope, however, supported open-ground mollusc communities. The alder woods of the valley bottom and 'old woodland' generally seem to have all but vanished. Robinson notes that this situation continues into the early part of the WF1b column but, curiously, the relevant beetle species increase higher in that column, perhaps a consequence of all the timber involved in the settlement structures.

Areas of reed swamp, inferred to have been present on a small scale during the later Neolithic, seem also to share in the expansion of more open conditions (pollen, insects and snails). Such habitats may well have been promoted by the alteration of the original anastomosed channel system discussed below (Chapter 11), which progressively left segments of the system cut off, to become standing water and ultimately marsh.

Late Bronze Age occupation

The extent of surviving Late Bronze Age occupation evidence is shown in Fig. 10.6, while the predicted extent of the contemporary island or peninsula available for occupation is in Fig. 11.9. Wherever Late Bronze Age occupation evidence has been located it has taken the form of many subsurface features accompanied by surface deposits,

relatively rich in artefacts and other debris. The upper part of the full depth of surface deposits was invariably flood-reworked, leaving beneath a variable thickness of in situ deposits, often of dark earth character and high in inorganic phosphate levels (Chapter 6). Only in Areas 27/28 do the surface deposits seem to have been entirely reworked by flooding; however, cut features survived immediately underneath, to demonstrate contemporary activity at this spot. The LBA occupation, therefore, presents a very clear marker horizon; it sits on top of parcels 2B, 3 and 4B, is incorporated within parcel 4A, is laterally cut by parcels 5 and 7, and is overlain by parcels 4C and 6 (Fig. 2.1). Although it should not be assumed that the detailed chronology of the deposits will be identical in all parts of the site, differences are unlikely to be significant in terms of the broad geomorphological history outlined here.

Most of the settled area overlies parcel 2B (Fig. 2.2), the surface of which had been stable long enough for a soil horizon to develop, but not for a mature B horizon to develop beneath the A horizon (Chapter 5). It is thought that aggradation had effectively ceased two or more centuries before occupation. Although difficult to quantify the rate at which these developments would take place, it may be significant that the much longer exposed land surface on top of parcel 1K did develop a mature soil profile during the Late Mesolithic and Early Neolithic periods.

The topography over most of the island was therefore the gently domed spine, aligned NW–SE, described above (parcel 2B). This was modified by the addition of parcel 4B along the northern margin, creating a trough dipping down to below 13.90 m OD (Area 21) and raising the adjacent 'levee' to around 14.30 m OD (Figs 2.2 and 10.5). The addition of land on the south-east flank, the result of parcel 3, also gave rise to a trough, this time descending to 13.40 m OD before rising to an outer rim of *c.* 14.10 m OD (Fig. 11.8). On the north-east edge of the island (Area 6) the levee remained at between 13.80 and 14.00 m OD, the top surface of parcel 2B.

Area 6: waterfront

The interpretation of the main sequence of 4A/4C, with its associated waterfront structures, has not altered. The chronology is barely modified by the addition of a few more ^{14}C measurements (Needham and Spence 1996, 230, table 63). In summary, the early silting yielded no indications of occupation on the adjacent bank; this first showed up as a substantial burnt horizon (unit 6.B). A succession of two waterfront palisade footings and a hard-standing followed amid more silting, before the growth of peat trapped further rubbish and an articulated dog skeleton. There was then a hiatus in occupation evidence (unit 6.H) before a modest concentration of refuse in layer 6.006, unit 6.I; this may represent flood-reworked material, rather than in situ dumping.

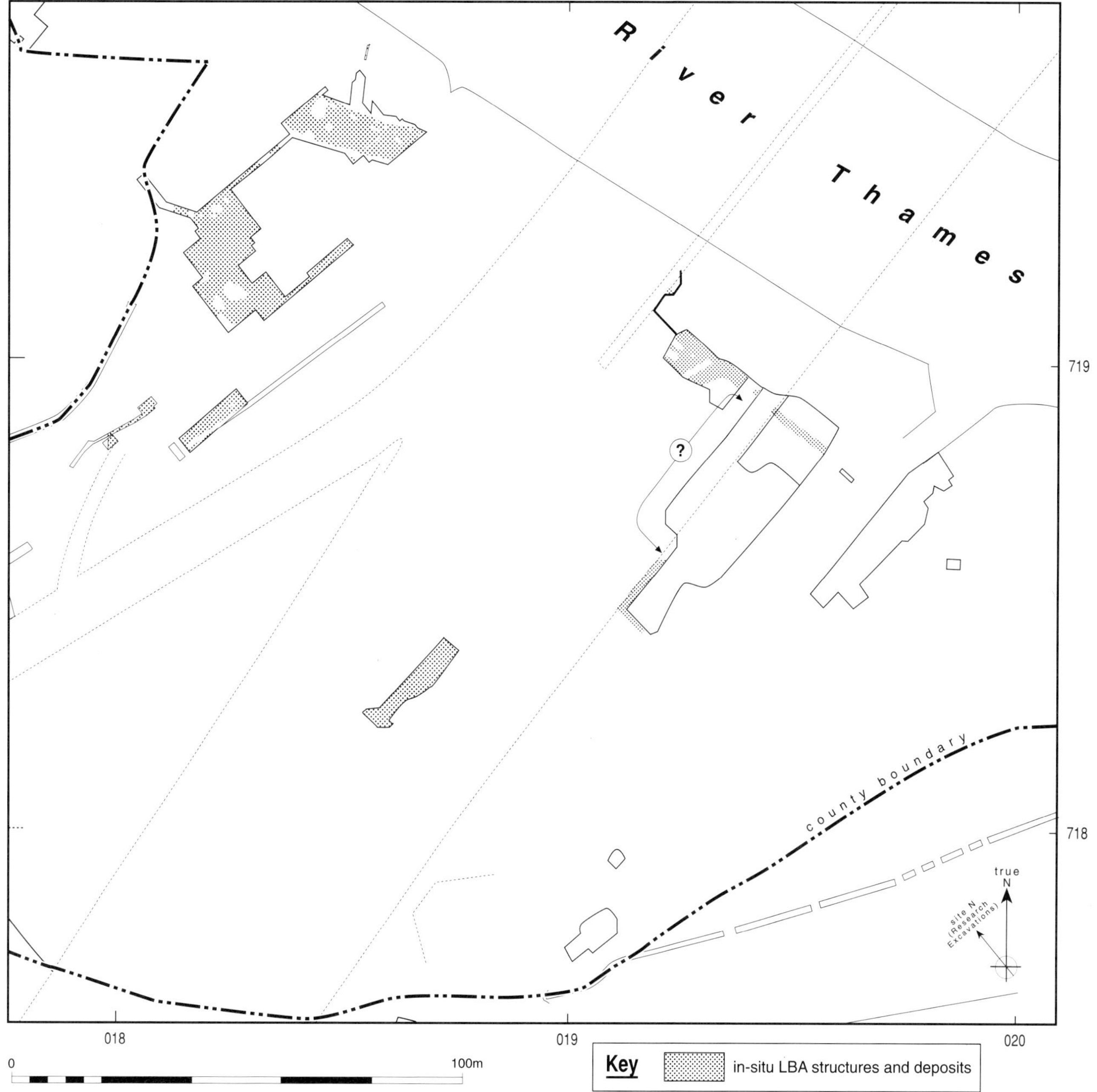

10.6 *Site plan showing the extent of excavated in situ Late Bronze Age occupation evidence. [Only large later intrusions into the occupied surface are shown.]*

The end points of the dated sequence have been re-evaluated by Janet Ambers using the CALIB programme to incorporate some of the known stratigraphic data. The best dating of the clearance horizon (unit 6.B) is early in the ninth century BC (68% confidence), whereas unit 6.G with its dog skeleton is best dated to around the end of the ninth century BC. The dates on the waterfront structures themselves are happily accommodated in the interim.

Riverside zone

Attempts to date a series of key contexts in the LBA deposits of this area using bone material have largely been unsuccessful because of poor protein survival. This may be due to regular fluctuations in the water table having affected these low-lying parts of the bank. A single radiocarbon date has been obtained on fresh cattle bones from the base of the

209

LBA deposits in Area 31 (BM-2659). This gave a calibrated range of 990–800 BC (at 95% confidence), not inconsistent with early levels in Area 6. However, it is not on its own sufficient evidence for exact contemporaneity and the ceramic assemblages may suggest a slightly later inception to refuse accumulation in this zone, though still probably before the end of the ninth century BC.

As always, the duration of occupation is difficult to estimate, but one noteworthy bronze ornament from fairly high in the riverside stratigraphy is a ring-headed pin, traditionally regarded as an Early Iron Age type. Its context (14.205) is grouped with unit I, interpreted as a 'conflation' horizon in which the uppermost in situ deposits were infiltrated by flood re-sorted material. The pin itself is in very good condition and has suffered little if any fluvial reworking. It is a typologically 'evolved' form and should indicate a post-Ewart date, i.e., later than 800 BC, perhaps significantly later.

Of considerable relevance to the nature of occupation is the question of the environmental conditions in the trough, since this is the lowest part of the occupied surface seen in the research excavations. The immediately preceding sandy silts at the top of parcel 4B are confirmed, on the strength of the molluscs, to derive from fast-flowing water, but there are also swamp and mudflat components to the assemblage, perhaps indicative of seasonal cycles (Chapter 7). Samples from the LBA material sitting in the deepest part of the trough (column 36) do not show any immediate discontinuity in terms of species composition; a very high proportion continue to be aquatic species at first, but by sample 71 this has reduced dramatically and one wonders whether this may have been a response to humanly constructed barriers, rather than simply to the further small increment of sediment. Lower levels of aquatics, generally well under 50%, were a feature of the nearby sequences on the flanks of the trough both high in 4B sediments (column 5) and within LBA dark earth deposits (column 11). The exception in column 5, sample 48, comes from very thin LBA deposits and may be affected by overlying flood silts of parcel 6.

The phosphate column through the deposits in the trough (column 36, Area 21) is of some note in that it peaks in sterile sandy sediments just below the dark earth deposits. This may be a localised feature, for elsewhere along the trough phosphate measurement gave high values within the dark earth (gridded phosphate sampling to be published in a future volume). In this extremely low-lying situation, this is unlikely to be a shift caused by worm sorting. It might, for example, be the result of the preferential accumulation of organics in this natural trap just before occupation. But another possibility to be considered is that organic phosphates from occupation, particularly things like cess, were leaching into the underlying sediment before conversion to inorganic, and thus less mobile, forms. It is perhaps significant that close to this horizon at the centre of the trough a marked concentration of mineral enrichment was encountered (section RH21, Fig. 2.16, context 21.223).

Area 16 East

It has already been noted that the soil profile immediately beneath LBA occupation material showed some woodland regeneration to have taken place and, while this might be a truncated profile, it could be seen to connect with the interpreted clearance horizon in the waterfront site (unit 6.B), which presupposes there was some scrub/wood vegetation requiring clearance. During occupation this part of the site, lying a little inland, evidently experienced very little flooding (Chapter 7).

The in situ occupation deposits survived relatively thickly in Area 16 East and contained enormous quantities of cultural remains. The sequence has been published in detail in Needham and Spence (1996). Dating by radiocarbon (Ambers and Leese 1996; Ambers, pers. comm.) favours the formation of the lowest LBA layers late in the ninth century BC. This ties in with the associated ceramics, which seem to be a little later than those at the base of the waterfront sequence. The ensuing sequence at Area 16 East was long enough for significant changes in ceramic style to occur.

Interior zone

This represents the highest ground on the island, still riding on the backbone of the early Holocene marl and silt (parcels 1C and 1K). It falls away gently to north-east and south-west (Fig. 2.2). It is unlikely that there was any significant increment to the soil profile on top of parcel 2B in the centuries leading up to occupation. The possibility of an earlier start to occupation, or land use, in the interior zone cannot therefore be excluded. As yet the ceramics have seen little study and no radiocarbon dating has been attempted. In broad terms the artefact assemblage does not suggest any radically different chronology from other parts of the site.

Area 2: the south-eastern side

The bulk of the pottery from Area 2 comes from layers overlying burnt timbers dated to 2750 ± 70 BP (HAR-1834) and 2620 ± 70 BP (HAR-1833) and its style again suggests a mid- to late stage. The lower ground to the northern end may have been subjected to some later scouring, but this did not remove all the cultural layers and cannot account for the sparsity of postholes. The depression here was subsequently filled with a layer (9) distinguishable from the main overlying alluvium (11), but it is not clear how much later than occupation these layers were (Longley 1980).

Alluvial parcels 4C and 5: mid- to late first millennium BC; post-Bronze Age channel fills

Parcel 5 was defined previously on the basis of a limited section through a channel edge in Area 8, section SA7, and its less certain reappearance in section SA8 (Fig. 2.29). Although little was visible, the channel had potentially important ramifications for the overall history of the site, since it had carved into the bank supporting Late Bronze Age occupation (8.053). A lens of greyer silt trailed away from the occupation material on the bank crest, and while it could conceivably have been contemporary interleaving, the appearance given was rather of later washing-out. But there was no way of establishing how much later. In order to gain more information on the alignment, structure and date of this channel, the research campaign investigated undisturbed areas to the south-east of the M25/A30 bridge embankment, Areas 17 and 23. The latter gave sections more or less transverse to the predicted alignment of the channel.

Considerable information was yielded on the parcel 5 channel, allowing deposits to be linked not only to Area 8, but also, via Areas 17, 4 and 7, to the main waterfront sequence in Area 6. These linkages suggest strongly that the later part of the latter sequence – that immediately following in situ occupation and the demise of waterfront 2 – is stratigraphically interleaved with parcel 5. At first sight the character of the relevant parts of the two parcels seemed utterly distinct, one dominated by gravel, the other by organic deposits and clayey silts, but the Area 8 sections in particular clearly show the interleaving of very different deposits (just as seen in the earlier parcel 4B). Area 17 proves to be the linchpin in the chain of correlations; it is reconstructed to have lain very close to the spit at the confluence of the parcel 4 and 5 channels (Fig. 10.7).

The sequence in parcel 5 is best described independently first. It is dominated by a thick gravel shoal, which can be reconstructed as a linear formation running inset from a concave palaeobank (Fig. 10.7). It seems likely to be the tail end of a point-bar extending from the south-west. It had a domed cross-section, peaking in Area 23 at 13.57 m OD, then dipping away towards the bank to a tail at 12.85 m OD in Area 8 (SA7); in section SA7 the gravel deposition, signifying high-energy fluvial conditions, can be seen to coincide temporally, and presumably causally, with the above-mentioned erosion deposit trailing from the occupation surface (8.047). Further north-east in section SA8 (Fig. 2.29) the gravel edge feathers into a light brown clay (8.147) with some shell-sand and charcoal at its base, perhaps again derived from the occupied bank. The top of the clay lay at about 13.20 m OD. Further north-east still, Area 17 yielded a layer at similar altitude and of similar character (17.005 and 006), charcoal again being present.

This gravel bar/erosion silt horizon, however, overlies some earlier deposits lying along the foot of the palaeobank, these showing in Areas 8 and 17. They are predominantly sandy lenses, but an organic-rich lens occurs at 12.70 m OD in Area 8 (8.051) and as a thicker layer between 12.72 and 12.99 m OD in Area 17 (17.007). Above the gravel bar/erosion silt horizon there is a return to a sandy deposition regime, a variable number of distinct lenses being

discernible in different places. In Area 23 they reach 14.00 m OD, but unlike the underlying gravel, the surface is more horizontal, 14.07 m OD being attained close to the palaeobank in Area 8. The molluscs from the channel fills in Area 23 indicate a fast-flowing, well-oxygenated river (Chapter 7).

The interrelationship of the parcel 4C sequence now needs to be established. Taking the three broad units of parcel 5 in turn from the bottom, the sand-dominant lower unit is not replicated in parcel 4C. However, the north-eastwards thickening organic silt 17.007 is matched by similar deposits at similar altitude in the Area 4 section just 6 m away (Needham 1991, fig. 13). The layers in question are 4.101, 102 and 103 and occupy a slight hollow which appears to have been enhanced by scouring after the deposition of 4.106. Part of layer 4.102 at least was noted to be sandy, conceivably coarser material washing up into this channel mouth from the main (parcel 5) channel and intermixing with quiet-water accumulation. These organic-rich layers in Area 4 have already been tentatively linked by way of Area 7 (7.054a) to a peat bed in Area 6 (6.009; Needham 1991, 52, table 3a). The upper surface of the latter was at 12.95–13.05 m OD, a little lower than that of the equivalents, but it had probably suffered erosion before the next deposit. This erosion would have re-exposed the dog skeleton, 6.F111, which originally was probably fully immersed in the peat.

The middle parcel 5 unit, the gravel bar/erosion silt horizon, seems to be entirely absent from parcel 4C. However, the tendency for erosion off the adjacent banks already noted, plus the high-energy depositional regime, could connect with a consistent erosional interface at this point in the parcel 4C sequence. There is the erosion of the Area 6 peat raised above, and a string of charcoal flecks and stones at the top of 4.101. The stones might be either part of the redeposited cultural debris, or elements of the gravel deposition swept in from the conjoining channel. Just into that channel, Area 17 also shows evidence for erosion: clasts of the underlying deposit were reincorporated into 17.004.

The upper unit with the resumption of sandy silting can be identified in Areas 4 and 6. Layer 4.100 is a dipping lens on the slope of the palaeobank (13.26–13.90 m OD), and might perhaps be a remnant left after later scouring in the trough. Layer 6.007 laps less far up the slope (highest recorded, 13.40 m OD) and survives as a fairly thick deposit sealing the erosional interface described. Lenses of charcoal and a charred branch were trapped within the layer, while sparse charcoal was also noted in the contemporary layers 17.002 and 003. There was a sandy component to at least some of the overlying layers in Area 6 (6.004, 005, 006) and the continued redeposition of eroded cultural debris.

Parcel 5 sits against the major erosion scarp in Area 8 already described and a scouring horizon in Area 6 parcel 4A/4C has been correlated with it. However, the possibility that there was any prolonged hiatus after LBA occupation is

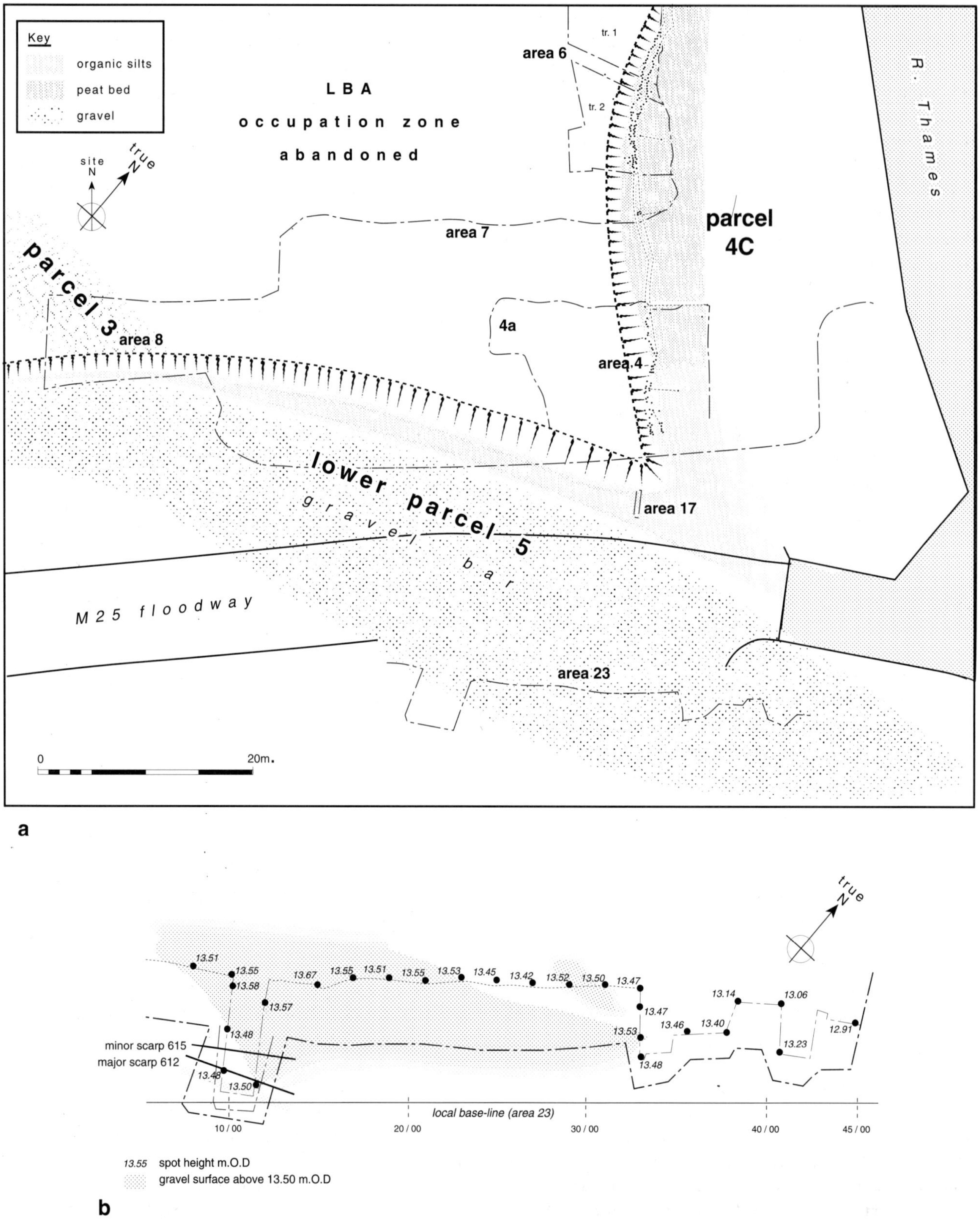

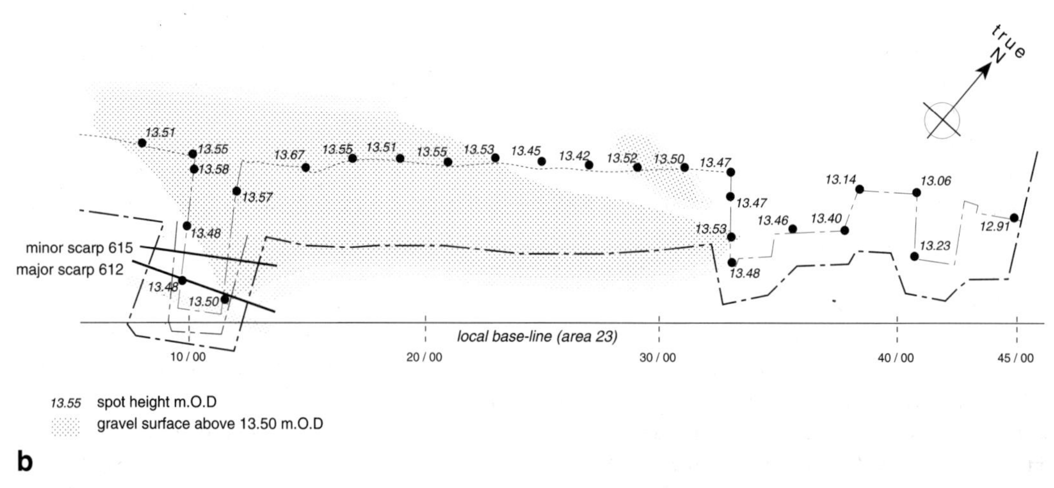

10.7   *a) Reconstruction of the channel confluence at the eastern end of the site with parcels 4C and 5 accumulating; b) detail of the surface of the gravel bar through Area 23.*

ruled out by the dating evidence from Area 6. Low in the 4C sequence an articulated dog skeleton was found (6.F111); it lay on top of the peat bed and, as mooted above, may have lain within it prior to another scouring episode. Radiocarbon dating of the skeleton has given the range 1000–790 BC (OxA-3368), but the dating of various underlying contexts would constrain it very much towards the end of this range (use of CALIB program, as discussed above). The overall correlation in the matrix (Fig. 2.38) suggests that a similar date would apply to the lowest unit in parcel 5. There is no dating evidence from parcel 5 itself; the only find from Area 23 was a flint-gritted body sherd from 23.613, in overlying parcel 6A.

The upshot of this set of correlations is striking. The earliest silting against the edge of the parcel 5 channel can barely date much later than 800 BC and may overlap with the later stages of the main occupation. This in turn means that the eastern side of the settled area (Area 8) was probably being eroded laterally before abandonment; furthermore, there is the suggestion that occupation soil was being eroded off the dry banks only a little later, particularly during the phase of gravel bar formation. The later stages of occupation and the abandonment of the site need now to be viewed against a backdrop of turbulent channel flow and increasing flood risk.

Alluvial parcel 7: mid- to late first millennium BC; post-Bronze Age channel fill

The final rescue excavation in 1980 had shown that intact LBA land surfaces with in situ refuse deposits were present in Area 10; indeed, there also proved to be an occupied Neolithic land surface beneath (excavations continued by the Surrey Archaeological Unit). However, the inspection of exposures in the walls of the drainage system some 50 m to the south-west failed to find any material and this suggested that another edge to the prehistoric deposits would be found in between. Consequently, in 1988 the long trench Area 27 was dug by machine, until artefacts were encountered. This succeeded in locating the edge of a river channel which seemed to truncate the deposits underneath the LBA horizon, and perhaps the LBA material itself. The following year a small trench, Area 28, was excavated to clarify better the nature of the land/channel interface here. One problem was that the LBA horizon appeared, at first sight, to be a continuous layer running across the interface. If indeed continuous and in situ, then the channel fill would obviously be earlier. However, the LBA horizon seemed generally rather thin, pale in colour and lacking in obvious in situ artefact groups. Although eventually some distinction appeared in the density of finds in the two zones (Fig. 10.8a), the overall conclusion was that the horizon had been reworked throughout the small trench. Nevertheless,

flooding had not severely truncated the underlying land surface for LBA cut features were found to survive (Fig. 10.8b).

Having drawn this conclusion on the nature of the occupation deposits, the chronology of the channel incision and fill relative to the bankside activity was still left open. All that could be said with assurance was that the incision post-dated the bank's make-up, i.e., parcel 2B. As outlined above, the aggradation of 2B would have ceased well before LBA occupation, and it is not inconceivable that a channel had already cut back to this position during the Late Bronze Age. Dating evidence for its fill is not strong. The horizon of artefacts immediately above parcel 7 (27.016, 28.323), being redeposited, serves only as a *terminus post quem*. The fill itself lacked obvious cultural debris, perhaps an indication that the great bulk was not contemporary with the occupation.

Independent dating is, however, provided by an archaeomagnetic column (65). Although tentative, the results do suggest sedimentation during the Late Bronze Age or Iron Age, but they do not really help with a finer chronology. Clark wondered whether high intensity values at samples 11, 14 and 15 might relate to contemporary on-site activity (Chapter 4).

In the absence of specific evidence for contemporaneity, it is assumed here that at least the infilling took place during the Iron Age. If the preceding eastward incision into the island also post-dated LBA settlement, this evidence would nevertheless still suggest the presence of a channel nearby during occupation. The only alternative would be that this was a recent breakthrough channel between two established branches of the river, but it is difficult to accommodate such an event within the broader sequence of channel changes reconstructed below.

Alluvial parcel 6: early first millennium AD; Roman period overbank alluviation

The fact that the LBA occupation levels were, in general, covered by a fairly thick layer of alluvium has been appreciated since David Longley's first excavation on the site (Longley 1980, layer 11). Later excavations have consistently revealed some covering alluvium, except very close to the Lutyens Bridge (Areas 5 and 6) where it had been largely removed during construction work (in 1959–60). The essential problems concerning parcel 6 are, firstly, how closely did alluviation follow the Late Bronze Age and, secondly, was it synchronous across the occupied surface; in other words, might some zones have remained unsealed and subject to further scouring for longer than others?

Alert to these possibilities, it has seemed prudent to label the covering alluvium in unconnected zones of the site differently – hence 6A/B on the east side, 6C on the south-

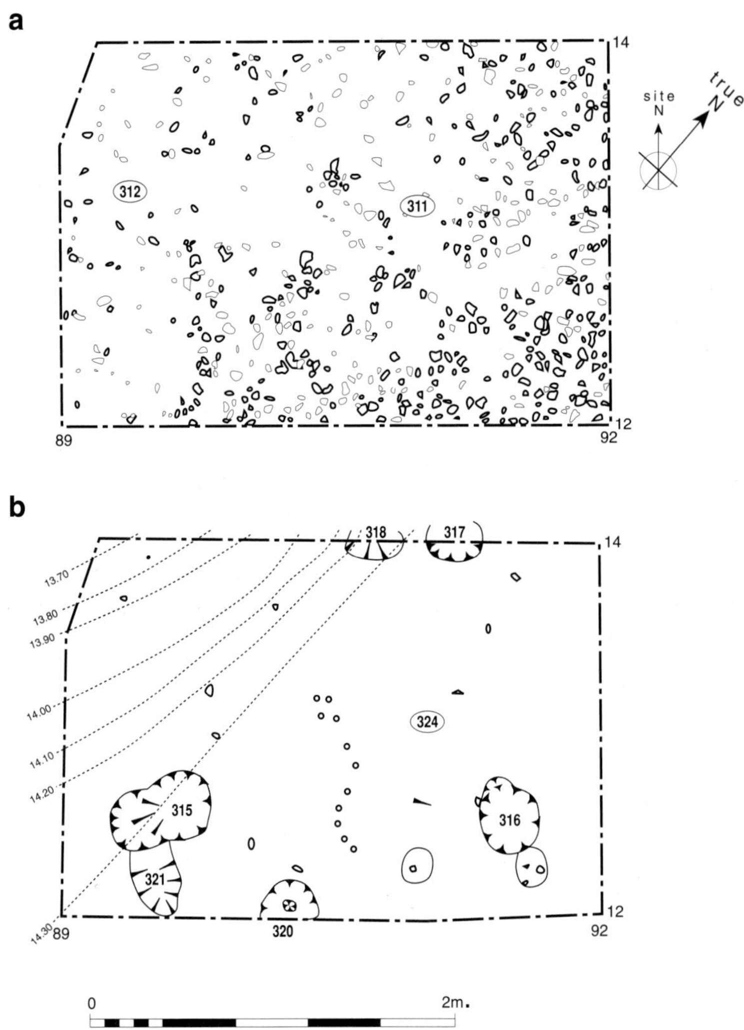

10.8 *Area 28 Late Bronze Age evidence: a) surface distribution of occupation debris, probably all reworked, but showing varying density; b) the underlying cut features in relation to interpolated palaeobank contours.*

west, and 6D on the north (riverside and interior zones). Identification of a major erosional interface in the eastern zone has enabled a vertical division, 6A below 6B.

The Area 8 section SA7 had already shown that on the south-east side of the site the covering alluvium (8.045) was very different in character from the underlying channel deposits of parcel 5. There was no sense of this being a gradation in an upwards-fining sequence, hence the separate parcel designation. This in itself suggested an interlude between abandonment and the sealing alluviation. The more recently garnered evidence from Area 23 gives a better picture of the sequence of alluviation on this flank. Layer 8.045 was around 0.6 m thick and perhaps encapsulated the similar silts in Area 23 that have been split into parcels 6A and 6B, namely the sequence from 23.614 up to 23.605 (Fig. 2.38). This, too, was around 0.6 m thick as overbank alluvium. However, the important additional evidence in Area 23 is the association of these overbank deposits with channel edges. A succession of two recuts are associated with

the parcel 6 silts, the first cutting into the top of parcel 5. The interval between abandonment and the start of covering alluviation can thus be seen roughly as the duration of the middle and upper units of parcel 5, plus any hiatus that might have been associated with the first erosion scarp (interface 23.615). There is, of course, no way of quantifying this period, other than to say it was not brief.

The sequence on the western side, as shown in the Area 27 section, is rather similar. If the attribution of the parcel 7 fill to a post-occupation phase is correct, then again there was a significant time interval prior to the inception of the covering alluvium. Interestingly, in this area the only clear horizon of eroded cultural material occurs around the parcel 7/6C interface (27.215, 216) and this might suggest that significant overbank scouring, leading to the reworking of the cultural deposits on land and their redeposition across the top of the channel, only occurred at a late stage here. It could in fact be seen to be the inception of the new alluvial regime marked by parcel 6.

214

Across the island itself, parcel 6 contains many Late Bronze Age artefacts. However, studies of their condition, along with the paler nature of the enveloping sediment and the lack of structural evidence, have all led to the conclusion that the artefacts are reworked. This has led to the definition of an 'upper zone' (containing LBA material) as opposed to a 'lower zone' featuring in situ material (Needham and Sørensen 1988). A vertical gradation is typical, with both darkness of colour and density of artefacts increasing downwards towards the upper/lower interface; phosphate levels also typically show a progressive increase downwards (Chapter 6). One implication is that prior to partial transformation, thicker in situ Late Bronze Age deposits would have been present in general across the site. Some of the sediment stripped and reworked could have contributed to the parcel 6 sediment, but it is clear from its thickness as well as the relative cleanness of its extensions into the flanking channels that the great majority must be newly in-washed sediment. In some parts of the site at least, Limbrey found parcel 6 deposits to present a marked break from those below. In column 15 there was a dramatic shift from silt to clay, although this might in part be due simply to the cessation of anthropic contributions to the sediment which tends to enhance silt fractions.

Aside from the reworked cultural material of the early first millennium BC, which can only act as the broadest of *termini post quem*, there are three sources of dating evidence for parcel 6. The first is the upper continuation of the archaeomagnetic column in Area 28, for which Clark suggested a match with the inclination variations of the Roman period (column 65, samples 1–7; Chapter 4). Nearby, a Potin coin dating to the early first century BC (Chapter 3) was recovered from a modern feature which cut to the bottom of parcel 6, but not below it at this point. It seems more than likely that the coin was redeposited from the parcel 6C silts, but obviously this cannot now be proven. In secure 6C contexts, however, are two sherds dating between the first century BC and the second century AD (Table 3.6; context 28.307).

Indeed, the main evidence to suggest a rather late covering of much of the site comes from a broader scatter of post-Bronze Age pot sherds. There are nineteen Roman ceramics and eleven certain or probable Middle to Late Iron Age sherds (Chapter 3). The great majority of these were stratified within the alluvium covering the island, and came from widespread locations throughout the excavated trenches (Fig. 3.7). Three sherds came from lower stratified positions and deserve individual comment. Most surprising is a Late Iron Age foot-ring from the base of in situ LBA deposits in the southern end of Area 13 (13.032). However, the in situ deposits here survived very thinly and it would not be unreasonable to suppose that a small sherd could have migrated downwards by means of faunal activity, especially if just through a soil profile rather than a thicker body of alluvium. A possible Late Iron Age bowl sherd came from

the uppermost in situ layer in Area 32 (32.002), while a similar stratigraphic position in Area 10 (10.010) yielded a first-century AD piece of Samian ware.

All of the recovered later Iron Age and Roman sherds are small and generally in poor condition and, given that no cut features or other structures have been found on the site for these periods (indeed, none between *c*. 700 BC and AD 1700), it would seem more than likely that their introduction to the site was incidental, as sherds, rather than deliberate, as vessels.

The most obvious explanation for the late pottery in parcel 6 contexts is their redeposition during the flooding which must have brought in the extra silt load. The pottery could either have derived from pre-existing contexts on the site in company with the quantities of Late Bronze Age artefacts, or have washed in from nearby. Whatever its initial context of deposition, such a thin scatter of pottery hardly seems indicative of erosion from a nearby settlement and a likely alternative would be their incidental deposition in the course of manuring. This raises the possibility that during the later Iron Age and Roman period the site itself or the immediate environs were being used for arable agriculture, even if there was some contemporary flood risk. The upper zone of the LBA-bearing deposits might in this scenario at some stage have been a plough soil, the plough presumably having bitten down to the top of the in situ deposits. No plough marks were ever detected at this interface, but this cannot be regarded as conclusive negation since subsequent homogenisation across the boundary could easily have obfuscated them. Roman ploughing is attested elsewhere in a waterlogged valley floor situation at Warren Villas on the River Ouse, Bedfordshire (Robinson 1992, 203). The Late Bronze Age occupation soil spread across the island would have been a very rich base for crop production before it was too deeply buried. The phosphate levels decline upwards from the in situ LBA deposits, sometimes rather rapidly (Chapter 6), which suggests an absence of any sustained manuring regime, but this would relate to the phase after alluviation had set in; any cultivation might have tailed away and might anyway have been relying heavily on the existing nutrient base.

The terrestrial molluscs from parcels 5, 6 and 7 could relate to the proximate land conditions during both the Iron Age and the Roman periods, but more specifically from the latter in the case of those sampled from parcel 6 alluvium. Consistent with the sediment's interpretation, the snails indicate overbank flooding and generally open conditions, but including 'some bare ground and/or mudflats' (Chapter 7). While these habitats might represent foreshores and patches of ground flood-stripped of turf, an alternative is that they reflect disturbance by cultivation.

Flood reworking with or without cultivation could give rise to the pottery becoming well distributed vertically through parcel 6; however, these mechanisms would seem less able to account for the three lower stratified sherds, had any thickness of this deposit already developed. On balance it may be suggested that the site, or at least parts of it, bore little alluvial covering until after the first century AD. Three of the sherds in contexts not far above in situ Late Bronze Age levels date to the third to fourth centuries AD; this suggests that a good deal of the alluvium dates to the third century at the earliest. If the alluviation was essentially Roman (and possibly continuing later), it would tie in with the archaeomagnetic result as well as with the significant interlude documented following the beginning of parcel 5. It could also tie in with an episode of alluviation seen at the Friends' Burial Ground in Staines which occurred in the third century AD, after the early phases of occupation (Crouch and Shanks 1984, 18, 128). The flood silt (phase XI) was up to 2.10 m deep where lying over the depressions on top of palaeochannels. It was then overlain by later Roman activity, after just half a century of abandonment of this area (*c*. AD 220–70). Crouch and Shanks wondered whether the third-century flood deposit at Staines might be fairly local and dependent on 'a combined surge of water from the Thames and Colne rivers, so affecting Staines far more than sites downstream' (1984, 128), but a regional regime of overbank alluviation is possible if Gibbard's brownish silty clays (mentioned above with parcel 2B) in general correlate with parcel 6. Indeed, more recent excavations at Horton Manor Farm, 3 km north of Runnymede Bridge, also yielded alluvium following first- to second-century AD activity (Steve Ford, pers. comm.). This regional picture may reflect a wider phenomenon of increased alluviation in the Roman period (Brown 1997, 226).

The switch in the recorded sequence from vigorous channel activity, seen in parcels 3, 4, 5 and 7, to steady overbank alluviation early in the first millennium AD (parcel 6), may be more apparent than real. Firstly, the parcel 6 phase of alluviation may not have been especially prolonged, as we have seen, and, secondly, the behaviour of the associated channel(s) may not have altered. The change from non-alluviating flooding, however, seems clear and would relate more to an increase in sediment supply than simply to hydrological factors. One obvious cause would be the large-scale conversion of a pasture-dominated valley landscape to arable; this question awaits evidence from appropriate sites in the middle Thames.

Alluvial parcel 8: second millennium AD; Medieval to Post-Medieval channel fills

This parcel covers a series of channel deposits investigated during rescue work in 1980, in Areas 11 and 12. Two sondages were dug by the contractors for the purpose of constructing oil traps; opportunity was given for inspection and recording of the sections. They lay to the south of the prehistoric settlement within the belt of land which remained marshy until at least the nineteenth century AD, a final vestige of the once major channel which has been dubbed the 'southern loop'. The fact that the county boundary followed this channel has been used to argue that it was a dominant channel, and perhaps the only channel, in early Medieval times when boundaries were being fixed (see further discussion in Chapter 11).

There is no obvious linkage between the layers recorded in the two areas and they are thus given different sub-parcels. The Area 12 sequence, parcel 8A, was overall more variegated and lensed than that in Area 11. The sediments were dipping to the south-east, i.e., in line with the drawn section (Fig. 2.30), and thus towards the county boundary whose position could represent either the midpoint of the early Medieval channel, or more likely the latest position of the open water before final silting. No datable material was retrieved and it is impossible to confirm the relationship to either the parcel 5 sequence, which it probably post-dates, or the 8B/C sequence.

The Area 11 sequence is broken by one interface which seems to cut through contexts 11.006 and 007. On this basis it is split into parcels 8B, pre-scarp, and 8C, post-scarp. Although only recorded in this one section, observations around the walls of Area 11 allow the scarp to be interpolated as running diagonally across the trench, between E–W and ENE–WSW, facing north. This feature, whether a late channel incision or an artificial cut, for example to enhance a small creek for mooring craft, is evidently aligned along the course of the southern loop. The deposits are otherwise essentially horizontally bedded, but 11.003 was seen to thicken from about 0.2 m in the drawn section to 0.5 m further along the north-west wall of the trench. In the latter spot 11.002 was intact below a turfline and about 0.3 m thick.

The sequence drawn in the south-west wall of Area 11 (11.009, 007, 005, 004, 003, 002) was later seen to continue for at least 13 m to the WSW when the contractor extended operations. Similarly, a bundle of apparently laid brushwood partially excavated (11.016; Fig. 2.31) from on top of 11.005 in the western corner of Area 11 was found to have extended westwards. These lay more or less horizontally at 12.92 m OD and it seems likely that they were deliberately laid at a stage when the channel was marshy but not normally flowing. They could have served to consolidate a crossing point to *Tyngeyt*, the island or near-island on which prehistoric deposits have survived. In fact on Lindley and Crosley's map of 1793, a small access lane is shown approaching *Tyngeyt* at just this point on the southern side (Fig. 11.15; Table 11.1).

The earlier parcel, 8B, included a deposit of gravel with shelly silt (11.007 and 006) above a clay layer. The gravel bar peaked at 12.81 m OD, which is a little below the current and probably the long-term water table. Its character argues against human dumping and it can probably be regarded as an accumulation while stream-power was still relatively strong in this channel. Within the deposit were bones and four tile fragments dated to the Post-Medieval period, the sixteenth to eighteenth centuries AD (Chapter 3), indicating that the channel remained active into Post-Medieval times.

Layer 11.005, within parcel 8C, also yielded tile and brick pieces, one of which is identified as a Late Medieval floor tile, and must, therefore, have been old at the time of deposition. In addition there was a copper alloy band, not closely datable.

As well as the brushwood already described, there were other structural components in the form of wooden stakes. Observed briefly in situ in the surface of the gravel (11.006) was a cluster of small and medium diameter stakes in the extension, some 9 m WSW of Area 11. Without investigation, it is impossible to say whether these were driven through the gravel or, perhaps more likely, had been engulfed by it, having been driven into lower deposits. There is some unstratified evidence that suggests the presence of earlier structural elements in Area 11; having been alerted that timbers had been excavated during the original excavation of Area 11 by the contractors, two piles and a pile tip were recovered from the spoil. Subsequently, one was radiocarbon dated to AD 685–965 (BM-2551). There is also an unstratified sherd of Saxo-Norman pottery, *c.* AD twelfth century. One of the workmen said that he had seen two rows of piles set in a 'V' formation, suggestive of a fish trap. It is most likely that the recovered piles belonged to this structure, but how this reconciles with the piles immediately to the west is not clear. If these were of the same age, it would appear that they did not become engulfed in the gravel until centuries after installation, or that the gravel deposit was a gradual or multi-stage accumulation with only the upper layers dating to the sixteenth century AD or later.

Combined, the gravel deposits, the county boundary and the possible fish trap suggest that the southern loop carried significant flow for much of the Medieval period, perhaps extending into Post-Medieval times. There was also a deep enough channel to bring craft to a long-standing wharf in this reach (see Chapter 11).

Alluvial parcels 9 and 11: late second millennium AD; Post-Medieval to modern channel fill and make-up

With the excavation of Area 14 in the first season of research excavations, it quickly became apparent that the Late Bronze Age occupied surface and all the associated dark earth deposits were truncated along a line inset from the modern Thames bank and slightly askew to it. This feature (14.018 *et al.*) turned out to run in a straight line in both directions and throughout the riverside trenches (Fig. 10.6). Some 10 m behind and parallel with it was a ditch (14.002 *et al.*), also cutting the Late Bronze Age surface, which seemed to be functionally related. These two late features had a major effect on deciding how to expand the excavated area as the research campaign developed.

The cut bank (14.018 *et al.*), chopped through to a depth of at least 0.65 m, was clearly of human artifice; the lower part is very steep, but above it grades into a more gentle slope, presumably due to erosion prior to alluvial covering. From its foot the cut extended as a surface gently dipping towards the river (from *c.* 13.63–13.22 m OD), thereby truncating in planar fashion the upper deposits of the underlying parcel 4B (Figs 2.18–2.19). Because of the nature of parcel 4B this created, in effect, a gravel foreshore presumably extending all the way to the contemporary river. It was clearly frequently under water since the silt layers 14.014, 012 *et al.* accumulated rapidly, before significant erosion or collapse of the lower bank.

These silt layers yielded a number of finds (Table 3.8), most diagnostic of which were clay pipe fragments and a sherd of Border ware, all datable to the seventeenth to eighteenth centuries AD. The stem and foot of a clay pipe bowl in a low context 14.012, however, is seventeenth century, giving the most specific *terminus post quem* for the early silting.

The fill of the parallel ditch also contained material dating to the seventeenth to eighteenth centuries AD, as well as one later piece of mid-nineteenth century from apparent primary fill (18.702.03). This could suggest that the ditch was maintained open for some time, or that it was dug a little later, but at a time when the line of the original cut bank was still discernible, albeit in a very weathered form. In fact the ditch and the 'corridor' it defines along the river bank can be identified on the Common Lands map of 1800 (and again on the Tithe map of 1840). By careful comparison with later maps it is seen to be on the skewed alignment relative to the modern bank, while the corridor even corresponds in width, being shown as about 10 m wide on the map. On balance it seems probable that ditch and cut bank were dug as part of a single plan to define a broad towpath at a time when river traffic was approaching its zenith in the later eighteenth century AD.

After silt layer 14.010, the character of the deposits change quite dramatically to varied layers dominated by gravels and sands (Fig. 2.17 – toned layers). While one cannot exclude the possibility that these are natural deposits stranded on the foreshore by high water, their overall character is suggestive of deliberate dumping and their chronology, within the last two hundred years, offers a source. The lock and weir complex (Bell Weir) lies 125 m upstream from the riverside zone and was constructed between 1817 and 1819. Its siting exploited a natural gravel island lying closer to the southern bank and shown in a detailed survey of 1811 by Z. Allnutt of Henley (Table 11.1; Pl. 11). (This survey was in fact commissioned to advocate a more ambitious scheme which would have carved a canal through the middle of *Tyngeyt* and continued upstream for 60 chains to the bend at Milson's Point (Ankerwyke bend)). The construction of the lock would inevitably have led to deep excavations and the resulting spoil, probably rich in gravel, seems to have been dumped nearby. Evidently, in April 1818 the lock had to be altered, having not met the specifications required: '... surveyor suggested the towing path be repaired, then to deepen and ballast new cuts. Contractors agreed to co-operate, but at the same time did not feel themselves bound by their contract to remove all the gravel in the cut above and below the pound lock accumulated by the late flooding' (quoted in Chaplin 1982, 181). This sloppy attitude would likely have led to spoil being dumped as locally as possible.

The river bank alongside the riverside zone had followed a gentle concave line, as illustrated in both the 1800 and 1811 maps. The dumping of much material on top of the early parcel 9 silt, however, had the result of encroaching this part of the bank into the river and probably creating the essentially straight bank seen in more recent times, and on maps since the Tithe map of 1840. Indeed, this map shows clearly not only the new 'towing path' along the new straight bank, but also the former plot boundary constituting the ditch 14.002, now seen at an acute angle to the bank and enclosing a narrow triangular piece of land between. It is probable that this made-up land also accounts for part at least of the sequence recorded in Area 15 (parcel 11; Figs 2.2 and 2.20), excavated into the modern Thames foreshore. The uppermost, silt-dominated deposits in that area seem likely to derive from renewed flood deposition on the flanking shore after the interpreted dumping.

Alluvial parcel 10: late second millennium AD; Post-Medieval to modern channel fill

This parcel refers to the upper fill of the western arm of the southern loop, dated to the Post-Medieval period. It was only examined in Area 13 North-west, where the presumed eastern part of the fill laps up against a palaeobank cut into the ancient spine with its prehistoric occupation surfaces. In broad terms deposition is likely to be contemporary with parcel 9, but the finer chronology and the depositional environment need not be the same in these different branches.

The lowest fills recorded in the sections are organic up to about 12.75 m OD, but sump excavation revealed that they overlay a dense gravel, presumably a fossilised shoal or bar after the cessation of high-energy flow through the southern loop. It would seem to be too early to be gravel dumped from the Bell Weir lock excavations of the nineteenth century, since the organic deposits yielded a clay pipe bowl datable to the turn of the seventeenth/eighteenth centuries (although this can only strictly serve as a *terminus post quem*). The thicker body of silt lying above the water table is dated by more clay pipe finds to the mid- to late nineteenth century (plus one residual from the seventeenth or eighteenth century), i.e., certainly later than the initial lock of 1819, a relationship interpreted as significant (discussed below).

The top of the excavated stratigraphic sequence is a ditch running obliquely through the trench (Fig. 2.9). This is identifiable with a ditch in exactly this position on maps from 1800 onwards (Fig. 11.15). However, the cut as excavated clearly post-dates the main nineteenth century fill and must, therefore, be a late recutting. Although no earlier cut lines were observed, it seems possible that the slightly different character of the underlying layers as they approach the palaeobank (thus directly beneath the discerned ditch cut) indicates the fills of earlier phases of this ditch (13.420/450, 13.506/464).

CHAPTER 11

River and bank: the changing riverscape

Topography of the region

The finer points of the valley floor topography around the Egham–Staines reach are conveniently brought out by the 50-ft contour, as plotted by the Ordnance Survey in the pre-metric era (Fig. 11.1). It equates to 15.24 m OD, significantly above the post-Boreal river level of around 13.00 m OD and also above the highest point of the prehistoric occupied island at *c.* 14.60 m OD. While the ground above 15.24 m OD was never entirely risk-free of floods, it would only have been inundated by the more severe ones.

In general the 50-ft contour can serve to indicate which parts of the valley floor were the highs in the undulating gravel surface left at the end of the Late Devensian period. In points of fine detail, of course, there may be differences. There will have been some accretion by alluviation: for example, at a site close to the church in Egham High Street (TQ 01357135; excavation by David Barker) 0.25 m of alluvium was found to cover a Late Bronze Age pit dug from about 14.90 m OD; the pit cut into a further 0.45 m of alluvium, an orange clayey silt underlain by yellow sandy silt. This whole sequence was truncated, then overlain by Post-Medieval layers. There will also have been cases where the complete infilling of former channels caused the linking of separate high banks to form larger blocks. This has been demonstrated by excavations, for example at Staines (Crouch and Shanks 1984). Another form of 'accretion' is the human building up of land, but this would probably be restricted to the towns and occasional roads and railways prior to the OS contour plot (1870). The opposite process, the loss of formerly high ground, will mainly have been due to the lateral incision of shifting channels, occasionally perhaps by the creation of new channels carving through higher ridges during floods. Early gravel quarrying obviously needs to be taken into account wherever possible.

Overall, however, it is unlikely that these various processes will have dramatically altered the general distribution of higher ground on the valley floor; the finer changes in the topography are of particular interest to environmental reconstruction, as will be illustrated in this chapter.

The topographic preconditions for the channel configuration were set at the end of the last glaciation when massive melt-water floods firstly scoured a new deeper channel and, secondly, filled it with gravel and sand. The gravel/sand in the main Thames valley is the 'floodplain' gravel, Gibbard's 'Shepperton' gravel (1985). Gibbard has also demonstrated that the tributary valleys in the Thames basin have similar gravel fills contiguous with the main fill. This includes the broad swathe of the Colne valley, although there is evidently a steep drop in the bedrock (London clay) at the point it met the Thames floodplain at its maximum width. This meeting point is some 3 km north of the present-day Thames through Runnymede Bridge and is obscured by the mass of overlying gravel and later alluvium. In fact, it would appear that the northern part of the Late Devensian floodplain was never occupied by the Thames during the post-glacial period. Gibbard notes (1985, 71) that the surface of the gravel fill falls by as much as 2.4 m from north to south; it is unlikely that this is simply due to subsequent removal by erosion, but instead the result of the mass of material being swept down the Colne corridor from the Chilterns in the Late Devensian. The River Thames would from a very early stage have sought out the lower ground, which in this reach was the southern part of the Late Devensian scour channel. Gibbard observes that the alluvium tends to be a thinner cover to the north of the present river than to the south (1985, 89). The land form left by the Late Devensian deposition was, however, a little more complex; as may be picked out from the modern topography, it seems that ridges of gravel were left transverse or oblique to the main valley. These presumably relate somehow to complex vicissitudes of flow and deposition in the confluence zone.

The main ridge comprises three blocks of land above 15.24 m OD strung out between Thorpe Lea in the west and Hythe in the east (Fig. 11.1). This forms something of a barrier linking the 'high' ground south of Egham to that at Staines. It encloses an area about 1 km² between Pooley Green and Hythe, before interruption by another ridge, a spur running east from Egham town centre, through Glanty and part way along the southern bank of the modern Thames towards Staines. This ridge is predominantly gravel, as seen by the author on a site stripped for redevelopment at Glanty (TQ 019716), on the factory site immediately west of the Victoria public house (TQ 016717), as well as at Petters Sports Field (TQ 016715; O'Connell 1986). The result of

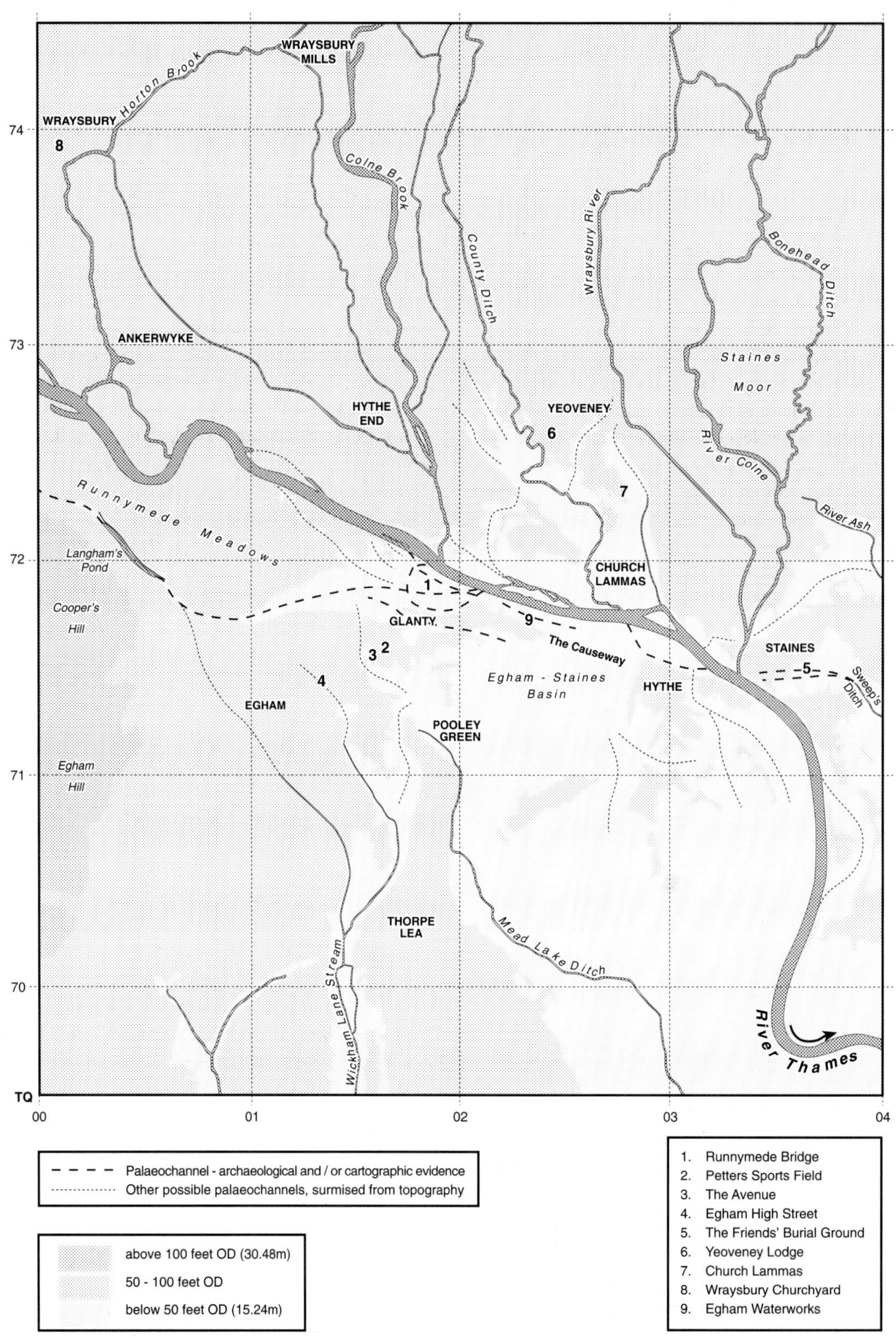

11.1 *The topography of the Egham–Staines region and the location of certain and probable palaeochannels.*

this spur is effectively to contain the valley bottom (land beneath 15.00 m OD) to an elongated strip running west from Runnymede Bridge into the Runnymede meadows, part way towards Langham's Pond and east towards Staines, where it opens into the main basin (Fig. 11.1). From this zone, low ground also runs northwards for almost a kilometre, being a valley occupied by two of the Colne streams – the County Ditch and probably the pre-canalisation course of the Wraysbury River.

Whether the Runnymede arm was ever in effect a separate basin would depend on the extent of modification to the topography since the Late Devensian; this is not easily determined on current evidence. However, it would seem that either this sub-basin or the larger Egham–Staines basin suffered impeded drainage early in the Holocene period, to account for the high water level of the Boreal (Chapter 10). Although poor drainage was a widespread problem in the early Holocene, the sudden cessation of ponding when marked incision took place implies a local topographic cause with the sudden clearing of blockages. The implication is that the gravel ridges were largely continuous; although some outlet channels must have flowed through, drainage was at first impeded. Further exposures of deposits dated to the early Boreal need to be identified in the locality. The palaeochannel fill at Mixnam's, intensively studied by Preece and Robinson (1982), is too far south, as is the recorded sequence south of Hythe (Gibbard 1985, 87).

Palaeochannels

There is evidence that this complex floodplain topography is laced with many palaeochannels which would have been active at one time or another during the Holocene. In addition to those established through excavations, some are shown by continuing streams or cut-off lakes (Langham's Pond) and yet others may be suggested on the basis of topography. They are shown in Figure 11.1 and selective observations are made below. Obviously much fieldwork and research would be required before all of these could be confirmed and worked into a wider picture of channel evolution for the area. Nevertheless, such judgements have had to be made in order to understand the very local, but critical, evolution of the configuration at Runnymede Bridge. It should go without saying that the reconstruction of channel alignments and periods of flow becomes more uncertain as one moves away from the excavated site.

Langham's channel. The ribbon pond running along the southern edge of Runnymede meadows, beneath the foot of Cooper's Hill, is obviously a palaeochannel relict. It may once have been a major branch of the Thames. The 50-ft contour shows the broad trend of its eastward passage towards Runnymede Bridge. This is confirmed by a borehole transect (nine boreholes, 88 m long) conducted in 1987 just west of the site (between TQ 01542/71800 and 01550/71888). This showed gravel to be present at 12.30 m OD in the middle,

rising to almost 13.00 m OD under the foot of the A30 embankment; it is not known whether this was Late Devensian gravel in situ, or the bed-load of the river. Overlying in sequence were grey-brown loamy silts with variable shell; orange to yellow-grey clay-loams, often with iron mottles; then light to mid-brown clay-loam. This last deposit dipped into a depression bottoming at 13.55 m OD in the middle of the transect, presumably the ultimate fill of the final channel position. A different sequence observed towards the northern end of the transect suggests the presence of a palaeobank around 40 m from the A308 Windsor Road. The results also hint at a migration northwards with time, though not necessarily by more than tens of metres. Charcoal was noted in certain samples from the lower channel fill.

A second transect, six boreholes, 100 m long, was made just beyond the south-east end of Langham's Pond (TQ 00695/71732 and 00730/71828). This may have caught the palaeochannel at a point of bifurcation, for there is a marshy plot running southwards shown on modern maps. If this is the last vestige of a channel, it would have to have once flowed somewhere through the west end of Egham; it is just conceivable it emerged again where a drainage ditch runs alongside Manorcroft Road (TQ 011709).

Starting at the northern end of the transect, the upper fill of a channel up to 50 m across could be identified; a mid-brown silt, underlain by a light brown one both dipped into a depression bottoming at 13.80 m OD. Beneath the middle was a dark brown fibrous silt with some shell and occasional gravel pebbles, whereas under the flanks there were buff to light grey-brown silts with variable shell content, not necessarily the same deposit on the two sides. Gravel was encountered below, at around 13.15 m OD, although at the northern end it was small gravel in a shelly silt matrix for 0.80 m depth. Towards the southern end of the transect the uppermost, brown silt deposits dipped away again from 14.45 to 14.10 m OD. These overlay an orange, very sandy loam; despite the sandiness, little shell was obvious. Gravelly deposits at the base here (13.25 m OD) rose to 14.00 m OD, apparently forming a shoal between the two possible channels.

Intermediate meadows channel. This is only deduced from the 50-ft contour plot. A string of low spots runs obliquely between the modern Thames and the A308 just upstream from the site. These seem to mark the top of a palaeochannel coming off the Thames below Ankerwyke bend and converging with the Langham's channel more or less where it meets the southern loop.

Southern loop. A pronounced meander looping round the west, south and south-east sides of the prehistoric settlement site. In its final form it is well documented by maps, topography, excavation and the position of the county boundary. However, it can be shown to have existed from at least the first millennium BC, the evidence discussed in detail below.

Wickham Lane stream and Mead Lake Ditch. Today, two small streams flow south from Pooley Green, the contours suggesting an interconnection at one time at this upstream end. They also indicate that Mead Lake Ditch reached back to the Avenue at TQ 01557160, immediately to the west of Petters Sports Field. A site excavated on the eastern edge by Steve Dyer seems to have identified channel deposits (pers. comm.). A re-entrant in the 50-ft contour just beyond the Avenue almost certainly shows that this channel interlinked with the Langham's channel. Thus at one time it had separated the Glanty ridge from the raised ground under Egham town. Whether a similar channel ran through this block to feed the Wickham Lane stream is not known, but the Egham High Street site mentioned above could lie on its path and did yield alluvium down to the level reached – 14.30 m OD. The Wickham Lane stream may once also have been fed by the channel hypothesised above as coming off Langham's channel on the north-west edge of Egham.

Colne Brook mouth. During the nineteenth century there was a ditch running parallel to the modern Thames between the lowest wiggle on the Colne Brook and Holm Island, where it runs into a spur creek. The contours would allow this once to have been an alternative mouth for the tributary. The reconstructions offered below work on the assumption that the Colne Brook was always present somewhere close to its current line, although there is only direct support for this from the second millennium BC onwards.

Glanty. Borehole 102 for the pre-M25 survey located a 3.8 m deep alluvial sequence which seems most likely to be the fill of a channel. It shows as such in Gibbard's plot (1985, 74, fig. 45) and lies a little to the south of the southern loop, presumably correlating with a linear east–west depression in the 50-ft contour.

Thames at Hythe. There used to be an island just upstream of the bridge at Staines opposite the extant Church Island; it is now landlocked on the southern, Hythe, shore. The southern channel was already merely a ditch by 1843, but it is shown in Stukeley's 1723 'Prospect of Staines' and, as was the case for the southern loop around *Tyngeyt*, the channel had earlier been sufficiently important to be used to define the county boundary.

Staines town/Sweep's Ditch. Two channels were identified in the excavations at the Friends' Burial Ground site, Staines (Crouch and Shanks 1984), which lies on the south-eastern edge of the high ground marked by the 50-ft contour. They ran more or less west–east and would presumably have come off the main Thames system and fed Sweep's Ditch, which rejoins the modern Thames at Penton Hook, 2 km downstream.

Historical settlement

The sitings of Egham and Staines town centres recognise the topography, each taking advantage of blocks of raised ground which stood close to the river system. In the case of Egham the core, as represented by the High Street, is some 500 m away from the southern loop at Runnymede Bridge and was linked with it from the beginning by the Avenue. This road also provided the driest access eastwards towards Staines, linking with the 'Causeway' running along the south side of the river. This long-standing route exploited the above-described tongue of land above 15.24 m extending halfway towards Staines, then ran up onto another raised terrace beneath the ancient settlement of Hythe, opposite Staines. The passage across lower ground was kept to the shortest possible distance, and this was presumably the part that needed building up the most.

The Causeway is renowned on account of the endless litigations in late Medieval times (thirteenth to fourteenth centuries) over who should bear responsibility for its upkeep (Turner 1926, 14 ff). The Abbots of Chertsey persistently refused to pay for repairs. All sorts of horror stories about deaths during flooding were propagated and, although there was doubtless a truth in the risks involved, one has the strong impression of a degree of embellishment to add persuasion to the litigation cases. The Causeway is said to have been built in the reign of Henry III (thirteenth century) by Thomas de Oxenford, a wool merchant (Turner 1926, 14), but it seems possible that in reality this was only the reconstruction of a pre-existing embankment, if a Roman road already existed on this alignment. The position of the road in this zone has long been uncertain, but now that its agger has been identified on Egham Hill, aligned directly on the centre of Egham town (Garner 1991), it becomes more than likely that the Roman engineers had also identified this as the best route. Directly between Staines and Egham lay the described basin of low-lying land 1 km across. To avoid it, the only 'island-hopping' alternative to the northern Causeway route was one skirting to the south through Frogs Island and Thorpe Lea. The latter involved crossing streams and associated floodplains and is not compatible with the alignment of the road beyond Egham Hill.

General features of the river system

The rich environmental records presented both in this volume and in Needham (1991) amply testify to the conditions of the water. Through much of the Holocene the river system would have carried well-oxygenated water, not too silt-laden and not languid. There is little doubt that the character of the channels changed substantially during this period in terms of both configuration and profiles, but this is not often easy to demonstrate because the full width of a channel only survives exceptionally when one branch of a multi-thread system dries up (Brown 1997, 251). No full channel widths have been revealed at Runnymede. An evolution from anastomosed systems in the early Holocene period to the effectively single-thread rivers of later times is now thought to be a common feature of lowland British rivers, but rarely has it been possible to demonstrate in detail the sequence of changes taking place within a limited part of a valley. The modifications to channels will be in response obviously to the volume of water carried, the seasonal variation and the number of active channels, but there are other important determining factors, notably the effect of containment due to the raising of alluvial banks and the variations in bank stability according to their particular material and vegetation cover.

Channel migration usually follows predictable patterns for given flow conditions and this is of great assistance in linking together the fragmentary observations that are an inevitable consequence of any valley that has undergone significant channel change. Nevertheless, one does have to bear in mind that bank erosion is sometimes more erratic, or that aggradation occasionally occurs on the unexpected, concave, bank (Brown 1997, 67). Sometimes there may be two or more possible explanations for observed bank migration, depending critically on what was happening in another channel, which might no longer exist for study. Overall, however, the strength of the sequence of channel changes proposed below lies in the long-time trajectory, because it may be supposed that anomalous patterns of erosion and deposition are unlikely to be long-lived. Furthermore, the fact that pieces of channels of many different ages have been recorded serves to constrain options in terms of their transformation through time.

The sequence of channel changes

Eighth to seventh millennia BC

It is suggested above that from the beginning of the Holocene period the channels of the Thames in the Egham–Staines reach were essentially confined to the southern part of the Late Devensian scour channel, i.e., close to the modern course. The high water table, in conjunction with the environmental evidence retrieved and the sediments accumulating at the time, suggests that up until the mid-Boreal the site suffered impeded drainage and was within a marsh and/or lake, although this would have had water flowing through it, perhaps preferentially in certain flow channels. Under this expanse of water, thick deposits of marl and probably other sediments formed (parcels 1C and 1J).

Lewis *et al.* (1992) argue that the occurrence of high water tables up the valleys is a direct product of the dramatic rise in the base level (i.e., at the Thames estuary) from –27 to –15 m OD during the late eighth millennium BC. However, local water tables do not all appear to fluctuate in synchrony and consequently their argument needs qualification. While there might have been some lessening of the overall river gradient, the rise in sea level would primarily have the effect of inundating low coastal zones and encroaching up river

mouths. More important for the formation of the ponds and marshes would have been the local topography and the resulting drainage potential. Early in the Holocene this would have been strongly conditioned by the land-form left at the end of the Late Devensian period, modified by any significant reworking of floodplain deposits. In our region the modern topography (Fig. 11.1) outlined above would seem to be significant in this respect. The land between Runnymede and Staines, to the east, could once have formed a basin almost entirely enclosed by ground above 15.24 m OD (the 50-ft contour); hence it is conceivable that quite large marshy areas were present in the locality. This environment would only have changed as the outflow channels were widened and deepened by erosion to allow freer drainage of the basin. This would account for the gaps now present in the encircling ring at Pooley Green, Frogs Island (Thorpe Lea) and Staines (where the present river flows).

It was only with the sudden fall in the local water table late in the Boreal that a channel system crystallised by down-cutting into the 1C/1J type deposits. There is evidence from the erosion profile of the marl that two channels would have run through, or alongside, the site (Fig. 2.2). The better documented, on the north side, had a southern bank running roughly west–east through Areas 13 and 22 (Fig. 10.1), and soon experienced aggradation with parcels 1E/1D during the Boreal/Atlantic transition. The more southerly channel can at present only be inferred from the fact that the marl has evidently been eroded away to the south-west of Areas 13/19. It is not impossible that other channels flowed further to the north and further to the south. As the 1D silts steadily built up, alder was establishing itself along the valley bottom, which might well have had some effect on the stability of the river banks as root systems began to bind it.

One effect of the drop in the river level initially to below 12.75 m OD would have been a corresponding incision in the lower reaches of the Colne tributaries in this zone. These smaller streams, individually with much less erosive power than the Thames, may to some degree have become entrenched in steeply cut channels and were consequently less mobile laterally. It is possible that signs of near-contemporary down-cutting may be seen in the sequence at Three Ways Wharf, Uxbridge, 13 km north of Runnymede Bridge. Peat growth during the Boreal period implies a high water table, but this ceased late in the period and thereafter no further deposition took place until much later in the Holocene (Lewis *et al.* 1992).

Sixth to mid-fourth millennia BC (Figs 11.2 and 11.3)

The channel reconstructions for the sixth and mid-fourth millennia need to be read together since the former is largely an extrapolation of the latter. By the sixth millennium BC (mid-Atlantic period), the southern bank of the northern channel was aligned WNW–ESE and abutted the silt bank

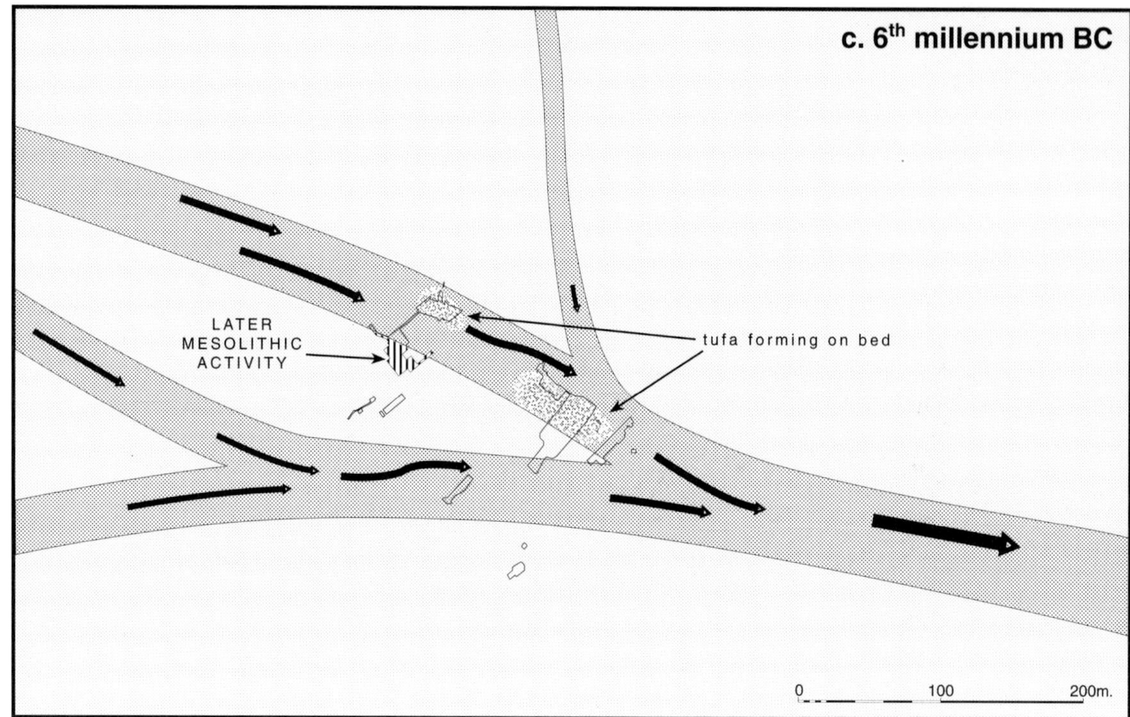

11.2   *The channel configuration and human activity in the sixth millennium BC.*

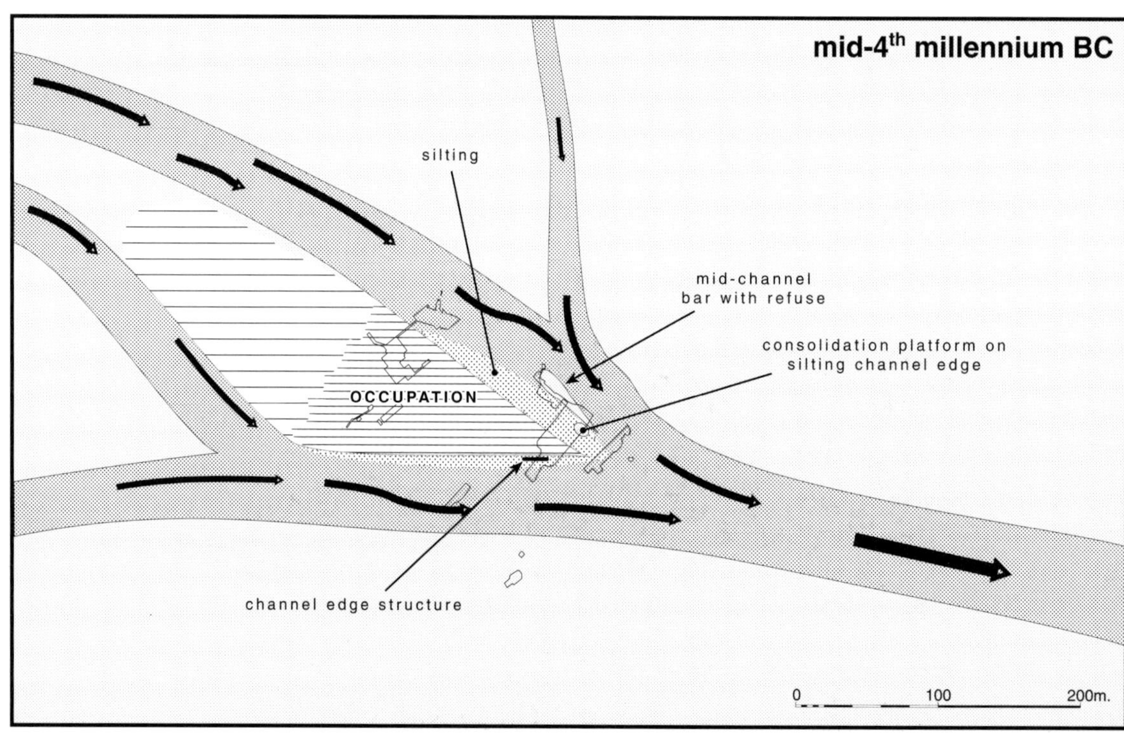

11.3   *The channel configuration and human activity in the mid-fourth millennium BC.*

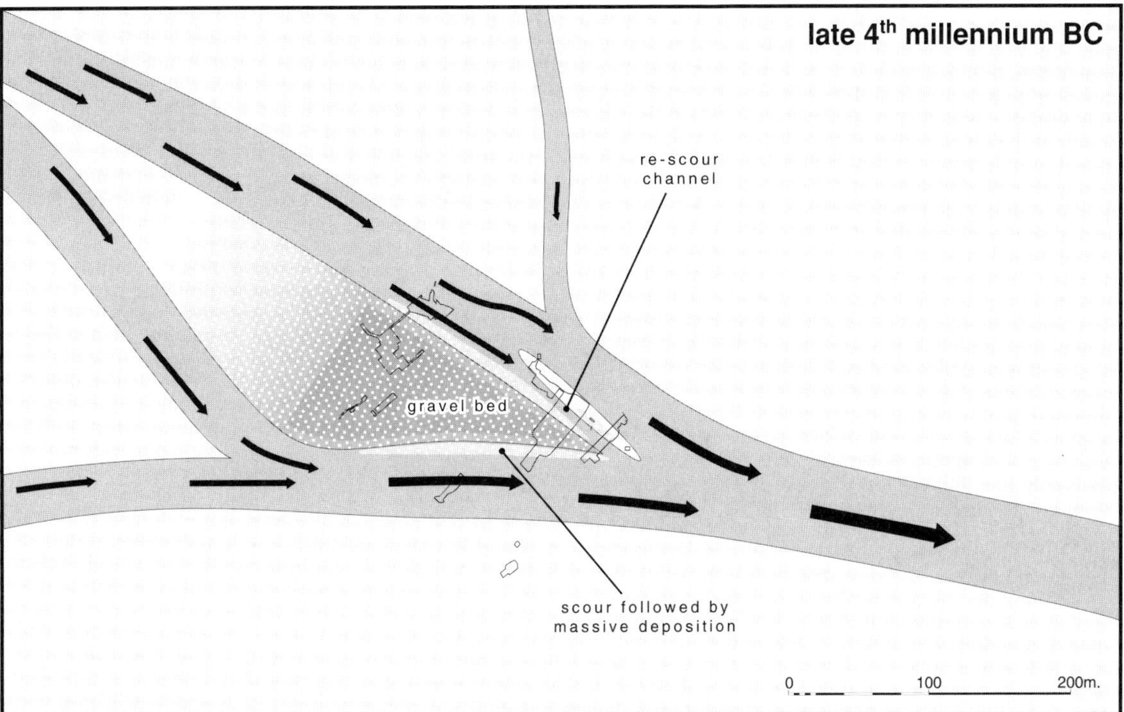

11.4   *The channel configuration and human activity in the late fourth millennium BC.*

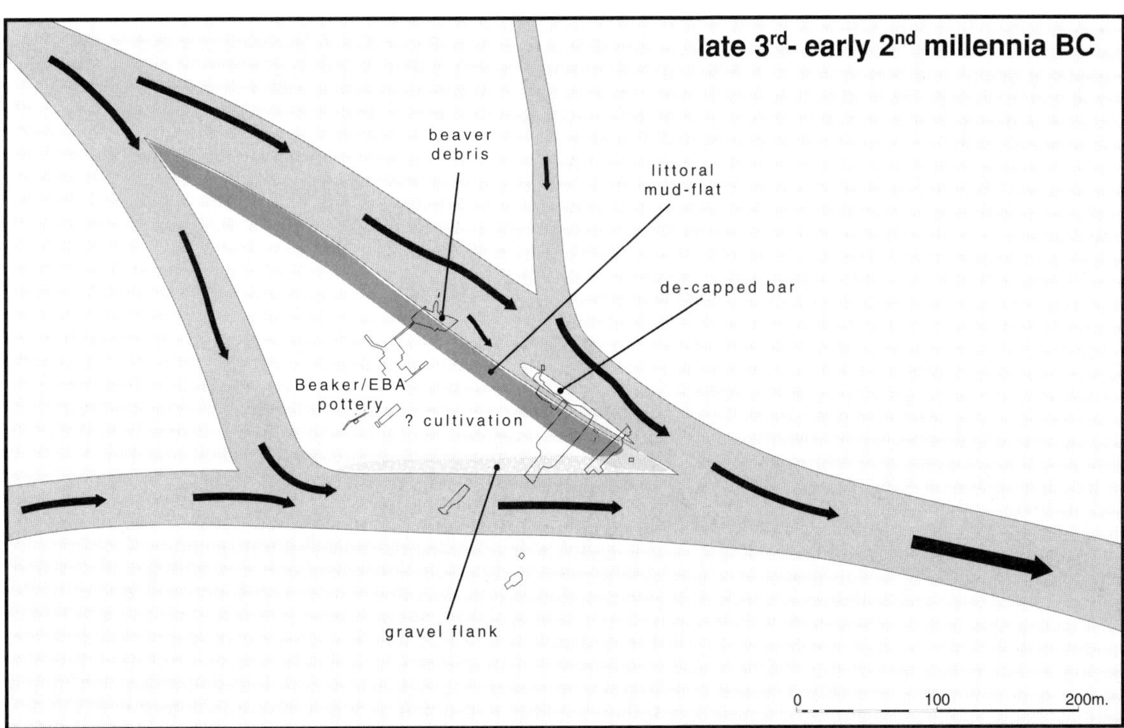

11.5   *The channel configuration and human activity in the late third to early second millennia BC.*

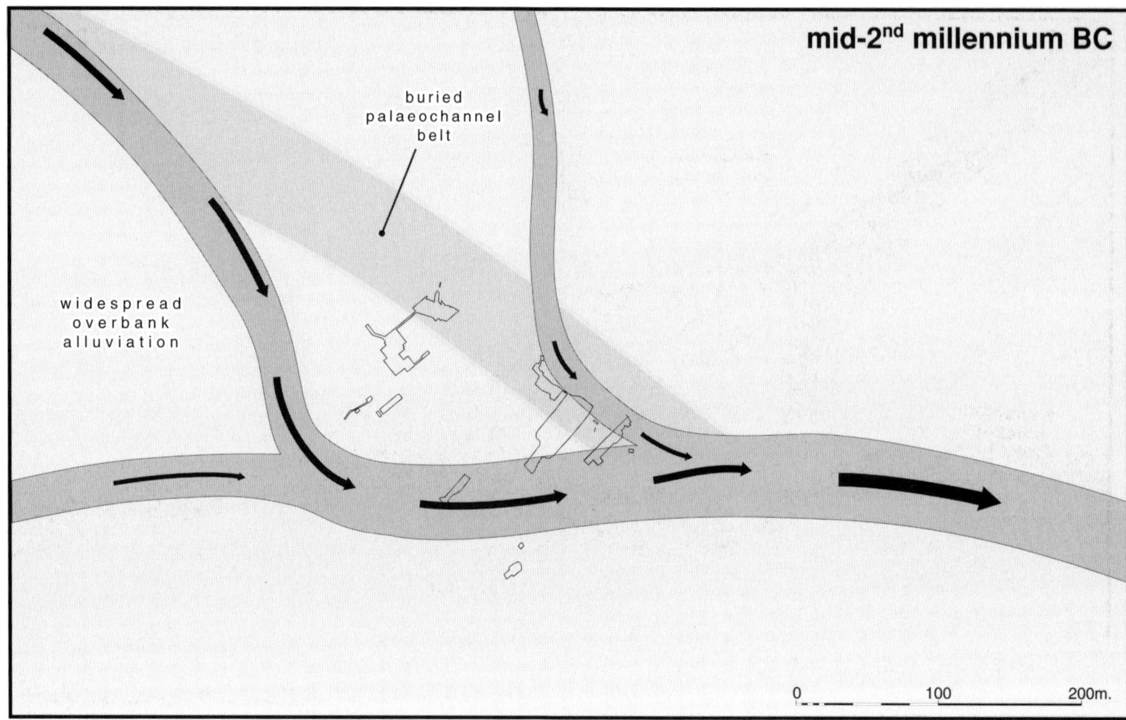

11.6   *The channel configuration and human activity in the mid-second millennium BC.*

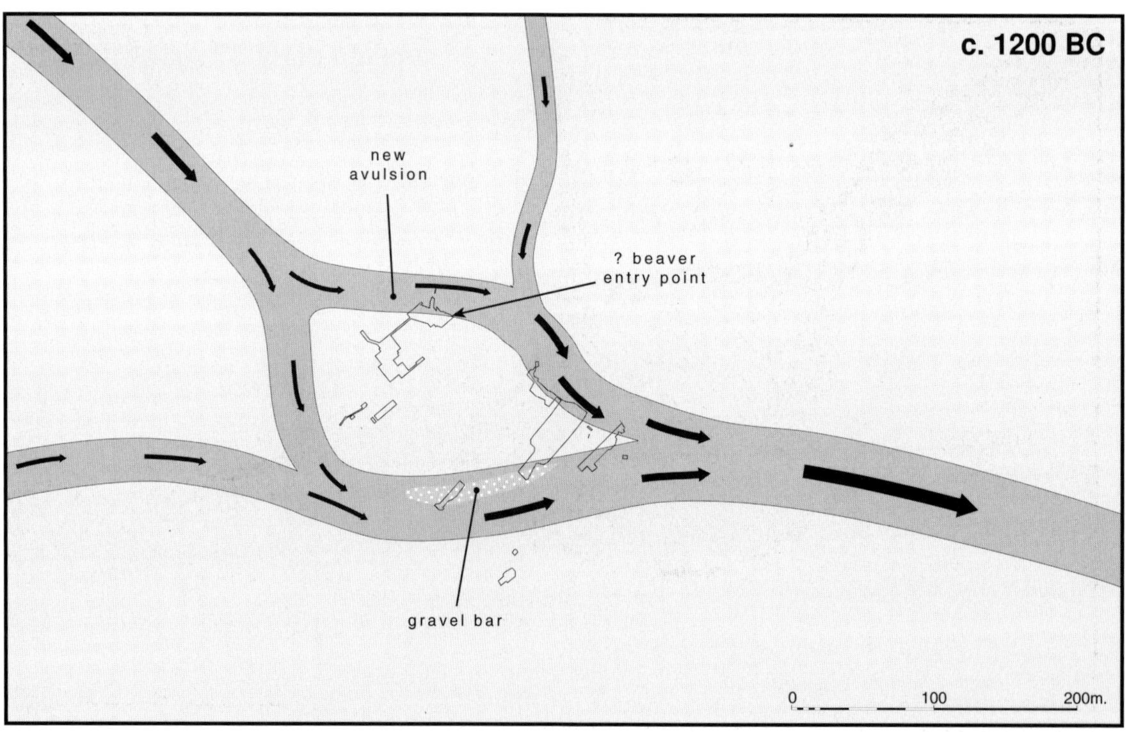

11.7   *The channel configuration and human activity c. 1200 BC.*

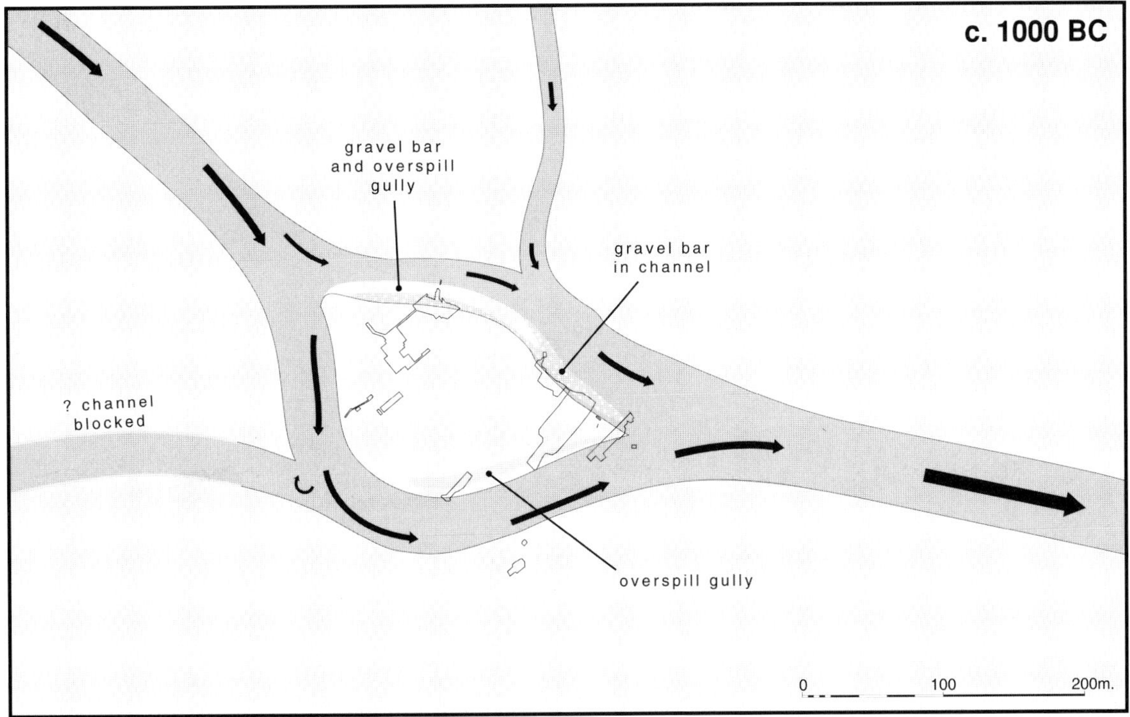

11.8   *The channel configuration and human activity* c. *1000 BC.*

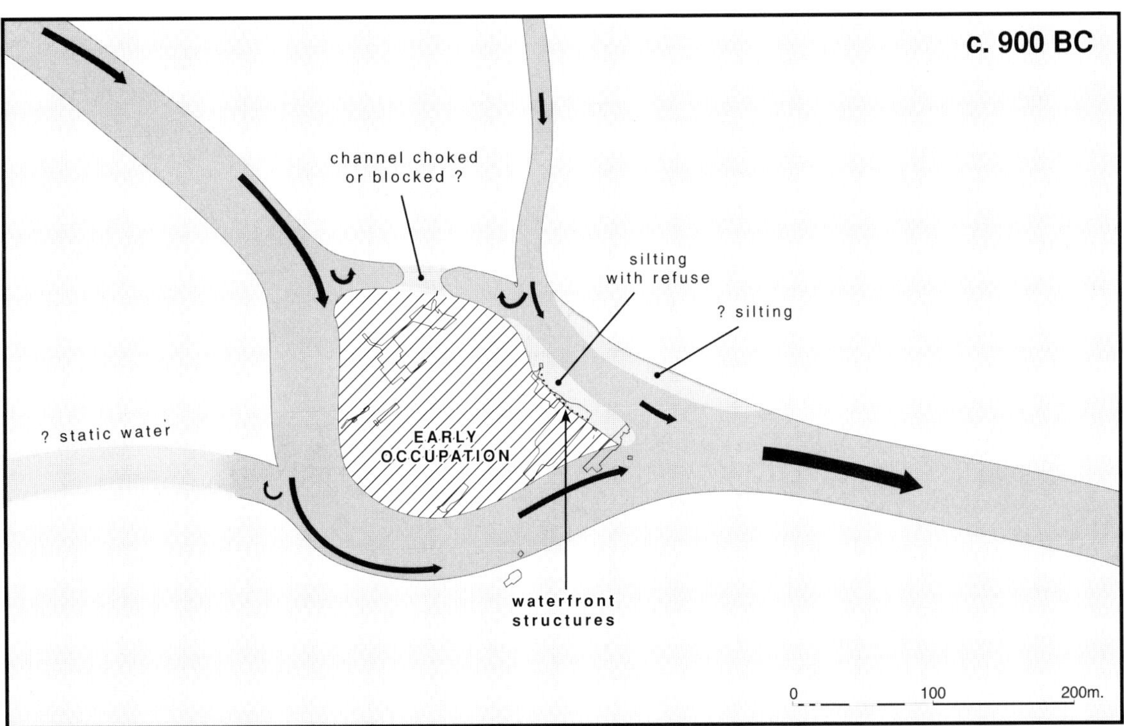

11.9   *The channel configuration and human activity* c. *900 BC.*

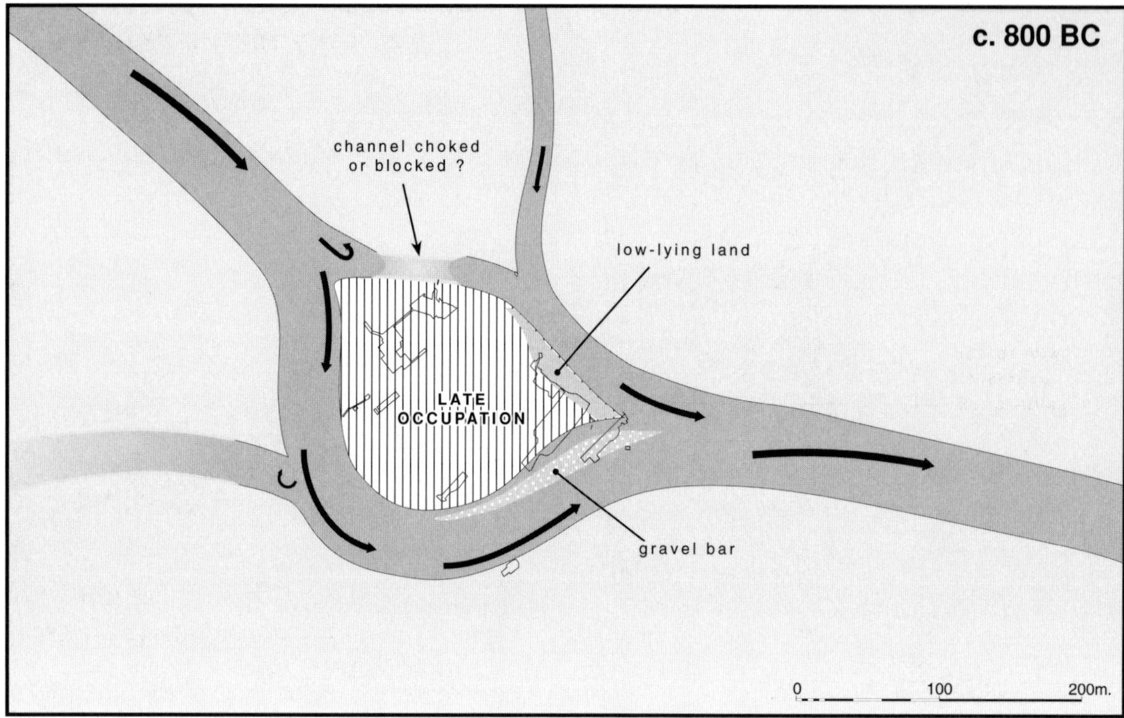

11.10 *The channel configuration and human activity* c. *800 BC.*

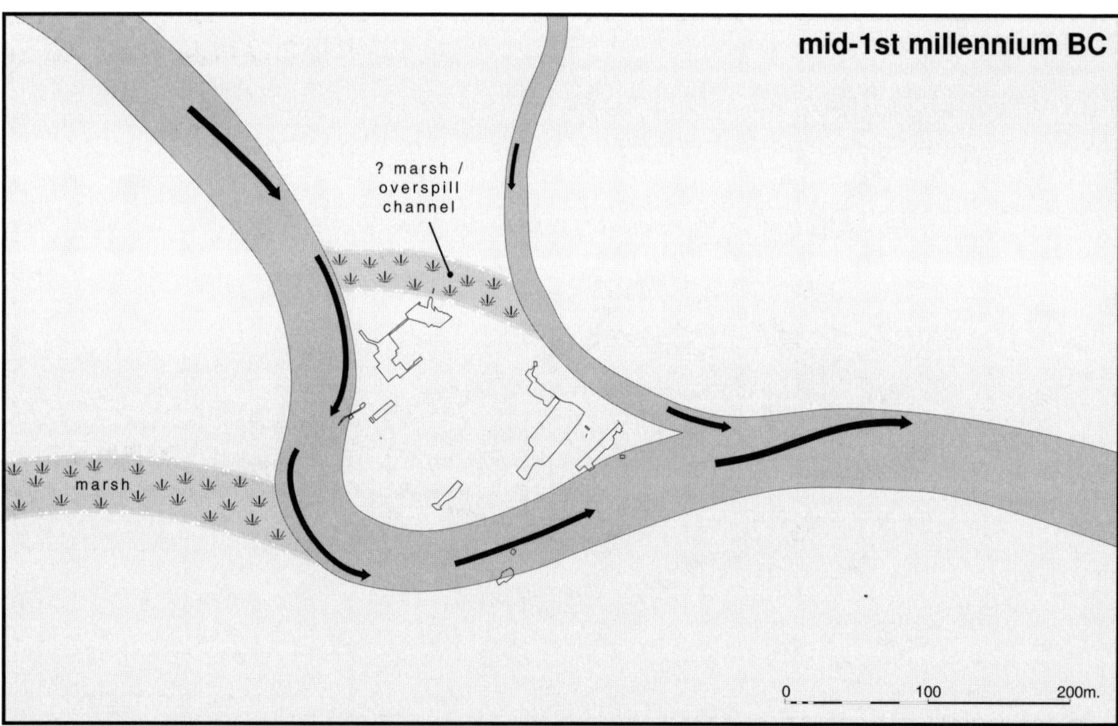

11.11 *The channel configuration and human activity in the mid-first millennium BC.*

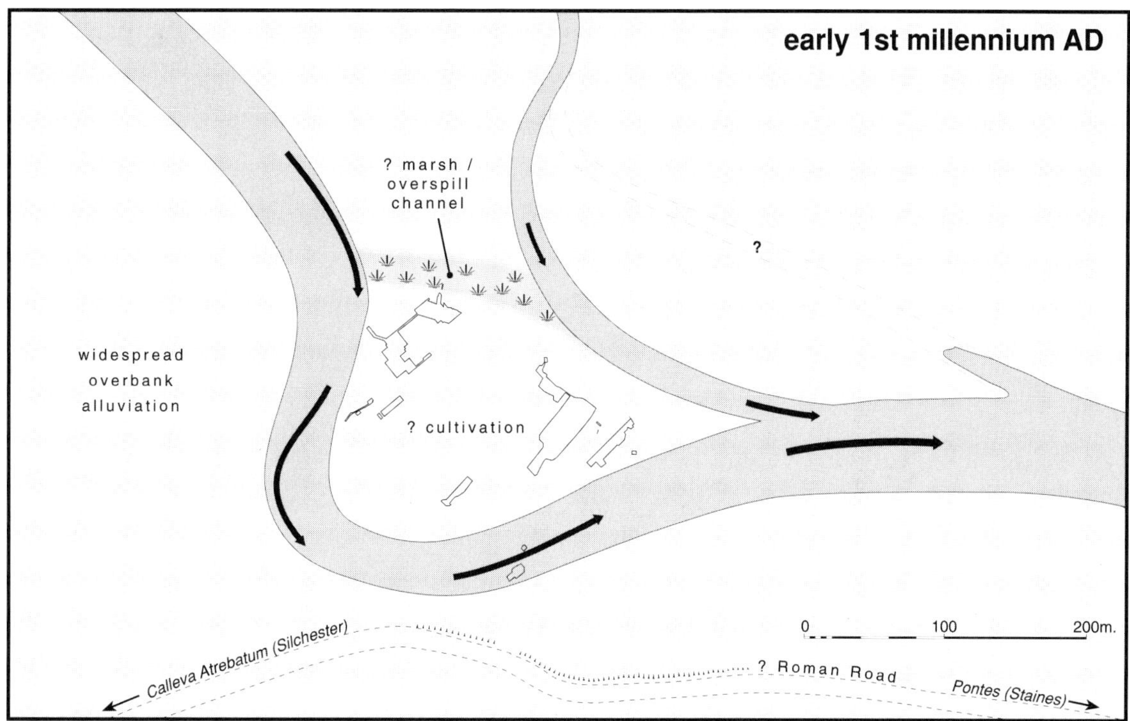

11.12   *The channel configuration and human activity in the early first millennium* AD.

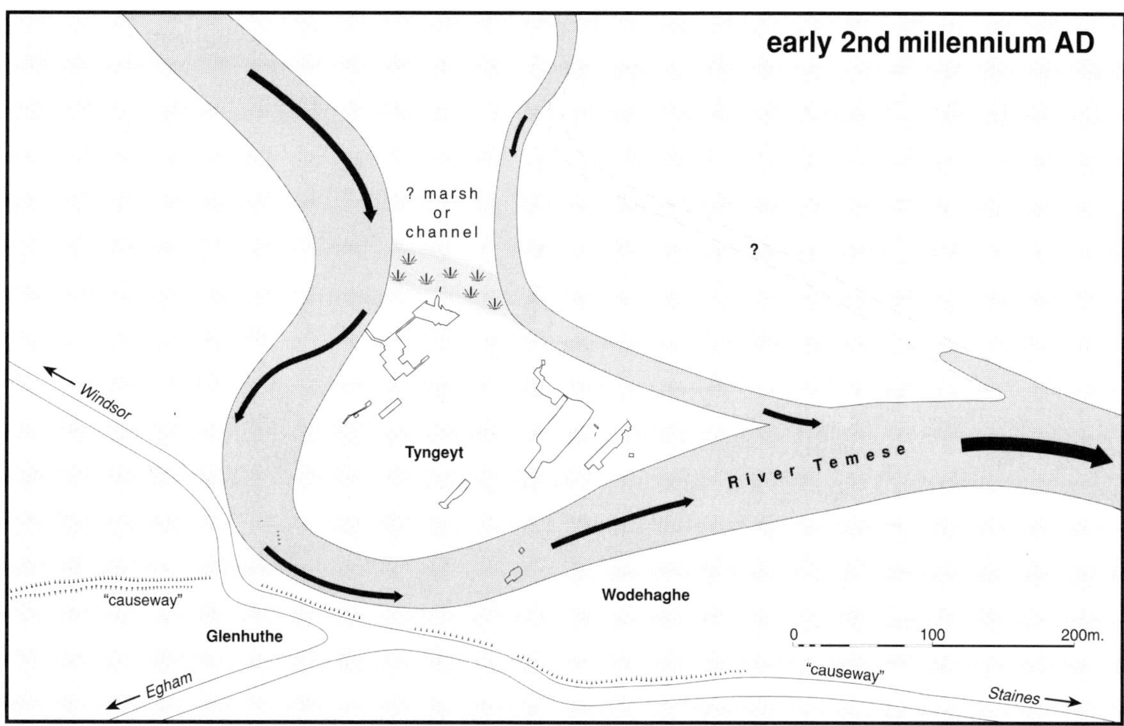

11.13   *The channel configuration and human activity in the early second millennium* AD.

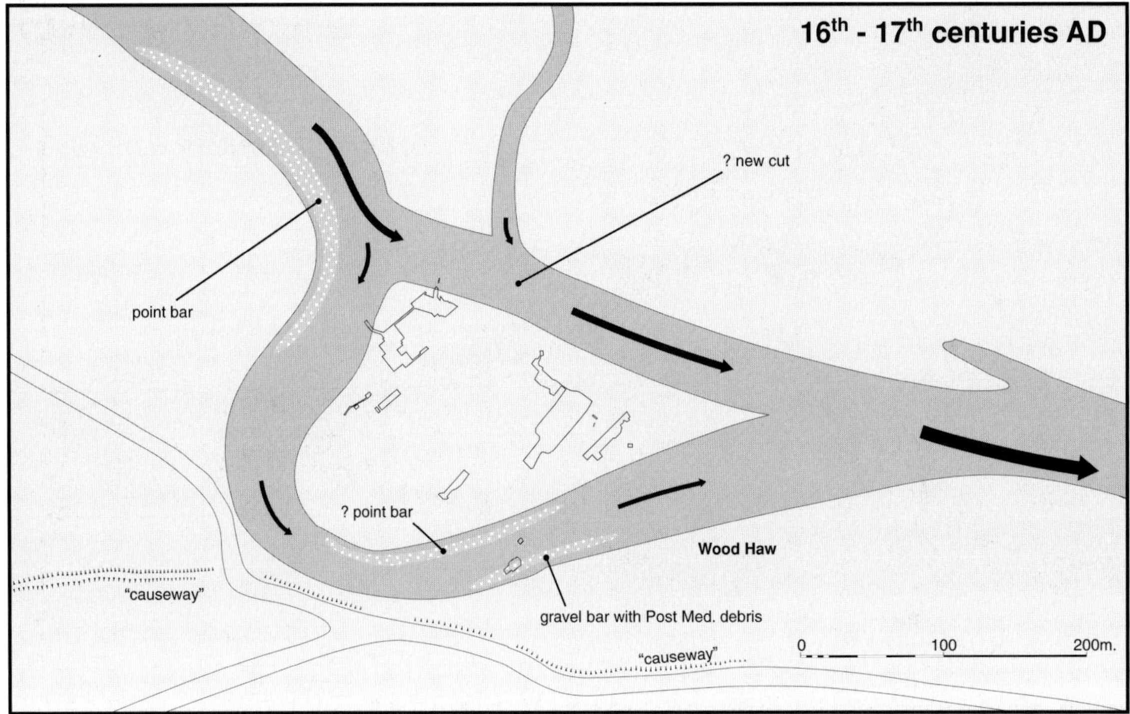

11.14   *The channel configuration and human activity in the sixteenth to seventeenth centuries AD.*

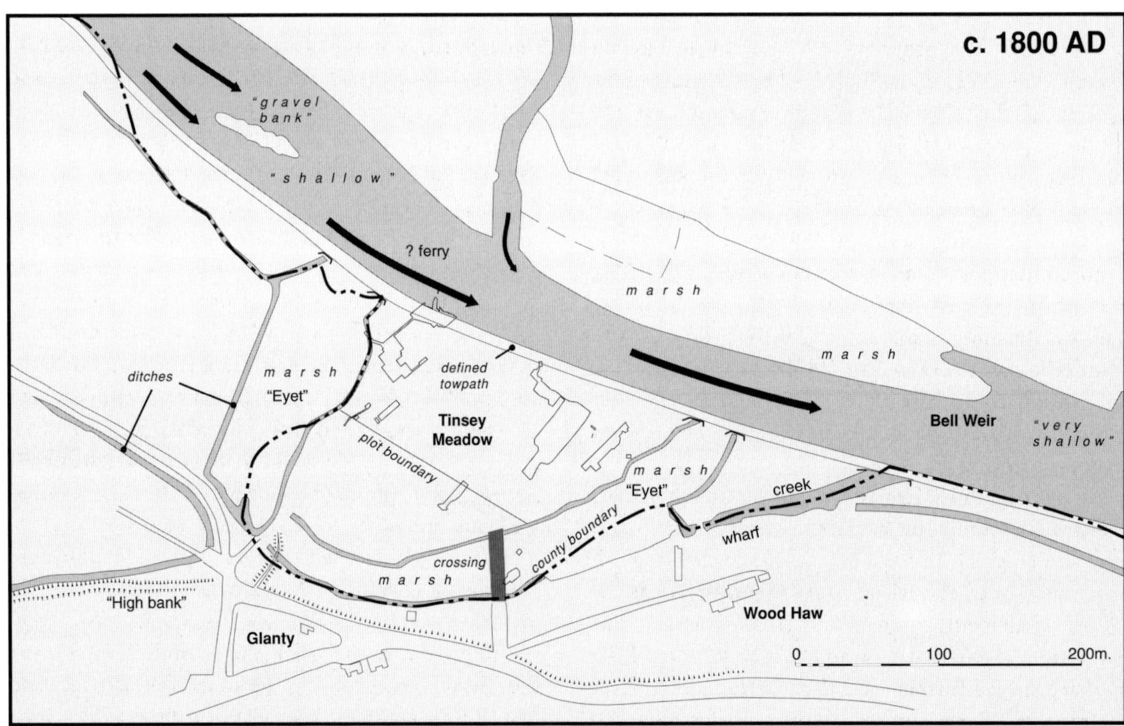

11.15   *The channel configuration and human activity c. AD 1800.*

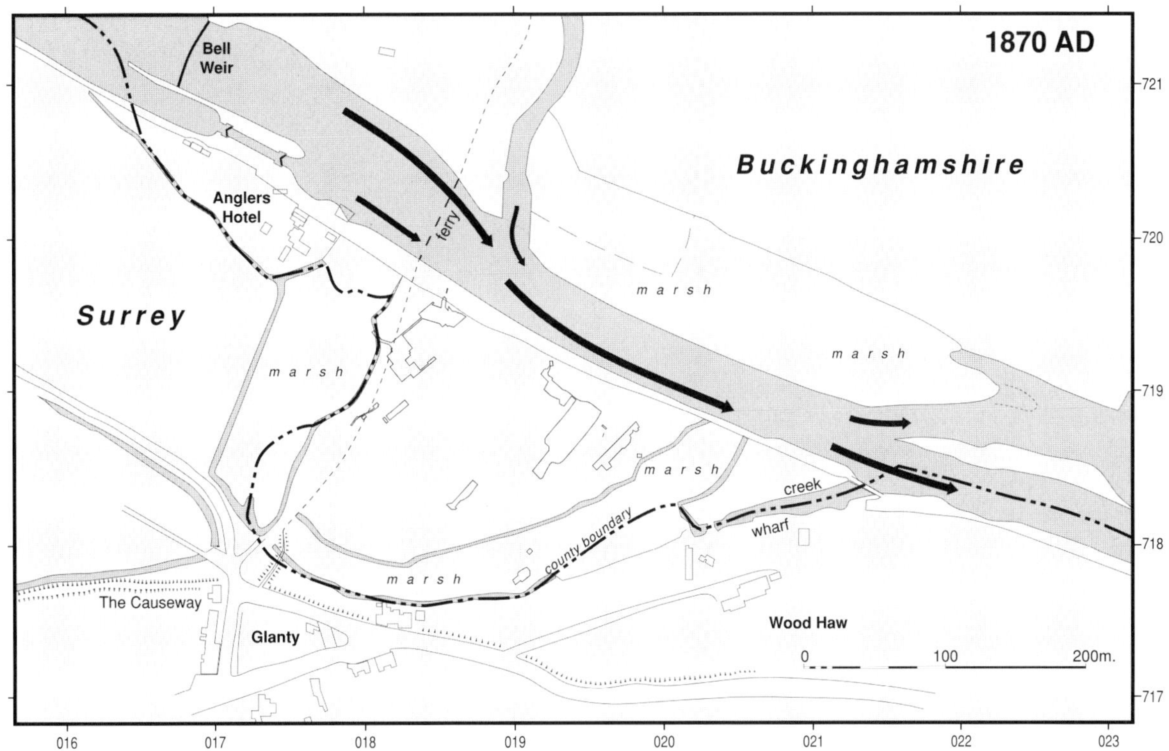

11.16   *The channel configuration and human activity in* AD *1870.*

topped by parcel 1K (Fig. 11.2). As the Atlantic period advanced this bank was progressively aggrading, moving 20 m through Areas 22, 26 and 32, with just temporary reversals seen in the erosion scarps that demarcate the different alluvial parcels (1F, 1G, 1H). These reversals need be due to little more than exceptional floods, sustained periods with higher than average flow, or the vagaries of current patterns according to shifting shoals in the channel. There is no contemporary migration showing in the long Area 7 section to the east (Needham 1991, fig. 10) where instead tufa was forming under flowing water; tufa was also noted to underlie parcel 2D in Area 14. It is of interest that channel migration could have taken place despite the alder root systems that presumably had a grip on most of the river banks (cf. Brown 1997, 210).

The existence of a third Thames branch (the intermediate meadows channel), joining the southern channel immediately west of the site (Fig. 11.2), is hypothetical for this stage. If not present, the size of the island, on whose eastern tip lay the prehistoric settlement, could have been much larger. The later Mesolithic and Neolithic occupation activity could in either case have spread further west than documented by the excavated trenches. Some support for this comes from the Neolithic pottery in the lower fill (parcel 4B) of the later second-millennium BC channel, which carved in part through virgin land which had not been eroded since the fourth millennium. This is a likely source for the redeposited cultural material which is argued not to have suffered more than brief fluvial action.

By the early fourth millennium BC, the northern channel had begun to silt along its southern shore towards the tip of the island (Fig. 11.3). Human interference cannot be ruled out, but the silting can fundamentally be put down to slacker water caught behind a mid-channel gravel bar recorded in Area 6 (its north-eastern side truncated by the LBA channel). It was high enough to poke above mean river level and might have been a more extensive feature upstream than shown in Fig. 11.3, but could have been kept clear downstream by the combined flow of the northern channel and the Colne Brook. Towards the middle of the millennium, when human activity reached its peak, a potentially wide belt (up to 20 m) of seasonally drier shore had been formed and opportunities were taken to extend activities onto it. This is witnessed in particular by the consolidation platform of Area 4. It also resulted in the accumulation of cultural debris on the shoal itself, although this was likely washed into its excavated positions during times of higher flow, as indicated by the channel-aligned, prostrate wooden stakes on its flanks. Working on the assumption that the northern channel was still carrying a significant portion of the Thames' flow during this period, that flow would have been concentrated to the north of the bar and potentially eroding the northern bank beyond.

The evidence for a southern channel at this stage has been outlined above (Chapter 10, parcel 1L), but nothing is known of its orientation. Upright Neolithic piles were found in Area 8, evidently the in situ foundations for a river's edge structure. This bank was also silting up from a date earlier

231

than this construction, and the embedded pile foundations escaped being dislodged by any scour associated with the large flood event which followed.

Late fourth to late third millennia BC (Figs 11.4 and 11.5)

There is some evidence that the northern channel saw noteworthy scouring along the silting edge around the time of the large gravel deposition, parcel 2F. This can be argued from the immediately underlying palaeobank in Area 32, as described above, but also from the erosion scarp at the parcel 2A/2C interface in Areas 4 and 7 which suggests the re-scouring of a narrow channel alongside the gravel bar. The top of the latter was truncated, but probably at a later stage in the Neolithic period. The large deposit of gravel seems to have been spread over virtually all of the dry land of the site, although on the highest ground the evenness of the bed might have been enhanced by later disturbance. Gravel was also left stranded along the channel edges, surviving differentially according to later erosion patterns. Along the southern channel edge a thick wedge survived (Area 8) and would have constricted the channel, possibly forcing some southerly migration or channel deepening. Subsequent erosion scarps seen in the northern channel in Area 4 indicate further periods of flow strong enough to keep reopening part of this silted zone, but in the longer term aggradation won out and eventually extended upstream to the riverside zone of the research excavations. By the mid-third millennium BC a broad strip at least 20 m wide had developed into a low-lying mudflat.

There was further lateral incision along the southern bank in the late third millennium as evidenced in Area 14 (parcel 2D/2E interface) and Area 4 (within 2C). Shortly after, there was beaver activity in the channel, but nothing suggests that the animals had any major impact on the river system in the immediate vicinity. There is little evidence to decide whether the mudflat development implies renewal of the northward migration pattern of the previous millennium, or was associated with reduced flow in this channel. The latter possibility is the preferred explanation, for it is argued below that during the second millennium BC the northern channel silted right up and it is therefore possible that a process of steady waning of the northern channel and consequent strengthening of the southern was beginning during the third millennium. It is not obvious that the major flood event played any part in initiating this pattern.

Second millennium BC (Figs 11.6 and 11.7)

Previous publications (especially Needham 1991) have dealt with channel reconstructions of this period and the Late Bronze Age in some depth. The overall picture has not altered, but more detail can now be added and all the key evidence will be reiterated here for the sake of continuity.

The second millennium BC saw a marked change in the hydrology of the valley around Runnymede Bridge. Thick deposits of slow overbank alluvium, parcel 2B, built up over a few centuries probably as part of a feedback cycle, whereby the raising of the floodplain tended to constrict channels between better-defined banks, and this led in turn to bank-full stage being reached more quickly, hence more flooding. As already discussed in Chapter 10, another significant factor in this new alluvial regime was the much more extensive clearance of woodland, allowing the more rapid run-off of rainfall into the river system, taking sediment from bare/arable ground with it – the ultimate source of the alluvium.

The excavated sections can only prove that parcel 2B filled the southern edge of the northern channel, covering the previous mudflats, but it seems more than likely that the 14 m OD altitude achieved extended right across the channel (Fig. 11.6). When next a palaeochannel enters the frame of the excavated areas, late in the millennium, it takes a distinct alignment from the previous system, a rotation anticlockwise of 30°, which is very difficult to accommodate in terms of a meander development given the course of the northern channel over the previous millennia (Fig. 11.7); one would need to invoke a very unusual migration locally due to some aberrant feature in the river or bank. The changed alignment is better explained as a wholly new cut after the complete silting of the previous channel. This is reinforced by the changes that took place in the Late Bronze Age channel immediately downstream. The concave sweep of the palaeobank established through Areas 5, 6, 7 and 4 by late in the second millennium can only be associated with a former course of the Colne Brook in very much the same position as today (see also Fig. 10.8). Initially this was clearly uninterrupted by the crossing flow of a Thames northern branch, but at some point towards the end of the millennium this eroding bank switched to become an aggrading one; the obvious explanation is that this was a direct result of restored flow on the new axis, which would have thrown current onto the convex bank of the tributary, the lowermost reach of which became part of the new northern Thames.

The second millennium BC thus saw a reconfiguration of the channel system, one that was to have long-lasting effects on its later evolution. Blockage of the Neolithic northern channel would obviously have diverted all flow around the southern one, thereby augmenting the erosive capabilities of this single branch (below the possible confluence with the Langham's channel, if this continued to flow – Fig. 11.7). Channel dimensions would almost certainly have increased, which neatly explains why the northern bank in Area 8 cut back into the pre-existing 2F and 2B deposits, being the palaeobank abutted by parcel 3. Accelerated meander development could also have been a response, giving rise to the situation where the new avulsion of late in the millennium only had to break through a narrow isthmus between the Thames and the Colne Brook. The course of the breakthrough is most likely to have been determined by the vagaries of scouring during a flood, although any undulations in the topography present might have been exploited. It occurred too early to be connected to the Late Bronze Age settlement phase and at a date when there was evidently no special interest in the site; no cultural material whatsoever can be attributed to the late second millennium.

With the breakthrough, renewed flow to the north would obviously have reversed the earlier switch of current to the south. Slackened current in the southern channel might well have precipitated the development of a gravel bar, probably a point-bar, close to the inner bank, which was to be followed by sand infill of the trough between (parcel 3). This all ties in with the relatively tight date bracketing of these deposits and it shows that the land along the southern margin of the LBA occupation area, sampled by Areas 1 and 2, was only recently created, just as was the case along the northern margin.

The aggradation along the southern bank of the breakthrough channel seems to have begun almost as soon as the channel had been established. This is inferred from the survival of fragile Neolithic pottery, argued to have been scoured out of primary Neolithic deposits very close by during the channel's formation (Chapter 10). The pottery could not have endured more than brief exposure to fluvial agencies.

The coarser elements in the channel fills could also result from the increased mobility of the channels and the destabilisation of material on adjacent banks and slopes; inevitably there would be places along its course where the river carved into bank deposits of sand and gravel. In addition, there is the possibility that the increased stream power simply allowed heavy bed material to be more dynamically redistributed, but it seems unlikely that this alone would account for the proportion of coarse material in these channels and the slightly later ones (parcels 5 and 7), if these are in any measure representative of the broader catchment.

By the turn of the millennium the newly formed island had thus already changed shape, with expansion to north and south and, probably, further loss to erosion on the west side. If it had not happened earlier, it seems likely that by this time the Langham's channel was essentially defunct in terms of flow, for it is otherwise difficult to comprehend the strong meander development that can be documented during the course of the first millennium BC. The nature of the development seems inconsistent with the continued input of a second feeder channel. From this stage on one can talk of a 'southern loop' around the site.

First millennium BC (Figs 11.8–11.11)

It has been argued for some time now that the Late Bronze Age inhabitants exploited an island or near-island, a plot later to be called *Tyngeyt*. The distinction between these two situations is the central issue regarding the channel system of the early first millennium BC. What can be deduced with some confidence is that over this half-millennium the meander of the southern loop became more accentuated with, first, the eastern reach cutting a new alignment back into Area 8, then probably a little later the western reach cutting back as far as Area 28. The onset of the wetter conditions of the Sub-Atlantic would have increased river flow generally and perhaps enhanced erosion, but it is not of course a precondition for the meander development itself.

Had the northern channel remained open during the whole of this period, as the more direct course, it would surely have taken more and more of the current. As postulated for the second millennium sequence of alternation, one would expect eventually to see slacker water in the southern channel, not really consistent with the meander amplification described. The preferred interpretation then is that the northern channel effectively became blocked to normal flow early in the period. If this had not already been achieved by the lateral extension of parcel 4B across the full width of the channel, then we need look no further than the occupants themselves as likely cause. There is evidence that the orientation of structures on the site was to the north, and it is possible that an entrance way is present on the northern periphery among a dense concentration of features. Passage across the remnant channel would have doubtless been desirable and could easily have led to blocking, deliberately or inadvertently. If this happened early in the occupation, then it would not be unreasonable to expect some effects of enhanced erosion around the southern loop a century later.

Soon after the eastern arm had cut back to its limit position in Area 8, it began to aggrade. Following the pattern of previous centuries in this reach of the valley, the sequence probably began with a point-bar emerging around the southern side of *Tyngeyt* and, at first, sandy and sometimes organic sediments in its lea in the concave sweep from Area 8 to Area 23 (Figs 10.8 and 11.10). These came to be covered by the downstream extension of the bar, an extension that might have continued further along the 'S' bend than normal because of the influence of the current entering from the Colne Brook. Overall, the channel in this segment was shifting to the south-east (Figs 11.10 and 11.11), and the trough left along the island shore steadily filled with sandy silts to create a new bank. There is no close dating, but this stage was probably reached long after the cessation of occupation, perhaps into the later first millennium BC. Meanwhile, the western reach had been developing a stronger 'S' curve in the predictable manner. This led first to its cutting back into Area 28, but then as the inflexion point shifted, to the aggradation of sandy silts. It appears that there was more rapid migration around the tighter southerly bend than the reverse bend upstream (Figs 11.11 and 11.12).

Despite the likelihood of increased river flow during this millennium, there is no evidence on the site for overbank alluviation. This certainly does not exclude the possibility of a flood regime that was carrying little silt load. Cultural debris found in various locations and stages of the parcel 5 accumulation is best interpreted as material having been swept off the island. In contemporary parcel 4C, Area 6, a largish group of pottery was evidently redeposited in a flood-silt in the channel edge (6.006).

First millennium AD (Fig. 11.12)

There is very little excavated evidence for the channel system during the first millennium AD. The re-cuts in the eastern arm of the southern loop preceding parcels 6A and 6B (Area 23) should belong early in the millennium, given the best dating of the overbank alluviation to the third century AD. These suggest some minor oscillations in bank positions against the general, but gradual south-easterly drift. Without further information it is difficult to identify any more specific significance. Of some passing interest, however, is that just as the southern loop nudged ever closer to the Glanty ridge to the south, this topographic feature was exploited by the Romans to carry the road from *Pontes* (Staines) to *Calleva Atrebatum* (Silchester), as argued above.

For the later part of the millennium there is the circumstantial evidence of a dated timber pile, sadly out of context (Area 11), but it may indicate the date of the in situ timber structure within the channel there. A hythe or hard was almost certainly installed on the southern loop to serve the Saxon town of Egham, while the southern loop itself was used to demarcate the border of both parish and county. Overall one can only assume that the meander continued to accentuate towards the shape documented for the later second millennium AD (Figs 11.14–11.16).

Second millennium AD (Figs 11.13–11.16)

Again a focal issue is the question of the reopening of the northern channel, the direct antecedent of the modern near-straight course of the Thames at Runnymede Bridge: when did it reopen to normal flow and how did this come about? By this time the meander was so pronounced that only a narrow neck of land between its upstream bend and the tributary Colne Brook had to be breached to re-create the northern channel (Fig. 11.13). This would have been the key factor controlling the abandonment of the southern loop; the diversion of the main current would allow the loop to silt up as any oxbow or other form of cut-off channel. Continued current from a second feeder channel, such as the Langham's channel, would alter the situation, but it has been argued above that this had ceased flowing much earlier.

The *Chertsey Foundation Charter*, believed to have been written down in the twelfth to thirteenth centuries (Turner 1926, 7 ff), describes the boundaries of Egham parish in a clockwise direction and states 'and so forth along Temese by midstream to Glenhuthe' (*ibid.*, 9). Glenhuthe evolved into Glanty, the place now occupied by the roundabout on the south bank of the southern loop. This corroborates the

deduction from the later mapped line of the county boundary, that the southern loop was regarded as the main channel. It is possible there was a bridge here at Glanty in the fourteenth century (Turner 1926, 19) and its *huthe* element indicates a hythe, but whether this refers to the wharf at Wood Haw immediately downstream, or an earlier location of it is uncertain. A mention of 'Woodhawick at Glenhuthe' in a 1373 document (Turner 1926, 50) suggests that at this time they were regarded as one and the same. *Le Wodehaghe* was obviously long established by the fourteenth century AD, at which date it was said to have taken a toll from river traffic 'from time immemorial', and it survived until the nineteenth century (*ibid.*, 19, 42). Turner says that the house of Wood Haw was very old and stood on the site of the old monastic wharf or landing-place (*ibid.*, 236).

On the evidence of the drainage ditches, which seemingly followed the channel's edges on the 1800 map, it would seem that the southern part of the loop became quite constricted in its late stage of evolution (Figs 11.14–11.15). This was probably due to resistance to erosion offered by the Glanty gravel ridge. Nevertheless, the current would have continued to scour the adjacent, concave bank allowing the inner one to aggrade and constrict channel width; it is likely that channel deepening resulted. The classic pattern of meander development (although not classic in origin) identified through the period 800 BC to AD 1500, is unlikely to have taken place while a straighter course of the river flowed, since this would have quickly deprived the longer course of any significant current, except perhaps during spate. This argues for a relatively late reopening of the straighter course, i.e., within the second millennium AD.

A *terminus ante quem* is provided by historical maps (Table 11.1); those accurate enough regarding the course of the Thames go back to John Senex's of 1729, where the *Tyngeyt* reach is shown as fairly straight. On the other hand the fixing of the county boundary along the line of the southern loop suggests that this was construed as the major branch, if not necessarily the only branch, when county boundaries were being defined in early Medieval times. The precise line followed by the county boundary would probably have moved with any later gradual channel migration and might have been subject to other modifications once marsh and drier land developed, allowing the apportionment of land plots. For instance, the wiggle just upstream of the site seems designed to take the complex of buildings associated with Bell Weir lock into a single parish (and county), presumably beneficial for administration. This complex did not exist before 1816. The boundary can, therefore, only give a broad indication of the river's course in its final phase.

The early maps, although they do not capture the open-water phase, do show the relict channel thus supporting the archaeological evidence (parcels 8 and 10) that final silting took place over the past three centuries. Senex's map is somewhat schematic and a little inaccurate; for example, the

Table 11.1 *Historical maps consulted.*

Date	Surveyor/publisher	Location (unpublished maps)
1729	John Senex	
1768	John Rocque	
1793	Lindley & Crosley	
1800	Common Lands map, Wraysbury parish	Buckinghamshire Record Office, Aylesbury: IR/111
1811	Z Allnutt, of Henley, on behalf of the Thames Navigation Commissioners	Buckinghamshire Record Office, Aylesbury: P/uA.25
1823	C. Greenwood & J. Greenwood	
1840/ 1843	Tithe map of Wraysbury	
1869	Ordnance Survey 1:2500, Surrey sheets V,5; V,9	
1872	Ordnance Survey, 6" sheet: Surrey V	
1881	Ordnance Survey, 6" sheet: Bucks XL	

Avenue does not reach the river opposite the Colne Brook confluence as it should. However, the downstream meander of Penton Hook is shown very prominently and there is every reason to have expected the southern loop to be depicted.

John Rocque's map of 1768 offers more detail. River islands are shown, both above Ankerwyke bend and at Staines, upstream of the bridge. The Tyngeyt plot is identifiable opposite the Colne Brook confluence because of the gentle loop expressed in the course of the road along the south bank from Staines to Windsor. Indeed, a shaded stroke here running into the river can be identified as the eastern arm of the southern loop. This shows more clearly in Lindley and Crosley's 1793 map which indicates a strip running right back to the road junction at the head of the Avenue. Judging from the succeeding maps of 1800, 1811 and 1840, it is likely that this strip was already a marshy belt laced with drainage ditches (see Fig. 11.15), rather than truly open water. The county boundary is shown continuing along the line of the former western arm. The marshy belt is depicted not only on the Greenwood and Greenwood's 1823 edition (a time when Runny Mead was occupied by the large circuit of Egham Race Course), but also on the first Ordnance Survey 6-in survey of 1870 (Fig. 11.16).

Other evidence must thus be sought to assess how much before 1723 the straight course might have existed. A further *terminus ante quem* should also be provided by the cessation

of significant current in the southern loop, and for this we can turn to the archaeological evidence. It can be a very lengthy process before a cut-off channel becomes merely a marsh; however, what would happen relatively rapidly would be the loss of stream power to move heavy bed material, such as the gravelly shoal 11.006, 007. The fossilisation of this shoal could therefore indicate quite closely the loss of previous stream power. This would be early in Post-Medieval times on the evidence of the entrapped finds, sixteenth to eighteenth centuries. This suggests that the critical change in flow conditions occurred not too much earlier than the eighteenth-century cartographic surveys; but we are still left with the problem of not knowing for how long strong currents might have rounded the loop after the re-creation of a straight course.

Further evidence comes from the placename of Late Medieval times. The possibility that the plot of land enclosed by the southern loop was already an island comes from the documentation that when Gilbert de Montfichet, lord of Wraysbury in the middle of the twelfth century, founded the small Benedictine house for women at Ankerwyke, he endowed it with land in that parish, while his son Richard gave 'an island in the Thames called Tyngeyt'; *Tyngeyt* can be identified as the plot we are concerned with, later mapped as Tinsey Mead (Turner 1926, 64). The plot was obviously recognisable as an 'island', as is clear from the *eyt* element, generally now transliterated as 'eyot' or 'ait', but it is possible that it was not at that time entirely encircled by active channels. Names meaning 'island' were frequently applied in Medieval times to raised patches of ground which were not necessarily entirely surrounded by open water, as classically seen in the Fens. The alternative would be that *Tyngeyt* was a narrow-necked peninsula, the short northern neck a marshy belt, perhaps partially a backwater, a relict of the late second-millennium BC breakthrough channel.

If *Tyngeyt* was a true island in the twelfth century AD, then it seems likely that the straighter course had only recently re-established itself, given the arguments presented above. Even so, there would have been a long period, some four or more centuries, during which the southern loop still carried a strong current. Whether it became a true island before the twelfth century, or somewhat later, the dating of the demise of the southern loop as an important current-carrying channel to as late as the sixteenth or seventeenth century, does beg the question as to whether the straight course was artificially created or enhanced at this time as a short-cut for the benefit of increasing river and canal traffic.

The seventeenth to eighteenth centuries was a period of growing interregional trade along British waterways, a trade which was further fuelled by the beginnings of the Industrial Revolution. The Thames itself had for some centuries already become a critical supply line for the developing metropolis of London, bringing downstream Cotswold stone, timber, malt and foodstuffs such as cheese, in particular. Staines was a thriving nodal point for waterborne and overland transportation by Late Medieval times; it became affluent and influential as a provincial town. The Great Fire of London in 1666 suddenly accelerated demands for building stone, this on top of the growth in supply needed to service the burgeoning population. The Reading to Newbury reach of the Kennet was dramatically improved from 1719 with 11 miles of new cuts and installation of twenty locks, while the opening of various new canals between 1790 and 1819 including the Kennet-and-Avon (Chaplin 1982, 80–2), opened up vast new trade potential by completing east- to west-coast waterways across southern England. By 1800 it is estimated that some 85,000 tons per annum were passing along the river at Staines, while it has been suggested that up to 95% of goods to and from Reading were carried by water (Wilson 1987, 69, 72–3). Every riverside community would have tried to take advantage of the passing trade (Chaplin 1982, 77), but by this stage the wharf at Wood Haw on the east side of Tyngeyt was sited on a creek, the last vestige of open water representing the southern loop (and surviving to this day), hence contributing to its demise.

During the early Post-Medieval period barges were being made larger and larger and there were continual problems with the navigation of shallows and the privately owned flashlocks that had been installed in an ad hoc manner over the previous centuries. There was growing pressure for rationalisation, better locks (and limits on tolls) and general river improvements (Wilson 1987). The first pound locks on the Thames were concentrated in the Oxford to Burcot reaches, notorious for a number of difficult passages, and these had been constructed by 1635. It was considerably later, in the early years of the nineteenth century, that the city of London finally accepted that locks would be beneficial in the reaches under their direct jurisdiction (upstream as far as Staines, marked by the 'London Stone' – Chaplin 1982, 23; strictly speaking the Corporation had had responsibility for the whole Thames since 1197, but beyond Staines it devolved that responsibility to various commissions). Bell Weir and Old Windsor locks came a few years after, in 1816–19, the former taking advantage of a gravel island shown in the 1811 map. It is possible that this island was a remnant of an earlier point-bar round the downstream bend (Figs 11.14 and 11.15). The lock was first built in wood, rebuilt in stone in 1867 (the plan seen in Fig. 11.16), and again rebuilt and enlarged in 1973–4. Previously, 'Bell Weir' seems to have referred to the reach just below the site, but there were shallows both here and where the lock/weir complex was finally built (Fig. 11.15).

In the preceding period a survey had been commissioned by the city of London in 1770 and, although its furthest-reaching recommendation was not accepted, it did stir the Commissioners into action to improve the river. Additionally, in 1795 an Act of Parliament '. . . gave powers to widen and consolidate the old haling ways for the use of horses'. This ties in well with the 1800 Common Lands map

which shows the well-defined towpath running through the riverside zone of the excavations (Fig. 11.15). In fact, it is possible that the improved towpath definition at Tyngeyt came a few years earlier; certainly a record for 1775 shows local activity: 'A path formed and gravelled over the ait ground above Staines Bridge being 483 yards long' (Chaplin 1982, 28). Chaplin also states that towpaths were almost non-existent until the eighteenth century (*ibid.*).

The detailed maps of 1800 to the present show that there have been some significant changes to the course of the modern channel, even over that relatively small timescale. These are largely attributable to lock and weir construction and their subsequent effects on flow patterns. It is hypothesised that the breakthrough channel may not initially have lined up with the linking reach of the Thames to the west (Fig. 11.14), but with time better continuity would be achieved by natural processes. If there had been any point-bar tendencies around the first tight bend, efforts may have been made to clear a channel alongside the southern bank with its towpath, potentially thus leaving some of the gravel bar stranded in mid-channel (i.e., the island which the lock took advantage of). Bridges would have been required across the entrances of the southern loop, in order for *halers* to benefit from the cut-through. By 1800, with the loop essentially reduced to marsh, the southern bank of the northern channel is presented as a broad concave sweep backed by defined towpath, albeit with some minor irregularities (especially in the 1811 map). The northern bank is more or less parallel.

One immediate effect of the construction of the lock and weir was to straighten the southern bank just downstream by dumping the upcast (i.e., in the riverside zone of excavations, as discussed in Chapter 10). Installation also gave rise to a continuous 'slalom' current from the weir which initially ate into the promontory of land on the west of the Colne Brook's mouth, but was then diverted to the south bank, adding to the regular flush from the lock. The first effect is seen in the bowed line of the northern bank by 1870, while the second resulted in the loss of a long plot of land/marsh, perhaps up to 20 m wide, further downstream on the southern bank, around the lower end of the former southern loop. Associated with this latter erosion was the creation of slack water in the embayment on the opposite bank. With time, the slack water was trapped behind two islands shown on the 1843 map, one probably having grown from a smaller gravel bank present in 1811 (the other is within the frame of Fig. 11.16). During the succeeding century these islands grew, becoming Holm Island of today, with little more than a creek against the former north bank.

One surprising aspect of the final, finer-sediment silting of the southern loop was its speed. It is assumed that normally the process of filling a channel would be rather prolonged after the initial cessation of any significant current. In particular, once the channel was truly cut off from the main flow, that is, blocked at its ends, the upper filling would be dependent on overbank flood sediment and detritus from vegetation growing nearby. Indeed, the rate of accumulation in Area 13 North is considerable; disregarding the lowest deposits encountered, the greater part of the depth (1.28 m) spans just 150 years, from the mid-nineteenth century onwards. This is a time for which there is little evidence for significant overbank accretion on the floodplain (although clearly many floods occurred, as historically documented).

The more likely explanation lies in the construction of the lock earlier in the nineteenth century. Locks are notorious for creating shoals a little downstream from the outflow, silt and gravel being drawn through each time they are filled and emptied (Wilson 1987, 62). At Runnymede it must be conjectured that the western arm of the southern loop was still a depression with shallow standing water opening onto the new straight Thames. The rush of water from each opening of the lock would cause a surge of silt-laden current, some of which would be forced round into the backwater where its energy would quickly dissipate as it spread, thus depositing its silt load. If this hypothesis is correct, it does imply that the southern loop was not blocked off from the Thames, artificially or otherwise, and that would presumably imply that the towpath was carried on a bridge here, despite the fact that no such bridge is depicted in nineteenth-century maps; bridges were certainly present across the creek and drainage ditches at the lower end of the southern loop.

The land and riverscape has thus been utterly transformed during the past two centuries, despite what was, until recently, a very rural plot of land of little special interest other than serving as a ferry point. It must be remembered that prior to the Lutyens Bridge of 1959–60, to carry the new A30, residents of Egham and Wraysbury on the opposite bank had no bridge crossing nearer than Staines downstream, or Windsor upstream. A ferry crossing from the site is documented in the later nineteenth century, but may have had earlier origins, conceivably respecting a long-standing route along the west bank of the Colne Brook towards the nearby mill. To speculate, such a route may originally have simply traversed a boggy belt of land before the modern Thames was re-established, although first the southern loop would have had to be negotiated. The latter-day crossing point of the modern Thames appears still to be marked by the survival of an old concrete setting on the foreshore alongside the site (Fig. 1.2). The setting forms a small angular funnel, perhaps the base of a flight of steps before the more recent retreat of the river bank by a few metres here.

The site had something of a local nodal position owing to the confluence with the Colne Brook and the deviation point of the Medieval and, perhaps, Roman highway from Staines to Egham and the West Country beyond, these in part explaining the long-standing wharfage. There is tentative documentary evidence for a Late Medieval bridge (Turner 1926, 19) which might link with the hypothesised routeway up the west bank of the Colne Brook. The early nineteenth century added to these features the lock and weir and adjacent complex, including a public house, the Angler's Hotel (Pl. 12), all natural magnets for those with time to idle, so inevitably a leisure focus emerged at this spot, as at many other points along the river. A boatyard offering varied craft for the pursuit of leisure (Pls 11 and 13) was added, being situated on the bank above the ferry point. Prior to the Second World War a row of five chalets had been erected along the southern bank to the east of the boatyard, in positions now obliterated by the combined A30/M25 embankment. The site also became the focus for camping holidays by the river, with caravans available. One lady visitor to the excavations, after enquiring in detail about what we were doing, finally divulged that she had lived on the site during the war as a child evacuee from London. She described a fan of caravans and little enclosing plots; doubtless some of the modern pits dotting the site can be attributed to this colony. She also remembered a flood immersing the wheels of the caravans. Faced today with the relentless pounding of the bridge traffic and the periodic bursts from the Heathrow jets, it is hard to imagine the relative idyll of the site even earlier this century. But the leisure connection has endured through the replacement of the Angler's Hotel with the Runnymede Hotel, and the seasonal, motley flotillas of passing pleasure craft.

There is still a further significant phase of activity on the site prior to archaeological works – the Bovis phase. Not only did this blight the site with a monumental construction, but it also left a legacy deriving from a workers' camp placed alongside the embankment on the north-west side (the main area of research excavations). A series of large pits were cut through the prehistoric land surfaces, leaving the usual frustrating holes in planned levels, but fortunately the initial laying of a thick raft of rubble has protected the earlier archaeological levels from other kinds of disturbance (Fig. 2.2). The swathe of the A30 and M25 bridges runs more or less centrally across *Tyngeyt* and the earlier, Late Bronze Age island (Fig. 1.2). In the gap between the town centres of Egham and Staines this crossing gave the nearest gravel abutments to the river, but between them was an ancient spine of silt entrapping the evidence of a long history of human interaction with a mobile valley-bottom environment.

CHAPTER 12

Conclusions

Several broader issues come out of the detailed topographic reconstructions and the associated settlement evidence. They relate to why the site was the repeated scene of human activity, the details of the topography utilised, phases of flooding and alluviation, the evolution of the channel system, and the contemporary perceptions of valley bottoms.

For a site to be the scene of separate occupations is not unusual, but one still is drawn to the question why, perhaps particularly so when there is the presumption of some persistent risk or discomfort. These themselves may be fallacies based on both modern attitudes and the character of the floodplain or valley bottom in recent history. It makes sense first, therefore, to tackle the questions relating to the nature of the ancient river and its flanking banks. It needs to be considered, for example, whether any features of the local reach of the river may have given it unusual geomorphological and vegetational properties, atypical of the middle Thames at large, properties that in turn might have given rise to distinct patterns of human exploitation.

Channel configuration

To begin with there is the circumstance of the Colne confluence, not a simple confluence but multiple confluences spread along 3.5 km of the Thames because of the multi-thread nature of the Colne system (Fig. 11.1). The survival of such a complex, and in part anastomosed, system until the present is a little unusual, and it may give a good impression of the complexity of other tributaries earlier in the Holocene period. The lacework of the Colne was in fact probably mirrored in the Thames itself in the local reaches to judge from the number of palaeochannels identified, or surmised (Fig. 11.1). It is by no means certain that all of these would have been contemporary, but it seems likely that early in the period a complex multi-thread Thames system would have flowed between the Colne confluence zone and the confluence of the River Mole, involving the River Ash, Mead Lake Ditch and the other extant streams in a broad floodplain corridor.

Nevertheless, there are grounds here for suggesting an unusual contribution to the Egham–Staines region, given the identification of a topographic basin in the surface of the Late Devensian gravel deposits. Irregularities in its surface,

while not remarkable, may often have been rather more local features.

Channel form and migration

Channel form is not easy to reconstruct when it is so rare to locate the second edge of a palaeochannel. However, it can be suggested that cross-sections changed during the Holocene from broader, shallower profiles to deeper, more contained ones. This is doubtless a simplification, and to some extent the predicted constriction effect of overbank silt accumulation from the mid-second millennium BC would have been countered by any reduction in active branches; with the same current being carried by fewer channels, narrowing is less likely, even given a deeper channel form.

During the Atlantic period channel beds seem to have been broad but slightly undulating and doubtless were associated with braided current patterns allowing the development of shifting shoals within the channel. They also encouraged the precipitation of calcium carbonate on the bed, as tufa. Even with this bed profile, however, there was a sudden steep bank at the river's edge rising to the early (Atlantic) island, as clearly seen in both Areas 7 and 32. It is possible that this particular water/land interface was still conditioned by the fact that the water table had dropped late in the Boreal, thus leaving relatively elevated blocks of land immediately alongside the channels. This is likely to be typical of the local reach and it had the effect of defining the channel and probably containing normal high flow. This is backed up by the evidence that flooding of the high silt spine was not very frequent during the Atlantic period. Seasonal variation within the channel, however, allowed not only the development of mudflat type habitats along the borders, but also the extension of human activity onto the damp but exposed shores.

Although it is abundantly clear that the river banks were in general densely vegetated with alder carr or woodland by this stage, it is noteworthy that the root systems did not create enough bank stability to inhibit lateral erosion entirely, for it would seem that both northern and southern channels were migrating away from the island. High flow, perhaps seasonally, was clearly of sufficient strength to penetrate and undermine the root systems. The alder trunk buried in parcel 2E (31.675; Table 3.9) may once have been one such casualty, albeit at a slightly later date. The process can be seen more strikingly in the powerful River Trent, where many fallen trees trapped in the river sediments have been recorded (Salisbury 1992).

With the opening up of the valley landscape, especially after the early second millennium BC, banks would become more prone to lateral erosion, other factors being equal. This is certainly borne out by the reconstructed evolution of the southern loop meander, which steadily became more and more accentuated over the course of the next two millennia, although eventually it was partly constrained on its southern edge because of butting up against a raised gravel ridge (Glanty). Of further interest concerning the southern loop, however, is the demonstration that it could not have evolved all the way from a gently sinuous channel. Where early land surfaces are shown not to have been subsequently carved away, they provide valuable negative evidence; no river migrated through the centre of the Runnymede Bridge site after the Boreal period. The protoform southern loop was in fact already a moderate meander at its inception; instead of having come about through classic meander development, it was a curve, or set of segments, inherited from the pre-existing anastomosed system. Its originating form was thus a fortuitous product of the pattern of channel redundancy as a multi-thread system evolved towards a single-thread one. Instant meanders formed in this way are likely to have been a recurring feature on rivers that followed this evolutionary trend.

The concentration of flow in a single channel by the late Holocene arises because of the tendency for some channels to get blocked at critical points during spate and then later to progressively choke with silt; this would be accompanied by the deepening of the remaining channel. It may be suggested that in this part of the Thames valley this transition occurred in the second millennium BC, a consequence of the blanket-like 2B alluvium laid during the early half of the millennium. Once these processes have taken place, it may take more exceptional flood conditions to modify the channel configuration by means of avulsions through ostensibly virgin floodplain. At Runnymede Bridge this happened twice in the form of a neck cut-off across the southern loop/Colne Brook confluence. The evidence is that on the first occasion, late in the second millennium, the breakthrough did not properly establish itself, remained subsidiary to the main channel around the south of the site, and fairly soon began to choke and dry up, perhaps aided and abetted by the Late Bronze Age occupants. It seems that there was then an interlude of at least 2,000 years before a similar avulsion or cut took place, this time with lasting effect – the establishment of the modern Thames course and the steady filling of the old meander.

Gravel and sand shoals clearly could develop in this reach of the middle Thames at virtually any stage of the Holocene period; that in Area 6 dates to the late fifth or early fourth millennium BC, but the evidence does point to a more systematic pattern in the evolved channel system of the later Holocene. The now essentially single, winding channel, combined with a good stream power and healthy coarse sediment supply, lent itself to the formation of point-bars of gravel and sand. These have been recorded in parcels 3, 4 (4A and 4B), 5 and possibly 7 (sand only), and there is a consistent formation of a trough along the nearside bank which filled up more gradually thereafter. The gravel shoals along the final course of the southern loop do not entirely respect this pattern and this is probably because they had built up in a slackening channel, rather than on one edge of a migrating channel.

Make-up of the islands

Any multi-thread river has islands and these would have been of varied and varying size as channels migrated. It is often supposed that islands would be founded on hummocks left in the underlying gravel at the end of the Late Devensian glaciation. In many cases, particularly for the larger islands, this remains valid, but it cannot be assumed to be the origin of all islands as has been pointed out in the past (e.g., Berry 1979). Berry recognised that the island at Windsor, a few kilometres upstream from Runnymede, comprised a body of silt, and it can now be shown that the changing island at Runnymede Bridge was built around a marl core with encircling silt, all sitting on a relatively flat Late Devensian surface. Paradoxically, the gravel component present higher up was added only at a later stage of its evolution and merely formed a thin capping. In the case of the mid-Holocene island postulated (Figs 11.2 and 11.3), current evidence does not exclude the possibility of a gravel core upstream, but the very means of its formation, by the down-cutting of channels, makes it more likely that its composition is entirely of the prior lake/marl sediments. The more general raising of floodplain surfaces with fine alluvium later in the Holocene period obviously gave rise to further possibilities for the creation of silt islands by dissection, for as long as an anastomosed system endured and avulsions took place.

Flooding and alluviation

At times when overbank alluviation takes place, flooding can be assumed, although predicting frequency is still fraught with difficulty. On the other hand, flooding regimes which deposited negligible silt are less easy to identify. In essence they can only be inferred from the composition of specialised organisms surviving on the floodplain land surfaces – in the case of Runnymede the Mollusca are the key. In relative terms the proportion of aquatic species should be sensitive to changes in the frequency and duration of flooding, but a residual problem is that, because there is no vertical accretion of sediment, one may obtain an 'average' situation representing the conflated records of a long period. There is only minimal scope to discern any temporal trends where worm sorting of the soil profiles has taken place.

Using just such evidence, Evans and Evans suggest (Chapter 7) that the stable land surface of the Atlantic period, atop parcel 1K, although relatively dry, was nevertheless subjected to some flooding. The cessation of the 1K silting regime and its transformation into a non-depositing regime early in the Atlantic, around the later seventh millennium BC, is probably down to sediment supply, in this case that of fine clayey sediments also present within the earlier marl deposit. The transformation can probably be connected with the vegetation cover of valley-floor deposits and the previous fall in the water table locally. The Boreal vegetation on the valley floors of the lower Thames basin seems not to have been very thickly wooded (Lewis *et al.* 1992; also Healy *et al.* 1992, for Kennet valley), while the incision of the river system in the Egham–Staines reach at least would have given rise to much newly eroded sediment from the underlying deposits, such as the clayey marls in parcel 1C, or other types. Allowing for the intermittent storage and reworking of such sediments and any initial adjustments of the incised channels, it may have taken some centuries before a new equilibrium was achieved (e.g., Brown 1987, 1995, 54; 1997, 200).

The main bodies of overbank alluvium encountered later in the sequence, those of parcels 2B and 6, have an entirely different background to parcel 1K. Now it may be argued with some force that it is the creation of arable landscapes in low-lying situations that gives the potential for an alluviating regime. What is noteworthy, however, is that these phases of alluviation are relatively restricted within the overall timescale of the late Holocene period: parcel 2B could have formed over several centuries at most; parcel 6 possibly over a much shorter period or periods. The vital question for future research then is whether this brevity is a product of shifting catch-points, such that at different periods different parts of the valley were trapping alluvium, or instead whether it was a broader regional pattern dependent on particular land-use patterns. Although there is as yet little comparanda for the 2B alluvium, the correspondence with the phase of substantial woodland clearance and the development of Deverel-Rimbury agricultural practices, could favour the latter option. A similar argument is favoured for parcel 6 of late Roman age, this time because other sites in the region show essentially synchronous alluviation.

Attention has been drawn before to the utterly exceptional nature of the gravel incursion, parcel 2F (Needham 1992). With the revised dating presented above it has been possible to argue that this particular fluvial event was triggered by the first significant destabilisation of the Atlantic woodland by Early to Middle Neolithic communities. This interpretation sees a major and sudden environmental impact from human activity which was still probably rather intermittent along the valley.

The nature and occurrence of occupation

During the Later Mesolithic and Early Neolithic periods, use of the island was not necessarily continuous; certainly it left only occasional features and material refuse. However, the absence of evidence for it becoming wholly closed in by trees has suggested that the site was being kept open against the backdrop of dense forest. Clearance may have been periodic, rather than persistent, but sufficient to maintain an inheritance of relatively open land binding the site through tradition to a seasonal cycle. The potential for marking of the site, its particular place in the landscape and cycle, offered by the seventh millennium BC post-pit, can be readily reconciled to such a picture. The later ditch is less easily understood, unless related to the seasonal occupation of the island, or serving as an aid to hunting. There might be many reasons why this site became one of those across the landscape chosen for particular use, but one possibility is that initially it was environmentally induced. The steady migrations of channels were leaving in their wake slowly accreting mudflats, which were also slow to become vegetated; if such tiny openings on the river banks had the effect of concentrating game coming to drink, then this would hold an obvious attraction for Mesolithic hunters. There is no real evidence from Runnymede at this stage for periodic clearance by fire, as for example has been argued up-river at Thatcham (Healy *et al.* 1992).

The Middle Neolithic dense occupation probably should not be read in terms of 'continuity' except insofar as the previous social and environmental legacy again gave preference to this site over others. If not necessarily all-year-round settlement, it was nevertheless very substantial and resulted in large numbers of dug and driven features, large quantities of food refuse and equipment, and hearths. If just a seasonal camp, it was evidently of some importance and durability and it would presumably have enjoyed an essentially flood-free environment during the summer months. Cattle were being grazed on the modest areas of grassland, but also probably alongside pig in the woods, continuing the practice of the Early Neolithic period deduced by Robinson (Chapter 9). The relationship of the site to the nearby causewayed enclosure at Yeoveney Lodge, Staines, is hard to articulate, but in terms of siting and potential functional differentiation, the latter offers complementarity, lying on the low bluff overlooking the Egham–Staines basin.

Real abandonment followed in that, although there is not evidence for full woodland regeneration in the locality, there do seem to be fluctuations in the vegetational cover and composition which suggest that the plot was utilised, subject to the changing fortunes and strategies of the communities of the territory, without being in any way of central importance. There were episodic visitations, but not obviously more than fleeting, and certainly by the end of the period (late third millennium BC) beavers were active in the reach. The presence of several later Neolithic and Early Bronze Age arrowheads invites speculation that hunting was one key reason for visiting the site and its environs, notwithstanding the fact that some are probably in secondary contexts.

Specific interest in the island may have resumed temporarily with a phase of cultivation, tentatively interpreted as having taken place in the Early Bronze Age, representative perhaps of widespread, but initially unbounded cultivation of the early second millennium BC on valley floors, among other locations. On this model, the presence of the Beaker/Early Bronze Age pottery would traditionally be linked with manuring. By the second half of the millennium, in mature Deverel-Rimbury times, the valley landscape had been utterly transformed, probably having consequences for land-use patterns. Regional occupation evidence, exiguous as ever in the former period, is more obvious and betrays, through its distribution, continued or growing concern for the newly raised floodplain. Whether much was turned to arable is unclear, but by the last quarter of the millennium the now extensive open land appears to have been predominantly grassland. This suggests an emphasis on grazing, a situation which continued into the early first millennium BC; the large faunal assemblage from the Late Bronze Age deposits shows the preferred livestock composition for meat production, a predominance of beef, followed by pork. This pattern of land use as grazing land recalls that interpreted for the Iron Age in the upper Thames valley (Lambrick and Robinson 1979, 1988).

The important question of the reasons for occupation here in the Late Bronze Age is returned to below, but the presumed pastoral pattern established later in the second millennium BC may largely have been followed for most of the succeeding millennia. One possible disruption to the ecological balance has been suggested in the form of Roman ploughing of low-lying ground in the catchment, including some very close to the site if not on it. Otherwise it was only with the marked intensification of commercial activity along the river in Post-Medieval times, followed by the rise of the pursuit of leisure, that renewed 'occupation' took place at and alongside *Tyngeyt*. In the long intervening period, any visitations have barely left a mark.

It is extremely unlikely that by the Late Bronze Age there was any folk memory or tangible evidence of former occupation of the island plot. The possible reasons for the siting of the new settlement need to be considered entirely in the context of the demands and preoccupations of the time. It is still difficult to identify a specific pattern of river-bank settlement of a nature comparable to that revealed at Runnymede, but there are a growing number of discoveries

pointing to activity on river banks and the setting of structures into and across contemporary channels. The site at Wallingford, Oxfordshire, just above the Goring gap into the upper Thames valley (Thomas *et al.* 1986) is a good parallel, being of the same date and occupying a terrestrial surface on top of floodplain alluvium immediately alongside a river channel, or perhaps two. It is now joined by another site just off the Thames, at Reading Business Park (Moore and Jennings 1992; and subsequent excavations), where a number of Late Bronze Age houses are set on a series of islands between the Foudry Brook and the Kennet. What is not apparently present there is any kind of refuse-rich deposit, a phenomenon that seems to bespeak specialised site formation and functions (Needham and Spence 1997). On the River Kennet, west of Reading, some occupation evidence and pile structures were recovered at Anslow's Cottages alongside a palaeochannel thought to be active at the time, but there was no indication of any very intensive occupation (Butterworth and Lobb 1992, 88–94). A single radiocarbon measurement and the pottery assemblage date the activity to around the LBA/EIA transition with perhaps an emphasis on the earliest Iron Age.

Looking further afield, one is drawn to the close comparison between Runnymede and a newly discovered site at Shinewater Marsh, Eastbourne, East Sussex. Apparently, substantial occupation activity took place on a timber platform set on peat and, although ostensibly located within a marshy embayment, boreholing suggests the presence of palaeochannels nearby (Chris Greatorex and Simon Jennings, pers. comm.). The activity is represented by a thick dark earth deposit with dense refuse of varied kinds; the range has much in common with the Runnymede assemblage and dating evidence so far, from metalwork, pottery and radiocarbon dates, suggests a focus on the ninth century BC.

Detailed exploration of channel deposits and associated structures and debris has been undertaken at Caldicot, on the small River Nedern, Gwent; although there is evidence for much activity over a long period of time, including during the Late Bronze Age, no occupied surface per se was discovered (Nayling and Caseldine 1997). There are further recent discoveries of later Bronze Age structures, for example bridges crossing a palaeochannel at Eton (Allen *et al.* 1997), just upstream of Runnymede, and at Test Valley Park on the River Test, Hampshire (A. Fitzpatrick – pers. comm.), where as yet no associated occupation can be demonstrated. Similarly, little occupation evidence per se has been recovered from the massive timber structure, the platform, at Flag Fen, near Peterborough (Pryor 1992). Here the comparison with Runnymede is poor in other respects, in its setting in a marshy embayment (albeit with at least one stream running through), and in its much earlier origins. It is hard to see in this site a closely comparable function to Runnymede, Shinewater and Wallingford. This discussion of sites bearing some comparison is not sufficiently exhaustive to cover those excavated settlement sites of the Late Bronze Age on floodplain terraces situated close to the river system.

It might be suggested that in many cases siting was determined by the desire to access certain resources. But absolute proximity to the river was not critical to enable the floodplain grasslands to be grazed, reeds and osiers to be harvested, or fish and fowl to be caught. It has also been argued that the concern of the occupants was not access to trade as such (Needham 1991, 383); rather, access to the social and political relations that went with exchange and the desire to demonstrate overtly those links, would have been more instrumental in determining river-bank site locations. This interpretation places the site in a nodal position, not so much in the geographical sense except insofar as this optimised maintenance of a social network. In this respect then, Runnymede Bridge could be seen to be a 'central place', because of its social nodality and the range of socially cohering activities it was engaged in (Needham and Spence 1996). The model envisages potentially large seasonal fluctuations in the population of the site and, on a very practical level, the geographical nodality would be beneficial to the transportation of supplies as well as the comings and goings of people.

Even so, this may not be the sole motivation. It is widely appreciated that wet places in general, and the River Thames in particular, had come to have a special meaning during the later Bronze Age. The river was the recipient of many offerings of metalwork, often lavish offerings (e.g., Bradley 1990). There is also the possibility that human remains were being placed, or were coming to rest in the river as a result of certain rituals concerning the dead or ancestral relics (Bradley and Gordon 1988). In this context the human skull buried in a pit in the riverside zone at Runnymede is worth recalling (Needham 1993, pl. 2). The Thames was thus a sacred place and perhaps a resource that it might be desirable to control, or at least associate with. It is interesting that, on the current evidence, the development of sites such as Runnymede only occurs at a very advanced stage of the Bronze Age, but this may be part of the process of taking into central sites the full range of affairs important to the reproduction of Late Bronze Age society (Needham 1993). The Thames, it might be ventured, would have become the centre of the cosmos for occupants of the valley; the siting of the settlement at Runnymede Bridge allowed it to host that vital cosmological organ. The risk of flooding from time to time (though not evidently very frequently in the Late Bronze Age) viewed against this conceptualisation becomes rather immaterial. Indeed, the hazard might even have been accepted as an unavoidable part of the relationship between the powerful deific force of the Thames (Isis to the romantics) and the people of the territories bound to it. On the subject of flood myths, Brown (1997, 288) suggests that 'whatever their contextual meaning, they play a significant part in many systems of belief, inculcating reverence for divine authority'.

To reflect back on the quotation at the beginning of this volume: between the vein of the land that a river is, and the artery of commerce it becomes, a river may also be the spiritual heart of society.

References

Alexandrowicz, S.W., Sniesko, Z. and Zajaczkowska, E. 1984. Stratigraphy and malacofauna of Holocene deposits in the Sancygniowka valley near Dzialoszyce. *Quaternary Studies in Poland*, 5, 5–28.

Allen, M.J. 1995a. Before Stonehenge. *In* R.M.J. Cleal, K.E. Walker and R. Montague (eds) 1995, 41–62.

Allen, M.J. 1995b. Before Stonehenge: Mesolithic human activity in a wild wood landscape. *In* R.M.J. Cleal, K.E. Walker and R. Montague (eds) 1995, 470–3.

Allen, T., Hey, G. and Miles, D. 1997. A line of time: approaches to archaeology in the Upper and Middle Thames valley, England. *World Archaeology*, 29, 114–29.

Allen, T.G. and Robinson, M.A. 1993. *The prehistoric landscape and Iron Age enclosed settlement at Mingie's Ditch, Hardwick-with-Yelford, Oxon.* Oxford University Committee for Archaeology.

Allison, J., Godwin, H. and Warren, S.H. 1952. Late-glacial deposits at Nazeing in the Lea Valley, north London. *Philosophical Transactions of the Royal Society*, B 236, 169–240.

Ambers, J. and Bowman, S. 1994. British Museum Natural Radiocarbon Measurements, XXIII, *Radiocarbon*, 36 (1), 95–111.

Ambers, J. and Leese, M. 1996. The radiocarbon results and their interpretation. *In* S.P. Needham and A.J. Spence 1996, 78–82.

Ambers, J., Matthews, K. and Bowman, S. 1989. British Museum Natural Radiocarbon Measurements, XXI, *Radiocarbon*, 31 (1), 15–32.

Andersen, S.T. 1967. Tree pollen rain in a mixed deciduous forest in South Jutland (Denmark). *Review of Palaeobotany and Palynology*, 3, 267–75.

Andersen, S.T. 1970. The relative pollen productivity and pollen representation of North European trees, and correction factors for tree pollen spectra. *Danmarks Geologiske Undersøgelse*, Series II, 96, 1–99.

Andersen, S.T. 1973. The differential pollen productivity of trees and its significance for the interpretation of a pollen diagram from a forested region. *In* H.J.B. Birks and R.G. West (eds) *Quaternary Plant Ecology*, 109–15. Oxford, Blackwell.

Astill, G.G. and Lobb, S.J. 1989. Excavation of prehistoric, Roman and Saxon deposits at Wraysbury, Berkshire. *Archaeological Journal*, 146, 68–134.

Avery, B.W. 1980. *Soil Classification for England and Wales (Higher Categories)*. Soil Survey Technical Monograph 14. Harpenden, Soil Survey of England and Wales.

Avery, B.W. and Bascombe, C.L. (eds) 1982. *Soil Survey Laboratory Methods*. Soil Survey Technical Monograph 6. Harpenden, Soil Survey of England and Wales.

Baker, C.A., Moxey P.A. and Oxford, M. 1978. Woodland Continuity and Change in Epping Forest. *Field Studies*, 4, 645–69.

Barker, G. and Webley, D. 1978. Causewayed Camps and early Neolithic economies in Central Southern England. *Proceedings of the Prehistoric Society*, 44, 161–86.

Bell, M. 1992. Archaeology under alluvium: human agency and environmental process. Some concluding thoughts. *In* S.P. Needham and M.G. Macklin (eds) 1992, 271–6.

Bennett, K.D. 1983. Devensian Late-glacial and Flandrian vegetational history at Hockham Mere, Norfolk, England. *New Phytologist*, 95, 457–87.

Bennett, K.D. 1984. The post-glacial history of *Pinus sylvestris* in the British Isles. *Quaternary Science Review*, 3, 133–55.

Bennett, K.D. and Birks, H.J.B. 1990. Postglacial history of Alder (*Alnus glutinosa* (L.) Gaertn.) in the British Isles. *Journal of Quaternary Science*, 5, 123–33.

Bennett, K.D., Whittington, G. and Edwards, K.J. 1994. Recent plant nomenclatural changes and pollen morphology in the British Isles. *Quaternary Newsletter*, 73, 1–6.

Bethell, P. and Máté I. 1989. The use of soil phosphate analysis in archaeology: a critique. In J. Henderson (ed.) *Scientific Analysis in Archaeology*, 1–29. Oxford University Committee for Archaeology, Monograph 19.

Bird, D.G., Crocker, G., McCracken, J.S. and Saich, D. 1994. Archaeology in Surrey 1991. *Surrey Archaeological Collections*, 82, 203–19.

Birks, H.J.B. 1989. Holocene isochrone maps and patterns of tree spreading in the British Isles. *Journal of Biogeography*, 16, 503–40.

Birks, H.J.B., Deacon, J. and Peglar, S. 1975. Pollen maps for the British Isles 5,000 years ago. *Proceedings of the Royal Society of London*, B 189, 87–105.

Bishop, M.J. 1981. Quantitative studies on some living British wetland mollusc faunas. *Biological Journal of the Linnean Society*, 15, 299–326.

Blytt, A. 1876. *Essay on the Immigration of the Norwegian Flora during Alternating Rainy and Dry Periods*. Kristiana, Cammermeyer.

Bowman, S. 1990. *Radiocarbon Dating*. London, British Museum Publications.

Bowman, S. 1994. Using radiocarbon: an update. *Antiquity*, 68, 838–43.

Bowman, S.G.E. and Huntley, D.J. 1984. A new proposal for the expression of alpha efficiency in TL dating. *Ancient TL*, 2, 6–8.

Boycott, A.E. 1936. The habitats of fresh-water Mollusca in Britain. *Journal of Animal Ecology*, 5, 116–86.

Bradley, R. 1990. *Passage of Arms: an Archaeological Analysis of Prehistoric Hoards and Votive Deposits*. Cambridge University Press.

Bradley, R. and Ellison, A. 1975. *Rams Hill: a Bronze Age Defended Enclosure and its Landscape*. Oxford, British Archaeological Reports 19.

Bradley, R. and Gordon, K. 1988. Human skulls from the River Thames, their dating and significance. *Antiquity*, 62, 503–9.

Briggs, D.J., Gilbertson, D.D. and Harris, A.L. 1985. Molluscan taphonomy in braided river environments: model development and application to the Summertown–Radley terrace gravels of the River Thames. *In* N.R.J. Fieller, D.D. Gilbertson and N.G.A. Ralph (eds) 1985, 67–89.

Bronk Ramsey, C. 1995. Radiocarbon calibration and analysis of stratigraphy: the OxCal program. In G.T. Cook, D.D. Harkness, B.F. Miller and E.M. Scott (eds) Proceedings of the 15th International Radiocarbon Conference. *Radiocarbon*, 37 (2), 425–30.

Brown, A.G. 1987. Long-term sediment storage in the Severn and Wye catchments. In K.J. Gregory, J. Lewin and J.B. Thornes (eds) *Palaeohydrology in Practice*, 307–32. Chichester, Wiley.

Brown, A.G. 1988. The palaeoecology of *Alnus* (alder) and the Postglacial history of floodplain vegetation. Pollen percentages and influx data from the West Midlands, United Kingdom. *New Phytologist*, 110, 425–36.

Brown, A.G. 1995. Holocene channel and floodplain change: a UK perspective. In A. Gurnell and G. Petts (eds) *Changing River Channels*. Chichester, Wiley.

Brown, A.G. 1997. *Alluvial Geoarchaeology: Floodplain Archaeology and Environmental Change*. Cambridge University Press.

Buckland, P.C. 1979. *Thorne Moors: a Palaeoecological Study of a Bronze Age Site*. Birmingham University Department of Geography, Occasional Publication 8.

Bullock, P., Federoff, N., Jongerius, A., Stoops, G. and Tursina, T. 1985. *Handbook for Soil Thin Section Description*. Albrighton, Waine Research Publications.

Burrin, P.J. and Scaife, R.G. 1984. Aspects of Holocene valley sedimentation and floodplain development in southern England. *Proceedings of the Geologists' Association*, 95, 81–96.

Bush, M.B. 1988. Early Mesolithic disturbance: a force on the landscape. *Journal of Archaeological Science*, 15, 453–62.

Bush, M.B. and Hall, A.R. 1987. Flandrian *Alnus*: expansion or immigration? *Journal of Biogeography*, 14, 479–81.

Butterworth, C.A. and Lobb, S.J. 1992. *Excavations in the Burghfield Area, Berkshire. Developments in the Bronze Age and Saxon Landscapes*. Salisbury, Wessex Archaeological Report no. 1.

Canti, M. 1997. An Investigation of microscopic calcareous spherulites from herbivore dung. *Journal of Archaeological Science*, 24, 219–31.

Case, H.J. and Whittle. A.W.R. (eds) 1982. *Settlement Patterns in the Oxford Region; Excavations at the Abingdon Causewayed Enclosure and Other Sites*. London, Council for British Archaeology Research Report 44.

Chambers, F.M. and Mighall, T. 1991. Palaeoecological investigations at Enfield Lock; pollen, pH, magnetic susceptibility and charcoal analysis of sediments. Unpublished report.

Chambers, F.M., Mighall, T.M. and Keen, D.H. 1996. Early Holocene pollen and molluscan records from Enfield Lock, Middlesex, UK. *Proceedings of the Geologists' Association*, 107, 1–14.

Chaplin, P.H. 1982. *The Thames from Source to Tideway*. Weybridge, Whittet.

Clapham, A.R. and Clapham, B.N. 1939. The valley fen at Cothill, Berkshire. Data for the study of post-glacial history II. *New Phytologist*, 38, 167–74.

Clapham, A.R., Tutin, T.G. and Moore, D.M. 1987. *Flora of the British Isles*. 3rd edition. Cambridge University Press.

Clark, A.J. 1977. Geophysical and chemical assessment of air photographic sites. *Archaeological Journal*, 134, 187–93.

Clark, A.J. 1991. The excavations: archaeomagnetic dating. *In* N. Sharples *Maiden Castle* (in microfiche). London, English Heritage.

Clark, A.J. 1992. Magnetic dating of alluvial deposits. *In* S.P. Needham and M.G. Macklin (eds) 1992, 37–42.

Clark, A.J., Tarling, D.H. and Noël, M. 1988. Developments in archaeomagnetic dating in Britain. *Journal of Archaeological Science*, 15, 645–67.

Cleal, R.M.J., Walker, K.E. and Montague, R. (eds) 1995. *Stonehenge in its Landscape: Twentieth-Century Excavations*. London, English Heritage Archaeological Report 10.

Coope, R. 1977. Fossil Coleopteran assemblages as sensitive indicators of climatic changes during the Devensian (last) cold stage. *Philosophical Transactions of the Royal Society of London*, B 280, 313–37.

Courty, M.A., Goldberg, P. and Macphail, R. 1989. *Soils and Micromorphology in Archaeology*. Cambridge University Press.

Craddock, P.T. 1984. The soil phosphate survey at the Cat's Water subsite, Fengate, 1973–77. *In* F. Pryor, *Excavations at Fengate Peterborough, England: The Fourth Report*, 234–41. Northamptonshire Archaeological Society Monograph 2.

Craddock, P.T., Gurney, D., Pryor, F. and Hughes, M.J. 1985. The application of phosphate analysis to the location and interpretation of archaeological sites. *Archaeological Journal*, 142, 361–76.

Crouch, K.R. and Shanks, S.A. 1984. *Excavations at Staines 1975–76; The Friends' Burial Ground Site*. London and Middlesex Archaeological Society/Surrey Archaeological Society Joint Publication 2.

Deacon, J. 1974. The location of refugia of *Corylus avellana* L. during the Weichselian glaciation. *New Phytologist*, 73, 1055–63.

Devoy, R.J.N. 1977. Flandrian sea level changes in the Thames estuary and the implications for land subsidence in England and Wales. *Nature* (London), 270, 712–5.

Devoy, R.J.N. 1979. Flandrian sea level changes and vegetational history of the Lower Thames estuary. *Philosophical Transactions of the Royal Society of London*, B 285, 355–407.

Devoy, R.J.N. 1980. Post-glacial environmental change and man in the Thames estuary: a synopsis. *In* Thompson, F.H. (ed.) *Archaeology and Coastal Change*, 134–48. Society of Antiquaries of London, Occasional Paper (new series) 1.

Donisthorpe, H. St J. K. 1939. *A preliminary list of the Coleoptera of Windsor Forest*. London, Nathaniel Lloyd.

Edwards, K.J. and Hirons, K.R. 1984. Analysis of pre-elm decline deposits: implications for the earliest agriculture in Britain and Ireland. *Journal of Archaeological Science*, 11, 71–80.

Ellison, A. and Harriss, J. 1972. Settlement and land use in the prehistory and early history of Southern England: a study based on locational models. *In* D.L. Clarke (ed.) *Models in Archaeology*, 911–62. London, Methuen.

Erdtman, G. 1928. Studies in the post-arctic history of the forests of North-West Europe; I, Investigation in the British Isles. *Geologiska Foreningens i Stockholm Forhandlingar*, 50, 123–92.

Evans, J.G. 1991a. The land and freshwater Mollusca. *In* S.P. Needham 1991, 262–74.

Evans, J.G. 1991b. An approach to the interpretation of dry-ground and wet-ground molluscan taxocenes from central-southern England. *In* D.R. Harris and K.D. Thomas (eds) *Modelling Ecological Change*, 75–89. London, Institute of Archaeology, University College.

Evans, J.G. 1992. Mollusca. *In* C.A. Butterworth and S.J. Lobb, 1992, 130–43.

Evans, J.G. 1995. Land- und Susswassermollusken. *In* B. Ottaway, *Ergolding, Fischergasse – eine Feuchtbodensiedlung der Altheimer Kultur in Niederbayern*, 193–202. Kallmunz, Michael Lassleben.

Evans, J.G., Davies, P., Mount, R. and Williams, D. 1992. Molluscan taxocenes from Holocene overbank alluvium in southern central England. *In* S.P. Needham and M.G. Macklin (eds) 1992, 65–74.

Evans, J.G., Limbrey, S., Máté, I. and Mount, R. 1993. An environmental history of the Upper Kennet Valley, Wiltshire, for the last 10,000 years. *Proceedings of the Prehistoric Society*, 59, 139–96.

Farley, M. and Leach, H. 1988. Medieval pottery production areas near Rush Green, Denham, Buckinghamshire. *Records of Buckinghamshire*, 30, 53–102.

Fieller, N.R.J., Gilbertson, D.D. and Ralph, N.G.A. (eds) 1985. *Palaeobiological Investigations: Research Design, Methods and Data Analysis*. Oxford, British Archaeological Reports (International Series), 266.

Fowler, W.W. 1890. *The Coleoptera of the British Islands*, 4. London, L. Reeve.

Fowler, W.W. 1891. *The Coleoptera of the British Islands*, 5. London, L. Reeve.

French, C.A.I. 1988. Further aspects of the buried prehistoric soils in the Fen margin, north-east of Peterborough, Cambridgeshire. *In* P. Murphy and C. French (eds) *The Exploitation of Wetlands*, 193–212. Oxford, British Archaeological Reports 186.

Gale, R. 1991. The identification of wood remains. *In* S.P. Needham 1991, 226–33.

Garner, B. 1991. The Roman Road – Colchester to Silchester through Windsor Park. *Surrey Archaeological Society Bulletin*, 260 (October 1991).

Gibbard, P.L. 1974. Pleistocene Stratigraphy and Vegetational History of Hertfordshire. PhD thesis, University of Cambridge.

Gibbard, P.L. 1985. *Pleistocene History of the Middle Thames Valley*. Cambridge University Press.

Gibbard, P.L., Coope, G.R., Hall, A.R., Preece, R.C. and Robinson, J.E. 1982. Middle Devensian deposits beneath the Upper Floodplain terrace of the River Thames at Kempton Park, Sunbury, England. *Proceedings of the Geologists' Association*, 93, 275–89.

Gibbard, P.L. and Hall, A.R. 1982. Late Devensian river deposits in the Lower Colne Valley, West London, England. *Proceedings of the Geologists' Association*, 93, 291–9.

Gibson, A. and Kinnes, I. 1997. On the urns of a dilemma: radiocarbon and the Peterborough problem. *Oxford Journal of Archaeology*, 16, 65–72.

Gilbertson, D.D. (ed.) 1984. *Late Quaternary Environments and Man in Holderness*. Oxford, British Archaeological Reports 134.

Girling, M.A. 1988. The bark beetle *Scolytus scolytus* (Fabricus) and the possible role of elm disease in the early Neolithic. In M.K. Jones (ed.) *Archaeology and the flora of the British Isles*, 34–8. Oxford University Committee for Archaeology Monograph 14.

Girling, M.A. and Greig, J.R.A. 1977. Palaeoecological investigations of a site at Hampstead Heath, London. *Nature*, 268, 45–7.

Girling, M.A. and Greig, J.R.A. 1985. A first fossil record for *Scolytus scolytus* (F.) (Elm Bark Beetle): its occurrence in Elm Decline deposits from London and the implication for the Neolithic Elm Decline. *Journal of Archaeological Science*, 12, 347–51.

Godwin, H. 1940. Pollen analysis and forest history of England and Wales. *New Phytologist*, 39, 370–400.

Godwin, H. 1943. *Rhamnus cathartica* L. *Journal of Ecology*, 31, 69–76.

Godwin, H. 1975a. *The history of the British flora*. 2nd edition. Cambridge University Press.

Godwin, H. 1975b. History of the natural forests of Britain: establishment, dominance and destruction. *Philosophical Transactions of the Royal Society of London*, B 271, 47–67.

Greig, J.R.A. 1982. Past and present lime woods of Europe. *In* M. Bell and S. Limbrey (eds) *Archaeological Aspects of Woodland Ecology*, 23–55. Oxford, British Archaeological Reports (International Series), 146.

Greig, J.R.A. 1989. From lime forest to heathland – five thousand years of change at West Heath Spa, Hampstead, as shown by the plant remains. *In* D. Collins and D. Lorimer (eds) *Excavations at the Mesolithic site on West Heath, Hampstead, 1976–1981*, 89–99. Oxford, British Archaeological Reports 217.

Greig, J.R.A. 1991. The botanical remains. *In* S.P. Needham 1991, 234–61.

Greig, J.R.A. 1992. The deforestation of London. *Review of Palaeobotany and Palynology*, 73, 71–86.

Grichuk, M.P. 1967. The study of pollen spectra from recent and ancient alluvium. *Review of Palaeobotany and Palynology*, 4, 107–12.

Hare, F.K. 1947. The geomorphology of a part of the Middle Thames. *Proceedings of the Geologists' Association*, 58, 294–339.

Healy, F., Heaton, M. and Lobb, S.J. 1992. Excavations of a Mesolithic site at Thatcham, Berkshire. *Proceedings of the Prehistoric Society*, 58, 41–76.

Hedges, R.E.M., Housley, R.A., Bronk Ramsey, C. and van Klinken, G.J. 1993. Radiocarbon dates from the Oxford AMS system: Archaeometry datelist 16. *Archaeometry*, 35, 147–68.

Hillman, G. 1986. Plant foods in ancient diet: the archaeological role of palaeofaeces in general and Lindow Man's gut contents in particular. *In* I.M. Stead, J.B. Bourke and D. Brothwell, 1991, 99–115.

Hinton, H.E. 1940–1. The Ptinidae of economic importance, *Bulletin of Entomological Research*, 31, 331–81.

Hobbs, R. 1996. *British Iron Age Coins in the British Museum*. London, British Museum Press.

Hodgson, J.M. 1976. *Soil Survey Field Handbook*. Soil Survey Technical Monograph 5. Harpenden, Soil Survey of England and Wales.

Hoffmann, A. 1954. *Coléoptères Curculionides, 2*. Faune de France, 59. Paris, Lechevalier.

Holyoak, D.T. 1982. Non-marine Mollusca of the last glacial period (Devensian) in Britain. *Malacologia*, 22, 727–30.

Holyoak, D.T. 1983. The colonisation of Berkshire, England, by land and freshwater Mollusca since the Late Devensian. *Journal of Biogeography*, 10, 483–98.

Holyoak, D.T. and Preece, R.C. 1985. Late Pleistocene interglacial deposits at Tattershall, Lincolnshire. *Philosophical Transactions of the Royal Society of London*, B 311, 193–236.

Iversen, J. 1941. Landnam i Danmarks Stenalder. *Danmarks Geologiske Undersøgelse*, Series II, 66, 1–67.

Iversen, J. 1949. The influence of prehistoric man on vegetation. *Danmarks Geologiske Undersøgelse*, Series IV, 3 (6), 1–25.

Jacobi, R.J. 1978. Population and landscape in Mesolithic lowland Britain. *In* S. Limbrey and J.G. Evans (eds) *The Effect of Man on the Landscape: the Lowland Zone*, 75–85. London, Council for British Archaeology, Research Report 21.

Janssen, C.R. 1959. *Alnus* as a disturbing factor in pollen diagrams. *Acta Botanica Neerelandica*, 8, 55–8.

Jones, R.L. and Keen, D.H. 1993. *Pleistocene Environments in the British Isles*. London, Chapman Hall.

Keef, P.A.M., Wymer, J.J. and Dimbleby, G.W. 1965. A Mesolithic site on Iping Common, Sussex, England. *Proceedings of the Prehistoric Society*, 31, 85–92.

Keen, D.H., Jones, R.L., Evans, R.A. and Robinson, J.E. 1988. Faunal and floral assemblages from Bingley Bog, West Yorkshire, and their significance for Late Devensian and early Flandrian environmental change. *Proceedings of the Yorkshire Geological Society*, 47, 125–38.

Kelly, M. and Osborne, P.J. 1963–4. Two faunas and floras from the alluvium at Shustoke, Warwickshire, *Proceedings of the Linnaean Society of London*, 176, 37–65.

Kerney, M.P. 1971. Interglacial deposits in Barnfield Pit, Swanscombe, and their molluscan fauna. *Journal of the Geological Society*, 127, 69–93.

Kerney, M.P., Gibbard, P.L., Hall, A.R. and Robinson, J.E. 1982. Middle Devensian river deposits beneath the 'Upper Floodplain' terrace of the River Thames at Isleworth, west London. *Proceedings of the Geologists' Association*, 93, 385–93.

Kerney, M.P., Preece, R.C. and Turner, C. 1980. Molluscan and plant biostratigraphy of some Late Devensian and Flandrian deposits in Kent. *Philosophical Transactions of the Royal Society of London*, B 291, 1–43.

Kinnes, I.A. 1991. The Neolithic pottery. *In* S.P. Needham 1991, 157–61.

Kloet, G.S. and Hincks, W.D. 1964. *A Checklist of British Insects: Small Orders and Hemiptera*. 2nd edition (revised). Royal Entomological Society of London handbook, 11, part 1.

Kloet, G.S. and Hincks, W.D. 1977. *A Checklist of British Insects: Coleoptera and Strepsiptera*. 2nd edition (revised). Royal Entomological Society of London handbook, 11, part 3.

Kloet, G.S. and Hincks, W.D. 1978. *A Checklist of British Insects: Hymenoptera*. 2nd edition (revised). Royal Entomological Society of London handbook, 11, part 4.

Koch, K. 1989a. *Die Käfer Mitteleuropas Ökologie, 1*. Krefeld, Goecke and Evers.

Koch, K. 1989b. *Die Käfer Mitteleuropas Ökologie, 2*. Krefeld, Goecke and Evers.

Koch, K. 1992. *Die Käfer Mitteleuropas Ökologie, 3*. Krefeld, Goecke and Evers.

Lacaille, A.D. 1963. Mesolithic industries beside Colne Waters in Iver and Denham, Buckinghamshire. *Records of Buckinghamshire*, 17, 143–81.

Lambrick, G.H. and Robinson, M.A. 1979. *Iron Age and Roman Riverside Settlements at Farmoor, Oxfordshire*. London, Council for British Archaeology, Research Report 32.

Lambrick, G.H. and Robinson, M.A. 1998. The development of floodplain grassland in the upper Thames valley. *In* M. Jones (ed.) *Archaeology and the Flora of the British Isles*, 55–75. Oxford University Committee for Archaeology, Monograph 14.

Lewis, J.S.C., Wiltshire, P.E.J. and Macphail, R.I. 1992. A Late Devensian/Early Flandrian site at Three Ways Wharf, Uxbridge: environmental implications. *In* S.P. Needham and M.G. Macklin (eds) 1992, 235–47.

Limbrey, S. 1991. The sediments in Area 6. *In* S.P. Needham, 1991, 218–22.

Limbrey, S. 1992. Micromorphological studies of buried soils and alluvial deposits in a Wiltshire river valley. *In* S.P. Needham and M.G. Macklin (eds) 1992, 27–36.

Longley, D. 1980. *Runnymede Bridge 1976: Excavations on the Site of a Late Bronze Age Settlement*. Guildford, Surrey Archaeological Society Research Volume 6.

Longworth, I.H. and Varndell, G. 1996. The Neolithic pottery. *In* S.P. Needham and A.J. Spence 1996, 100–5.

Lowe, J.J. and Walker, M.J.C. 1984. *Reconstructing Quaternary Environments*. London, Longman.

Mangerud, J., Andersen, S.T., Berglund, B.E. and Donner, J.J. 1974. Quaternary stratigraphy of Norden, a proposal for terminology and classification. *Boreas*, 3, 109–28.

Mangerud, J. and Berglund, B.E. 1978. The subdivision of the Quaternary of Norden: a discussion. *Boreas*, 7, 179–81.

McVean, D.N. 1953. Biological flora of the British Isles. *Alnus glutinosa* (L.) Gaertn. *Journal of Ecology*, 41, 447–66.

McVean, D.N. 1956. Ecology of *Alnus glutinosa* (L.) Gaertn. VI Post-glacial history. *Journal of Ecology*, 44, 331–3.

Merriman, N. 1992. Predicting the unexpected: prehistoric sites recently discovered under alluvium in central London. *In* S.P. Needham and M.G. Macklin (eds) 1992, 261–7.

Mook, W.G. 1986. Business meeting: recommendations/resolutions adopted by the 12th International Radiocarbon Conference. *Radiocarbon*, 28 (2B), 799.

Moore, J. and Jennings, D. 1992. *Reading Business Park: a Bronze Age Landscape*. Oxford University Committee for Archaeology; Thames Valley Landscapes: The Kennet Valley, Volume 1.

Moore, P.D. 1977. Ancient distribution of lime trees in Britain. *Nature*, 268, 13–14.

Moore, P.D. and Webb, J.A. 1978. *An Illustrated Guide to Pollen Analysis*. London, Hodder and Stoughton.

Moore, P.D., Webb, J.A. and Collinson, M.E. 1991. *Pollen Analysis*. 2nd edition. Oxford, Blackwell Scientific.

Murphy, J. and Riley, J.P. 1962. A modified solution method for the determination of phosphate in natural waters. *Analytica Chemica Acta*, 27, 31–6.

Nayling N. and Caseldine, A. 1997. *Excavations at Caldicot, Gwent: Bronze Age Palaeochannels in the Lower Nedern Valley*. York, Council for British Archaeology, Research Report 108.

Needham, S.P. 1989. River valleys as wetlands: the archaeological prospects. *In* J.M. Coles and B.J. Coles (eds) *The Archaeology of Rural Wetlands in England and Wales*, 29–34. University of Exeter, Wetland Archaeological Research Project.

Needham, S.P. 1991. *Excavation and Salvage at Runnymede Bridge, 1978: The Late Bronze Age Waterfront Site*. London, British Museum Press.

Needham, S.P. 1992. Holocene alluviation and interstratified settlement evidence in the Thames valley at Runnymede Bridge. *In* S.P. Needham and M.G. Macklin (eds) 1992, 249–60.

Needham, S.P. 1993. The structure of settlement and ritual in the Late Bronze Age of south-eastern Britain. *In* C. Mordant and A. Richard (eds) *L'Habitat et l'Occupation du Sol à l'Age du Bronze en Europe*, 49–69. Paris, Comité des Travaux Historiques et Scientifiques; Documents Préhistoriques 4.

Needham, S.P. and Evans, J. 1987. Honey and dripping. Neolithic food residues from Runnymede Bridge. *Oxford Journal of Archaeology*, 6, 21–8.

Needham, S.P. and Macklin, M.G. (eds) 1992. *Alluvial Archaeology in Britain: Proceedings of a Conference Sponsored by the RMC Group plc, 3–5 January 1991, British Museum*. Oxford, Oxbow Monograph 27.

Needham, S.P. and Sørensen, M.L.S. 1988. Runnymede refuse tip: a consideration of midden deposits and their formation. *In* J.C. Barrett and I.A. Kinnes (eds), *The Archaeology of Context in the Neolithic and Bronze Age: Recent Trends*, 113–26. University of Sheffield, Department of Archaeology.

Needham, S.P. and Spence, A.J. 1996. *Refuse and Disposal at Area 16 East, Runnymede*. London, British Museum Press. Runnymede Bridge Research Excavations, Volume 2.

Needham, S.P. and Spence, A.J. 1997. Refuse and the formation of middens. *Antiquity*, 71, 77–90.

Needham, S.P. and Trott, M.R. 1987. Structure and sequence in the Neolithic deposits at Runnymede. *Proceedings of the Prehistoric Society*, 53, 479–83.

Pearce, J. 1992. *Border Wares: Post-Medieval Pottery in London, 1500–1700. Vol. 1*. London, HMSO.

Pearson, G.W., Becker, B. and Qua, F. 1993. High-precision [14]C Measurement of German and Irish Oaks to show the Natural [14]C Variations from 7890 to 5000 BC. *Radiocarbon*, 35 (1), 93–104.

Pearson, G.W., Pilcher, J.R., Baillie, M.G.L., Corbett, D.M. and Qua, F. 1986. High-precision [14]C Measurement of Irish Oaks to show the Natural [14]C Variations from AD 1840–5210 BC. *Radiocarbon*, 28 (2B), 911–34.

Pearson, G.W. and Stuiver, M. 1986. High-precision calibration of the radiocarbon time scale, 500–2500 BC. *Radiocarbon*, 28 (2B), 839–62.

Poole, H.F. 1936. Outline of the Mesolithic flint cultures of the Isle of Wight. *Proceedings of the Isle of Wight Natural History and Archaeological Society*, 2, 551–81.

Preece, R.C. and Robinson, J.E. 1982. Mollusc, ostracod and plant remains from early Postglacial deposits near Staines. *The London Naturalist*, 61, 6–15.

Pryor, F. 1992. Discussion: the Fengate/Northey landscape. *Antiquity*, 66, 518–31.

Rankine, W.F. 1961. The Mesolithic age in Dorset and adjacent area. *Proceedings of the Dorset Natural History and Archaeological Society*, 83, 91–9.

Rigby, V. 1995. The pottery. *In* A. Hunn and M. Farley, The Chalfont St Peter Roman coin hoard, 1989. *Records of Buckinghamshire*, 37, 113–26.

Robinson, M.A. 1988. Molluscan evidence for pasture and meadowland on the floodplain of the upper Thames basin. *In* P. Murphy and C. French (eds) *The Exploitation of Wetlands*, 101–12. Oxford, British Archaeological Reports 186.

Robinson, M.A. 1991. Neolithic and Late Bronze Age insect assemblages. *In* S.P. Needham 1991, 277–326.

Robinson, M.A. 1992. Environment, archaeology and alluvium on the river gravels of the south Midlands. *In* S.P. Needham and M.G. Macklin (eds) 1992, 197–208.

Robinson, M.A. and Lambrick, G.H. 1984. Holocene alluviation and hydrology in the upper Thames Basin. *Nature*, 308, 809–14.

Rose, F. 1974. The epiphytes of Oak. In M.G. Morris and F.H. Perring (eds) *The British Oak; its History and Natural History*, 250–73. Farringdon, Classey.

Saville, A. 1991. The flintwork: Mesolithic, Neolithic and Bronze Age. *In* S.P. Needham 1991, 125–31.

Scaife, R.G. 1980. Late Devensian and Flandrian palaeoecological studies in the Isle of Wight. Unpublished PhD thesis, University of London, King's College.

Scaife, R.G. 1982. Late Devensian and early Flandrian vegetation changes in southern England. *In* S. Limbrey and M. Bell (eds) *Archaeological Aspects of Woodland Ecology*, 57–74. Oxford, British Archaeological Reports (International Series), 146.

Scaife, R.G. 1987. The Late Devensian and Flandrian vegetation of the Isle of Wight. *In* K.E. Barber (ed.) *Wessex and the Isle of Wight, Field Guide*, 156–80. Cambridge, Quaternary Research Association.

Scaife, R.G. 1988. Pollen analysis of the Mar Dyke sediments. *In* T.J. Wilkinson, *Archaeology and Environment in South Essex*, 109–14. East Anglian Archaeology, 42.

Scaife, R.G. 1992. Plant macrofossils and pollen analysis. *In* F. Healy, M. Heaton and S.J. Lobb, 1992, 64–70.

Scaife, R.G. and Burrin, P.J. 1992. Archaeological inferences from alluvial sediments: some findings from southern England. *In* S.P. Needham and M.G. Macklin (eds) 1992, 75–91.

Scott, E.M., Harkness, D.D. and Cook, G.T. 1998. Interlaboratory comparisons: lessons learned. Proceedings of the 16th International Conference. *Radiocarbon*, 40 (1), 331–40.

Serjeantson, D. 1996. The animal bones. *In* S.P. Needham and A.J. Spence 1996, 194–223.

Serjeantson, D., Field, D., Penn, J. and Shipley, M. 1991–2. Excavations at Eden Walk II Kingston: environmental reconstruction and prehistoric finds. *Surrey Archaeological Collections*, 81, 71–90.

Sernander, R. 1908. On evidence of post-glacial changes of climate furnished by peat mosses of northern Europe. *Geologiska Foreningens i Stockholm Forhandlingar*, 30, 365–478.

Sidell, E.J., Scaife, R.G., Tucker, S., and Wilkinson, K.N. 1995. Palaeoenvironmental investigations at Bryan Road, Rotherhithe. *London Archaeologist*, 7, 279–85.

Sieveking, G. de G., Longworth, I.H., Hughes, M.J., Clark, A.J. and Millett, A. 1973. A new survey of Grimes Graves, Norfolk. *Proceedings of the Prehistoric Society*, 39, 182–218.

Simmons, I.G. 1975a. Towards an ecology of Mesolithic man in the uplands of Great Britain. *Journal of Archaeological Science*, 2, 1–15.

Simmons, I.G. 1975b. The ecological setting of Mesolithic man in the highland zone. In J.G. Evans and S. Limbrey (eds) *The Effect of Man on the Landscape: the Highland Zone*, 57–63. London, Council for British Archaeology, Research Report 11.

Simmons, I.G. and Tooley, M. (eds) 1981. *The Environment in British Prehistory*. London, Duckworth.

Sims, R.E. 1973. The anthropogenic factor in East Anglian vegetational history: an approach using APF techniques. *In* H.J.B. Birks and R.G. West (eds) *Quaternary Plant Ecology*, 223–36. Oxford, Blackwell.

Sims, R.E. 1978. Man and vegetation in Norfolk. *In* S. Limbrey and J.G. Evans (eds) *The Effect of Man on the Landscape: the Lowland Zone*, 35–43. London, Council for British Archaeology, Research Report 21.

Smith, A.G. 1970. The influence of Mesolithic and Neolithic man on British vegetation: a discussion. *In* D. Walker and R.G. West (eds) *Studies in the Vegetational History of the British Isles*, 81–96. Cambridge University Press.

Smith, A.G. and Pilcher, J.R. 1973. Radiocarbon dates and the vegetational history of the British Isles. *New Phytologist*, 72, 903–14.

Sparks, B.W. 1961. The ecological interpretation of Quaternary non-marine Mollusca. *Proceedings of the Linnean Society of London*, 172, 72–80.

Sparks, B.W. 1962. Post-glacial Mollusca from Hawes Water, Lancashire, illustrating some difficulties of interpretation. *Journal of Conchology*, 25, 78–82.

Sparks, B.W. and Lambert, C.A. 1961. The post-glacial deposits at Apethorpe, Northamptonshire. *Proceedings of the Malacological Society*, 34, 302–15.

Sparks, B.W. and West, R.G. 1970. Late Pleistocene deposits at Wretton, Norfolk; 1. Ipswichian interglacial deposits. *Philosophical Transactions of the Royal Society of London*, B 258, 1–30.

Stace, C. 1991. *New flora of the British Isles*. Cambridge University Press.

Stead, I.M., Bourke, J.B. and Brothwell, D. 1991. *Lindow Man: the Body in the Bog*. London, British Museum Publications.

Stockmarr, J. 1971. Tablets with spores used in absolute pollen analysis. *Pollen et Spores*, 13, 614–21.

Stuiver, M. and Pearson, G.W. 1986. High-precision calibration of the radiocarbon time scale, AD 1950–500 BC. *Radiocarbon*, 28 (2B), 805–38.

Stuiver, M. and Pearson, G.W. 1993. High-precision bidecadal calibration of the radiocarbon time scale, AD 1950–500 BC and 2500–6000 BC. *Radiocarbon*, 35 (1), 1–24.

Stuiver, M. and Polach, H.A. 1977. Discussion: reporting of ^{14}C data. *Radiocarbon*, 19 (3), 355–63.

Stuiver, M. and Reimer, P.J. 1987. *Users Guide to the Programs CALIB and DISPLAY, rev 2.1*. University of Washington, Quaternary Isotope Laboratory.

Tansley, A.G. 1949. *The British Isles and their Vegetation*. Cambridge University Press.

Tauber, H. 1965. Differential pollen dispersion and the interpretation of pollen diagrams. *Danmarks Geologiske Undersogelse*, Series II, 89, 1–69.

Tauber, H. 1967a. Investigation of the mode of pollen transfer in forested areas. *Review of Palaeobotany and Palynology*, 3, 277–87.

Tauber, H. 1967b. Differential pollen dispersion and filtration. *Proceedings of Congress, International Association of Quaternary Research*, 7, 131–4.

Thomas, K.D. 1985. Land snail analysis in archaeology: theory and practice. *In* N.R.J. Fieller, D.D. Gilbertson and N.G.A. Ralph (eds) 1985, 131–56.

Thomas, R., Robinson, M., Barrett, J. and Wilson, B. 1986. A Late Bronze Age riverside settlement at Wallingford, Oxfordshire. *Archaeological Journal*, 143, 174–200.

Thorley, A. 1981. Pollen analytical evidence relating to the vegetation history of the chalk. *Journal of Biogeography*, 8, 93–106.

Troels-Smith, J. 1956. The Neolithic period in Switzerland and Denmark. *Science* (New York), 124, 876–9.

Turner, F. 1926. *Egham, Surrey: the History of the Manor under Church and Crown*. Egham.

Turner, G.M. and Thompson, R. 1982. Detransformation of the British geomagnetic secular variation record for Holocene times. *Geophysical Journal of the Royal Astronomical Society*, 70, 789–92.

Turner, J. 1962. The *Tilia* decline: an anthropogenic interpretation. *New Phytologist*, 61, 328–41.

Turner, J. 1970. Post-Neolithic disturbance of British vegetation. *In* D. Walker and R.G. West (eds) *Studies in the Vegetational History of the British Isles*, 97–116. Cambridge University Press.

Tyers, I. 1988. The prehistoric peat layers (Tilbury IV). *In* P. Hinton (ed.) *Excavations in Southwark 1973–76 and Lambeth 1973–79*, 5–12. London and Middlesex Archaeological Society and Surrey Archaeological Society Joint Publication 3.

Van Arsdell, R.D. 1989. *Celtic Coinage in Britain*. London, Spink.

Walker, M.J.C., Griffiths, H.I., Ringwood, V. and Evans, J.G. 1993. An early-Holocene pollen, mollusc and ostracod sequence from lake marl at Llangorse Lake, south Wales, UK. *Holocene*, 3, 138–49.

Waller, M. 1993. Flandrian vegetational history of south-eastern England: pollen data from Pannel Bridge, East Sussex. *New Phytologist*, 124, 345–69.

Waller, M. 1994. Paludification and pollen representation; the influence of wetland size on *Tilia* representation in pollen diagrams. *Holocene*, 4, 430–4.

Waton, P.V. 1982a. Man's impact on the chalklands: some new pollen evidence. *In* M. Bell and S. Limbrey (eds) *Archaeological Aspects of Woodland Ecology*, 75–91. Oxford, British Archaeological Reports (International Series), 146.

Waton, P.V. 1982b. A palynological study of the impact of man on the landscape of central southern England, with special reference to the chalklands. Unpublished PhD thesis, University of Southampton, Department of Geography.

Williams, D.F. 1996. A note on the petrology of three sherds of Early Saxon pottery. *In* P. Andrews and A. Crockett, *Three Excavations along the Thames and its Tributaries, 1994*, 38–9. Salisbury, Wessex Archaeology Report no. 10.

Wilson, D.G. 1987. *The Thames: Record of a Working Waterway*. London, Batsford.

Wymer, J.J. 1962. Excavations at the Maglemosian Sites at Thatcham, Berkshire, England. *Proceedings of the Prehistoric Society*, 28, 329–61.

Yates, D.T. 1999. Bronze Age field systems in the Thames Valley. *Oxford Journal of Archaeology*, 18, 157–70.